Computational Approaches to Novel Condensed Matter Systems

Applications to Classical and Quantum Systems

Computational Approaches to Novel Condensed Matter Systems

Applications to Classical and Quantum Systems

Edited by

D. Neilson
The University of New South Wales
Sydney, Australia

and

M. P. Das
The Australian National University
Canberra, Australia

Plenum Press • New York and London

Library of Congress Cataloging-in-Publication Data

On file

Proceedings of the Third Gordon Godfrey International Workshop on Computational Approaches to Novel Condensed Matter Systems, held July 12–17, 1993, in Sydney, Australia

ISBN 0-306-44986-2

© 1995 Plenum Press, New York
A Division of Plenum Publishing Corporation
233 Spring Street, New York, N. Y. 10013

10 9 8 7 6 5 4 3 2 1

All rights reserved

No part of this book may be reproduced, stored in a retrieval system, or transmitted in any form or by any means, electronic, mechanical, photocopying, microfilming, recording, or otherwise, without written permission from the Publisher

Printed in the United States of America

PREFACE

This volume contains the lectures given at the Third Gordon Godfrey International Workshop on **Computational Approaches to Novel Condensed Matter Systems** which was held at The University of New South Wales July 12-17, 1993. Lecturers from Asia, Australia, Europe and North America gave a total of twenty-nine lectures which were spread over the five days. Unfortunately we were not able to include in this volume the lectures of S. Das Sarma from the University of Maryland on "Non-Equilibrium Growth as a Self-Organised Phenomenon" due to constraints of time.

The workshops have been held annually since 1991 in Sydney, each covering a novel research area in condensed matter physics that is of topical interest. Australia has a strong tradition of research in condensed matter physics. The workshops are jointly organised by the School of Physics at the University of New South Wales (Sydney) and the Department of Theoretical Physics, Research School of Physical Sciences and Engineering at the Australian National University (Canberra). The late Gordon Godfrey was an Associate Professor of Physics at the University of New South Wales. He bequeathed his estate for the promotion and teaching of theoretical physics within the university.

The primary purpose of each workshop is to expose post-graduate students in physics to both informal interaction and formal lectures from recognised international leaders in topical research areas. Past experience has demonstrated again and again that to be informed about a new field there is no substitute for personal contact and interaction. Australian students have been disadvantaged compared to their counterparts in other Western countries by the lack of opportunity to travel to overseas research centres. Each set of lectures started from a level understandable to post-graduate students in physics. Seventy participants officially registered for the workshop including post-graduate students from five states. The convenors want to thank the lecturers who all cooperated with us to give a broad and coherent coverage of the subject material at a level and in a style which made it interesting and informative for the audience.

Thanks go also the heads of both physics departments involved, Jaan Oitmaa (University of New South Wales) and Brian Robson (Australian National University). We want to particularly thank Anne Merton who took care of every detail of organisation and to thank all the individuals who helped make the meeting the success it was. Financial support for the workshop was provided by The University of New South Wales Gordon Godfrey Bequest, The University of New South Wales School of Physics, and the Department of Theoretical Physics, Research School of Physical Sciences and Engineering, The Australian National University.

D. Neilson
M.P. Das

August 15, 1995

CONTENTS

Computational Approaches to Novel Condensed Matter Systems: An Overview M. P. Das and D. Neilson	1
Introduction to Quantum Monte Carlo Simulations of Electronic Systems R. M. Martin and V. D. Natoli	7
Density Functionals, Molecular Dynamics, and More R. O. Jones	37
Large-Scale Electronic Structure Calculations in Solids P. Giannozzi	67
Computer Simulation of Materials Using Parallel Architectures P. Vashishta, R. Kalia, A. Nakano, W. Jin and J. Yu	87
Molecular Dynamics on a Massively Parallel Computer for Application to Surface Systems S. Pickering and I. Snook	125
Friedel Oscillations in Condensed Matter Calculations J. F. Dobson	139
Collective Electronic Oscillations on C_{60} M. Michalewicz and M. P. Das	163
Theoretical Studies of Semiconductor Surfaces with Particular Reference to Fluorine and Chlorine Chemisorption on Si(001) P. V. Smith, M. W. Radny and A. J. Dyson	175
Functional Integral Techniques in Condensed Matter Physics N. Van Hieu	191
Disordered Electronic Materials and Spin Glasses D. J. W. Geldart	235
Freezing: Density Functional Theory A. D. J. Haymet	255

Application of the Local Chemical Potential to the Quantum Hall Effect in a
　　　Ballistic Quantum Wire 261
　P. N. Butcher and D. P. Chu

Finite Lattice Calculations for Magnetic Systems 269
　J. Oitmaa

Index 279

COMPUTATIONAL APPROACHES TO NOVEL CONDENSED MATTER SYSTEMS: AN OVERVIEW

M.P. Das[*] and D. Neilson[†]

[*]Department of Theoretical Physics, R S Phys S & E
The Australian National University, Canberra 0200, Australia

[†]School of Physics, The University of New South Wales
Sydney 2052, Australia

A major challenge in modern condensed matter theory is the bridging of the gap between those quantum systems which consist of just a single body and quantum systems which are made up of many particles. Exact analytic solutions for many-particle systems exist only for highly simplified Hamiltonians or if major approximations are first introduced into the analytic expressions and these approximations are frequently not tightly controllable. Perturbation methods are often not suitable for many-body systems of condensed matter because of the interactions between the constituent particles can be far too strong to be treated as a small parameter in a perturbation expansion.

Large-scale computational calculations of condensed matter systems can be of tremendous assistance in understanding these complex and fascinating systems. The primary objectives of such computational calculations are

- the carrying out of *ab initio* calculations which avoid the use of uncontrolled analytic approximations;

- the prediction of results which would prove difficult to attain in a real laboratory experiment.

There is a tremendous variety of different computational approaches which have been used. These include,

- Density functional theory,

- Molecular dynamics,

- Car-Parrinell technique,

- Quantum Monte Carlo approach,

and we briefly discuss each of these in turn.

In Density functional theory the ground state energy of electrons moving in an ionic lattice in the presence of an arbitrary external potential $V_{\text{ext}}(\mathbf{r})$ can be written in terms of a functional which depends only on the inhomogeneous density distribution $n(\mathbf{r})$, $E_v[n] = \int V_{\text{ext}}(\mathbf{r})n(\mathbf{r})d\mathbf{r} + F[n]$. The formally exact universal but unknown functional $F[n]$ is written as $F[n] \equiv T[n] + U_H[n] + U_{xc}[n]$ where $T[n]$ is the kinetic energy of the corresponding non-interacting system with the same density distribution, $U_H[n]$ is the Hartree potential and $U_{xc}[n]$ is the so-called the exchange-correlation potential which is not *a priori* known.

Two solution methods are commonly used in Density functional theory. One is the direct minimisation approach which gives the well-known Euler-Lagrange equations. The other is the Kohn-Sham method, which is a generalisation of the one-particle self-consistent Hartree scheme. Both approaches are variational and contain a one-particle effective potential. Many body effects are contained within the effective potential in a mean field sense.

The exchange-correlation potential is commonly described by the local density approximation (LDA). In spite of the clear limitations of the local density approximation for realistic systems, for which it is often the case that the electron density is not really slowly varying, this approximation has led to impressive results for electronic properties of systems ranging from atoms and molecules to solids and surfaces.

In the molecular dynamics approach atomic motion is simulated numerically. The trajectories of motion for each particle I are explicitly calculated using the equations of classical dynamics, $M_I \ddot{\mathbf{R}}_I = -\delta E_I/\delta \mathbf{R}_I(t)$, and "observations" are then made paralleling what one might measure in an actual experiment. For this reason molecular dynamical calculations are often referred to as numerical experiments.

Molecular dynamical techniques are of interest in a wide variety of problems such as

- crystal growth,
- amorphous and glassy structures,
- melting,
- defect motion,
- ion implantation,

and Molecular dynamics has been used in simulations of

- the reconstruction of surfaces,
- the making and breaking of chemical bonds,
- defect and vacancy formation,
- the structure and bonding of silicon.

In studies of systems with covalent bonding such as silicon a significant limitation of the molecular dynamical approach becomes apparent due to the fact that the chosen form of the interaction remains the same during the entire simulation. In reality each alteration in the position of the atoms should change the electronic structure. This in turn affects the interatomic potential, thus affecting the atomic position, and so on.

The Car-Parrinello method unifies density functional theory and molecular dynamics, and takes into account this readjustment of the empirical potential which is missing from the molecular dynamical approach. In the Car-Parrinello method after each step in the molecular dynamical calculation the resulting adjustments in the electronic structure due to the changed positions of the nuclei are determined using density functional theory. The ionic degrees of freedom are treated by molecular dynamical trajectories $\mathbf{R}_I(t)$, and the electronic degrees of freedom by single particle Kohn-Sham orbitals. For this reason the Car-Parrinello method is also referred to as dynamical simulated annealing.

In Quantum Molecular Dynamics the exchange correlation potential V_{xc} in the Kohn-Sham equation is replaced by a time dependent potential to take into account the motion of the ions. The time dependent equation is then solved. However the ionic motion is still treated classically.

During past ten years there has been a lot of progress in *ab-initio* studies of this type on the electronic and geometrical structure of molecules, of clusters and of ordered and disordered solids. Detailed results are now available not only for individual systems, but also there have been a number of trends predicted across the periodic table, trends of local structure, bonding, atomic dynamics and similar properties. Such trends can be a great help understanding electronic and atomic dynamical phenomena even in the absence of large scale calculations.

The Quantum Monte Carlo (QMC) method is a numerical approach mainly used to evaluate expectation values of physical observables such as energy, density, pair correlation function etc. It is particularly useful for extended systems, reproducing the proper scaling of properties with the size of the system. The method commonly uses one of the following techniques,

- Variational Monte Carlo,

- Green function or diffusion Monte Carlo,

- Auxiliary field Monte Carlo or Path integral Monte Carlo method.

The first two of these are used mainly for calculating ground state averages, whereas the last technique can be used to determine finite temperature properties and thermodynamical properties.

In the variational Monte Carlo method one starts with a many body correlated trial wave function which has some variational parameters. A statistical sampling is carried out to obtain the expectation value of the operators by optimising these parameters. With a good choice for the parametrised form of the trial wave function one can directly visualise the effect of the correlations. Since the method is variational good estimates of ground state properties can be obtained.

The Green function or Diffusion Monte Carlo deals with the time development of the equation of motion for some fixed starting condition. It is an imaginary time formulation. The time development equation is numerically solved iteratively with small time increments. For Bosons the ground state is nodeless, and Quantum Monte Carlo gives exact energy and ground state properties. For Fermions however there is a sign problem due to anti-symmetry of the wave-function under exchange. The fixed node approximation is used to overcome this and seems to give results of good accuracy. This approximation has been heavily used in the recent past.

The auxiliary field or path integral method is a powerful technique for strongly interacting systems. In the path integral formalism an additional real or imaginary time dimension is introduced to the d-dimensional real space. One then defines the path integral as an integral over $(d+1)$-dimensional space. It should be noted that this does not avoid the Fermion change-of-sign problem. With the Hubbard Hamiltonian, for example, an operator identity for the interaction term together with a transformation for the Gaussian integral identity permits us to write individual time-slice Hamiltonians in terms of one electron operators interacting with auxiliary fields. From this point on it is then straightforward to carry out the calculations exactly. Monte Carlo computations involving these methods have been performed in the continuum limit as well as for small size lattice systems.

The breathtaking advances in computer technology show little sign of abating and there is not much doubt that this will bring even greater opportunities for improving our quantitative understanding of condensed matter systems. With our excitement over the possibilities with numerical experiments it is important to keep in mind that what we wish to gain is better insight into these fascinating systems, in particular by studying various trends. It is not sufficient simply to produce results showing good agreement with real experiments. Volker Heine has said,

> How often I have read a paper about a piece of computational physics which finishes with the words ... *and we obtain good agreement with experiment.*
>
> If you know the answer from experiment, I want to cry, why are you wasting so much of time calculating it?

Bibliography

[1] W. Kohn and P. Vashishta in *Theory of the Inhomogeneous Electron Gas*, (Ed) S. Lundqvist and N. H.March, Plenum (1983).

[2] R.O. Jones and O. Gunnarson, Rev. Mod. Phys. **61**, 689 (1989).

[3] R.M. Dreizler and E.K.U. Gross, *Density Functional Theory*, Springer-Verlag, Berlin (1990).

[4] D.J.W. Geldart in *Strongly Correlated Electron Systems*, (Ed) M.P. Das and D. Neilson, Nova Sc., New York (1992).

[5] M.P. Das in *Condensed Matter Physics*, (Ed) J. Mahanty and M.P. Das, World Scientific, Singapore (1989).

[6] A. Rahman, Phys. Rev. **136**A, 405 (1964).

[7] G. Ciccotti and W.G. Hoover (Ed) *Simulation of Statistical Mechanical Systems*, North Holland, Amsterdam (1986)

[8] R. Car and M. Parrinello, Phys. Rev. Lett. **55**, 2471 (1985).

[9] M.P. Allen and D.J. Tildesley, *Computer Simulation of Liquids*, Oxford Univ. Press, Oxford (1990).

[10] M.C. Payne, M.P. Teter, D.C. Allen, T.A. Arias and J.D. Joannopoulos, Rev. Mod. Phys. **64**, 1045 (1992).

[11] P. Vashishta, R. K. Kalia, S.W. de Leeuw, D. L. Greenwell, A.Nakano, W. Jin, J. Yu, L. Bi and W. Li in *Topics in Condensed Matter Physics*, (Ed) M. P. Das, Nova Sc, New York (1994).

[12] N. Metropolis and S. Ulam , J. Am. Stat. Assoc. **44**, 335 (1949).

[13] D.R. Hamann, M. Schlüter, and C. Chiang, Phys. Rev. Lett. **43**, 1494 (1979).

[14] D.M. Ceperley and B.J. Alder, Phys. Rev. Lett. **45**, 566 (1980).

[15] G. Sugiyama and S.E. Koonin, Ann. Phys.(USA) **168**, 1 (1986).

[16] D.J. Scalapino in *Modern Perspectives in Many Body Physics*, (Ed) M.P. Das and J. Mahanty, World Scientific, Singapore (1994).

[17] J.E. Hirsch, Phys. Rev. B **31**, 4403 (1985).

[18] K. Binder, *Application of Monte Carlo Methods in Statistical Physics*, Springer, Berlin (1984)

INTRODUCTION TO QUANTUM MONTE CARLO SIMULATIONS OF ELECTRONIC SYSTEMS

Richard M. Martin[a,b] and Vincent D. Natoli[a]

Department of Physics[a] and Materials Research Laboratory[b]
University of Illinois at Urbana-Champaign
1110 W. Green Street, Urbana, IL 61801

ABSTRACT

Monte Carlo statistical methods have a unique role in computational physics because random sampling can be used to carry out exact or nearly exact calculations for many-body systems. The subject of this paper is a brief outline of some of the current developments in applying Monte Carlo methods to quantum problems involving interacting electrons in condensed matter. We will discuss methods and examples, in particular, homogeneous electron systems, the metal-insulator transition in hydrogen at high pressure, and a brief introduction to *ab initio* calculations on general materials.

INTRODUCTION

All forms of matter - atoms, molecules, ordinary condensed matter, nuclear matter, and dense matter which occur in astrophysical situations - are composed of interacting particles. Except for two-body systems such as the hydrogen atom, and special one-dimensional problems, there are no exact analytic solutions to any of these many-body problems. Of course, there are approximate mean-field independent-particle solutions, which are very useful and often appear to describe certain properties of real systems; however, a complete theoretical description must deal with the full many-body problem of correlated particles. The subject of these lectures is an introduction to methods which can make possible exact or nearly exact calculations on large systems of many interacting particles.[1] We will consider only the first three states of matter in which the electrons and nuclei together form the many-body system of interest. Although the interactions among the particles and the quantum equations have been known for a long time, the difficulty of the many-body problems has precluded theory from making definitive predictions on any but the simplest systems.

Here we will emphasize the quantum system of interacting electrons which is ubiquitous in all condensed matter. Qualitative effects which are caused by electron-electron correlation include magnetism, superconductivity, and certain metal-insulator transitions.[2] [3] A famous recent example is the 2-dimensional electron system in the CuO_2 layers in the high temperature superconductor materials: The widely-used independent-electron local density functional approximation (LDA) fails to give the insulating anti-ferromagnetic ground state.[4] On the other hand, the Hartree-Fock approximation (HFA), which treats exchange properly while omitting all correlation, tends to favor magnetism too much. Of course, the HFA has other striking failures - such as a zero density of states near the Fermi surface in metals. In many other problems, the independent-electron approximations are qualitatively correct but cannot describe quantitatively important properties, such as the cohesive energies of molecules and crystals which are crucial for quantitative theories of binding, phase transitions between different states, etc.

There is currently a great effort to develop theoretical computational methods for accurate *ab initio* calculations of interesting properties of such "real" materials.[5] Independent-electron methods have an important role to play and it has been established that, for a great many systems, they are quite accurate for properties such as structures and energy differences between *similar* phases. However, truly *ab initio* methods must deal with the full many-body problem. This is the subject of the present paper.

In recent years, many methods have been developed to treat the effects of correlation. These methods can be broadly grouped into two categories. One type is designed to treat excitations, i.e., the energy differences due to an excitation in an extended system. These are usually derived from perturbation or diagrammatic expansions which involve analytic approximations, but often the resulting expressions are evaluated numerically.[6] The other category of methods treats interactions nonperturbatively. These methods, which are able to treat the full many-body problem, are generally either restricted to very few particles or are able to treat large systems because they are intrinsically designed to describe the ground state, i.e., the lowest energy state. One of these approaches is density functional theory [7] which in principle is exact for the ground state, but which has not been possible to make exact in practice.

Monte Carlo methods have an important role in the theory of many-body systems because statistical sampling methods can simulate exactly (or nearly exactly) large many-body systems.[1] Quantum Monte Carlo (QMC) denotes the set of methods which use random sampling to simulate the many-body Schroedinger equation. [8] [9] Configuration interaction methods [10] which are widely used in chemistry, can also find exact ground and excited states; however, because they require computational time which scales as $N!$ where N is the number of particles, they are limited to problems involving only a very few particles, such as light atoms and small molecules. In contrast, Monte Carlo simulations of the ground state of a many-body Fermion system scale as N^3. Thus QMC methods can be applied to large finite systems or, by considering different numbers of particles N, the results can be extrapolated to the large N bulk limit. In this paper we will consider QMC as a method to determine only ground state properties (even though certain excited states can also be treated by QMC and projection techniques are beginning to be developed for excitations [11] [12]).

Methods that are applicable to electronic ground states have an important role in calculations on materials because the electronic ground state determines the structure of the material, phase transitions between different structures, elastic properties, defect energies, phonon frequencies, magnetic order at T=0, and a host of other properties. In addition to the intrinsic interest of such ground state properties, determination of the atomic scale structure is a crucial step for *ab initio* calculations of excited electronic states.

Monte Carlo methods are now an important part of computational methods in many areas of science. For an introduction to the methods, two textbooks which contain both general formulations and specific examples of programs are "Computational Physics" by Koonin [13] and "An Introduction to Computer Simulation Methods - Applications to Physical Systems", by Gould and Tobochnik.[14] Each gives examples pertinent to our emphasis on quantum problems, and Koonin even gives actual programs which provide *exact* solutions of the helium atom and hydrogen molecule.

In these lectures we first discuss basic ideas of Monte Carlo sampling methods applied to electronic problems. The description is based upon the thesis of one of us [15] and we attempt to give sufficient detail to help a reader to understand all the important ingredients. Then we proceed to a review of some work, mainly recent work in the group at the University of Illinois. We make no pretense to be complete and only attempt to present examples which illustrate the types of problems at the forefront of current research. Previous reviews of work in our group have been given in other proceedings.[16] [17]

MONTE CARLO METHODS

The term Monte Carlo has been applied to a myriad of methods and techniques which in one form or another use random numbers and stochastic processes. The phrase was perhaps first used in connection with calculations on neutron transport in fissionable materials that were suggested by Fermi in the 1930's and carried out by von Neuman, Ulam and Metropolis in the early 1940's. The efficacy of stochastic methods in the solution of certain differential equations had already been demonstrated by Courant, Friedrichs and Lewy [18]. Metropolis and Ulam[19] and Donsker and Kac[20] demonstrated how the new Monte Carlo methods could be applied to the Schroedinger equation. With the advent of advanced computing power in the latter half of this century, Monte Carlo methods have flourished with applications in physics, chemistry, economics and many other disciplines.[1] [13] [14]

Quantum Monte Carlo (QMC) refers to a set of numerical techniques used to evaluate expectation values of quantum many-body physical systems. Through sampling, the generation of particle configurations with a probability determined by the quantum wavefunction, one can evaluate interesting physical observables such as the energy or particle pair correlation functions. Quantum Monte Carlo is a general term which covers any of several commonly used techniques, such as variational(VMC), diffusion(DMC), Green's function Monte Carlo(GFMC), and path integral Monte Carlo(PIMC). The VMC, DMC and GFMC methods are all ground state techniques, whereas PIMC [21] [22] [23] [24] is a method to calculate thermodynamic properties at finite temperature. Although, this technique is widely used, especially for model lattice hamiltonians,[25] it will not be discussed here because a useful description would require developing a number of additional ideas, as well as a discussion of the difficulties encountered in reaching temperatures sufficiently low to be of interest for condensed matter.

Among the many different types of QMC methods,[1] we will focus upon Variational Monte Carlo (VMC) and Green's Function (or Diffusion[26]) Monte Carlo (DMC), which are capable of treating real electron problems with the accuracy needed in condensed matter problems. VMC uses statistical sampling methods to evaluate expectation values of any operator with a correlated many-body trial wavefunction. By optimizing parameters in the wavefunction, one can find an upper bound to the exact energy and determine which types of correlation are most important among the electrons. The two great advantages of VMC are: 1) its computational and conceptual simplicity, and 2) the explicit analytic representation for the wavefunction, which manifestly includes some correlations but not others. The primary disadvantages are: 1) the desired correlations must be put in, and 2) there is no intrinsic way to check the assumptions. Since the energy is variational, one can often find good estimates of the energy; conversely, it is difficult to use energy minimization alone to find the wavefunctions since the energy is insensitive to the errors in the wavefunction.

The first application of Monte Carlo methods to a quantum many body problem is attributed to McMillan[27], who calculated properties of the ground state of liquid ^4He using a variational trial wavefunction. Refinements in the trial wavefunction led to further improvements in the calculations on liquid ^4He by Schiff and Verlet [28], and Chang and Campbell [29], and on solid ^4He by Hansen et. al. [30]. Variational Monte Carlo methods have made valuable contributions to many other Bose systems in condensed matter physics such as hard sphere fluids and solids by Hansen et. al. [31], and by Kalos et. al. [32], and the Bose one-component plasma (i.e., Bosons interacting with a Coulomb potential in a neutralizing background) by Hansen and Mazighi[33]. The seminal work of Ceperley et. al. [34] paved the way for variational calculations on Fermi systems. Amongst the Fermi systems that have been studied by Ceperley and others are the two and three-dimensional electron gas (Fermion one-component plasmas) [35], ^3He[34] and hydrogen[36][37].

More recently, Fahy, Louie, and their co-workers [38] have carried out the first calculations for real materials, namely carbon and silicon, using VMC together with pseudopotentials. This is an important advance and these authors have shown that full many-body calculations can be carried out on materials with resulting ground state energies which are potentially accurate enough to make accurate *ab initio* predictions for real condensed matter. Further steps that make the VMC wavefunctions more accurate have been made by Umrigar [40] and Mitas [41] [42] for atomic-like systems, and by Kwon, et.al.,[43] for the homogeneous electron gas.

The other methods, DMC and GFMC, are closely related and are based upon use of Monte Carlo methods to iterate the many-body Schroedinger equation in imaginary time to project out the best possible ground state energy and correlation functions.[1, 44, 45] For Boson systems, because the ground state has no nodes, QMC methods can find the exact ground state energy and all other ground state properties. For Fermions, however, the famous "sign problem" [1] - due to the antisymmetry of the particles under exchange - has prevented the formulation of an exact method that is feasible for more than a few particles.[46] Nevertheless, the "fixed node approximation" [48] [49] is remarkably accurate if one has a trial function with appropriate nodes. The accuracy has been established by carrying out "release node" calculations for many systems, including the homogeneous electron gas, small molecules, and solid hydrogen. [49, 45, 36] The primary advantages of DMC are: 1) its ability to find exact (or nearly exact) ground state energies and correlations beyond those included in the trial wavefunction, and 2) a greatly decreased sensitivity to the choice of the trial function. Its primary disadvantages are: 1) the wavefunction is represented only implicitly in the sampling and 2) difficulties in using the fixed-node method in conjunction with non-local pseudopotentials which arise in treating atoms with cores, as will be discussed below.

The first applications of GFMC[39][50][51] were few-body nuclear problems[52] and the Helium atom[53]. The method was improved substantially by Kalos et. al. [32] in its application to "helium-like" systems with hard sphere interactions. Applications to many-Fermion systems would wait until 1980 when Ceperley and Alder[49] introduced the DMC method for Fermions in a calculation of the ground state of the three dimensional electron gas in the fixed-node approximation. Further DMC and GFMC work ensued on the electron gas[54][43] and other fermion systems like ^3He[55][56] and hydrogen[36]. Recent work using this method includes a study of hydrogen at high pressures, [57] [15] calculations of energies of metallic surfaces, [58] and the first DMC calculation for a material like silicon. [59] Recently, Mitas and coworkers have developed of a new hybrid approach [41] [42] to allow fixed-node DMC calculations to be done with pseudopotentials, which is essential for calculations on atoms with cores.

Monte Carlo Integration

Monte Carlo methods are important in many-body problems in physics and other fields because they make it possible to carry out complex high-dimensional integrals. Consider the following integral

$$I = \int_0^1 dr f(r) g(r) \tag{1}$$

where $\int_0^1 dr f(r) = 1$. Here $f(r)$ behaves like a probability distribution for the variable r. With this in mind, we may estimate the integral I as follows,

$$I_N = \frac{1}{N} \sum_{\substack{i=1 \\ r_i \in f(r)}}^{N} g(r_i), \tag{2}$$

where $r_i \in f(r)$ means r_i chosen according to the probability distribution $f(r)$. I_N is now an estimate for I, and we wish to estimate the expected error in this value. Each specific set of N points will give different values for I_N, and the expected error may be defined in terms of the average variation over many repeated calculations. If we define an average over repeated calculations by $\langle ... \rangle$, the variance on I_N can be expressed as[14][60]

$$var[I_N] = \langle I_N^2 \rangle - \langle I_N \rangle^2. \tag{3}$$

Using a property of the variance for large numbers of randomly chosen samples, $var\{cx\} = c^2 var\{x\}$,[60] we have the relation,

$$var[I_N] = var[\frac{1}{N} \sum_{i=1}^{N} g(r_i)] = \frac{1}{N^2} \sum_{i=1}^{N} var[g] = \frac{1}{N} var[g]. \tag{4}$$

Here the variance in g may be estimated from the variance over the N points used in finding the integral I_N and is related to the variation in the function g weighted by the probability f. The expected error in the integral I_N is the square root of the $var[I_N]$, which is here defined to be σ. We see from these equations that σ is proportional to the root mean square variation in the function g divided by the square root of the number of sampling points $\sqrt{N}$.[14][60] A stronger statement about the distribution of I_N in the asymptotic limit is given by the central limit theorem[32]. The central limit theorem states that in the limit of many samples, the observed values of I_N will be normally distributed with mean $\langle g \rangle$ and variance proportional to $1/N$.

Herein lies the advantage of Monte Carlo, since this result is independent of the dimensionality of the original integral. This is not true of more traditional quadrature methods. The error when using the trapezoidal rule[61] for a uniform grid in d dimensions, for example, is,

$$\epsilon \approx N^{1/d} \left(\frac{\Omega}{N}\right)^{3/d} \langle f'' \rangle, \tag{5}$$

where Ω is the volume concerned, N is the total of quadrature points, $N^{1/d}$ is the number of grid points in each dimension, $\langle f'' \rangle$ is an averaged value of the second derivative of f and d is the dimension. The most critical factor to consider is the way error scales with computation time. Since computation time will be proportional to N, the size of the estimated error will scale as $T^{-2/d}$, while for Monte Carlo integration it would scale as $T^{-1/2}$. Thus in general, for integrals of dimension greater than four, Monte Carlo integration is more efficient than the trapezoidal rule. Similar analysis may be applied to other quadrature methods to show the efficiency of Monte Carlo integration in high dimensions.

Another important concept in Monte Carlo methods which will arise in the discussion of both VMC and DMC is importance sampling. Consider multiplying and dividing the integrand of Eq (1) by a function $\tilde{f}$, so that its value is not altered.

$$I = \int_0^1 dr f(r) \tilde{f}(r) \frac{g(r)}{\tilde{f}(r)} \tag{6}$$

Now we consider $f\tilde{f}$ as the probability distribution, and the new estimator for I is,

$$I_N = \frac{1}{N} \sum_{\substack{i=1 \\ r_i \in f(r)\tilde{f}(r)}}^{N} \frac{g(r_i)}{\tilde{f}(r_i)} \tag{7}$$

The term $\tilde{f}$ is the importance sampling function. The advantage of importance sampling is seen by noting that the variance of I_N is now proportional to the variance of $g/\tilde{f}$ not g. This means that if we can devise a function $\tilde{f}$ which approximates g such that $g/\tilde{f}$ is smoothly varying everywhere, we can significantly reduce the variance on I_N.

Variational Monte Carlo (VMC)

Variational Monte Carlo (VMC) is a method which provides an upper bound to the ground state energy of a quantum many-body system, as well as estimates of other interesting physical quantities. As with any variational method, it requires a trial wavefunction, obtained from other methods which usually rely heavily on physical intuition and analytic calculation. Typically, a trial wavefunction will have several adjustable parameters which are optimized to minimize the energy. The Monte Carlo aspect of VMC is in the particular way the integral which represents the energy is evaluated. We begin with a ground state trial wavefunction $\Psi_T(R, \lambda_1, \lambda_2, ...)$ which depends on the parameters λ_i and the collective coordinates of the particle $R \equiv \{r_i\}$. The energy of the ground state is the expectation value of the Hamiltonian operator $\hat{H}$,

$$E(\{\lambda_i\}) = \int_\Omega d\text{R} \, \Psi_T^*(\text{R}, \{\lambda_i\}) \, \hat{H} \, \Psi_T(\text{R}, \{\lambda_i\}). \tag{8}$$

The energy, E may now be minimized as a function of the λ_i and by the variational principle it represents an upper bound to the ground state energy. In problems of interest, the integral (8) is highly dimensional (typically $d \approx 100$) and is therefore an excellent candidate for Monte Carlo evaluation. Let us recast Eq.(8), using the idea of importance sampling, in the following form, which is reminiscent of (1),

$$E(\{\lambda_i\}) = \int_\Omega d\mathrm{R}\, \Psi_T^*(\mathrm{R},\{\lambda_i\}) \Psi_T(\mathrm{R},\{\lambda_i\}) \frac{\hat{H}\Psi_T(\mathrm{R},\{\lambda_i\})}{\Psi_T(\mathrm{R},\{\lambda_i\})} \qquad (9)$$

where, $\Psi_T^*(\mathrm{R})\Psi_T(\mathrm{R})$ properly normalized is the probability distribution of configurations R and takes the place of $f(r)$ in equation (1). The expression, $H\Psi_T/\Psi_T$, commonly referred to as the local energy, takes the place of $g(r)$. Following the previous prescription, E can be estimated as follows,

$$E_N(\{\lambda_i\}) = \frac{1}{N} \sum_{\substack{j=1 \\ \mathrm{R}_j \in \Psi_T^* \Psi_T}}^{N} \frac{\hat{H}\Psi_T(\mathrm{R}_j,\{\lambda_i\})}{\Psi_T(\mathrm{R}_j,\{\lambda_i\})} = \frac{1}{N} \sum_{\substack{j=1 \\ \mathrm{R} \in \Psi_T^* \Psi_T}}^{N} E_{local}(\mathrm{R}_j,\{\lambda_i\}), \qquad (10)$$

which involves evaluation of the local energy at N configurations R_j, generated with probability $\Psi_T^*(\mathrm{R}_j)\Psi_T(\mathrm{R}_j)$. Evaluating $E_{local}(\mathrm{R}_j)$ is presumably straightforward, if tedious, since we know $\hat{H}$ and $\Psi_T(\mathrm{R}_j)$. The question which remains, however, is how do we generate configurations R with the correct distribution $\Psi_T^*(\mathrm{R})\Psi_T(\mathrm{R})$.

For certain cases such as Gaussian distributions of a variable, there are methods to generate points which are completely uncorrelated and have the given distribution, as described in many places.[14][13] However, in general, there are no such methods to generate configurations R_j with the desired distribution $\Psi_T^*(\mathrm{R})\Psi_T(\mathrm{R})$. This is accomplished by the $\mathrm{MR}^2\mathrm{T}^2$ or Metropolis algorithm[62] which is outlined below.

At this point we can summarize that for variational Monte Carlo (VMC), any method which can generate points with this distribution can be used to evaluate the high dimensional integrals and thus find the many-body expectation values of the hamiltonian, as well as other operators. The only input needed is the wavefunction and the accuracy will depend upon the extent to which this wavefunction contains the correlation among the particles. There are methods to efficiently use the sampling to vary any parameters inside the wavefunction to optimize it and produce the lowest energy within the degrees of freedom allowed. The variational theorem shows that the resulting energy must always be above the true ground state energy; however, the quality of the results is always limited by the chosen form of the wavefunction.[34][38]

The Metropolis Algorithm

The Metropolis algorithm[62] generates configurations $\{\ldots, \mathcal{R}_i, \mathcal{R}_{i+1}, \ldots\}$ according to a given distribution $P(\mathcal{R})$, by making successive moves whereby configuration $\mathcal{R}_{i+1}$ is generated starting at the configuration $\mathcal{R}_i$. The algorithm involves a transition amplitude $T(\mathcal{R}_i \to \mathcal{T})$ connecting any $\mathcal{R}_i$ to any other point $\mathcal{T}$); however, a crucial point is that, in the limit of many steps, the algorithm generates points with the desired distribution for *any* transition amplitude T, subject only to the condition that T be positive for any move. The algorithm is exceedingly simple and proceeds as follows:

1. Choose a new trial configuration $\mathcal{T}$ according to $T(\mathcal{R}_i \to \mathcal{T})$, where $T(\mathcal{R}_i \to \mathcal{T})$ is the probability of choosing configuration $\mathcal{T}$, given $\mathcal{R}_i$.

2. If the "acceptance" function, $A(\mathcal{R}_i \to \mathcal{T})$, is greater than ζ, a random number between zero and one, then accept the trial configuration $\mathcal{T}$ as the next configuration $\mathcal{R}_{i+1}$. Otherwise reject the configuration so $\mathcal{R}_{i+1}$ is $\mathcal{R}_i$

The generalized Metropolis choice for $A(\mathcal{R}_i \to \mathcal{T})$ is,

$$A(\mathcal{R}_i \to \mathcal{T}) = min\left[1, \frac{P(\mathcal{R}_i)\, T(\mathcal{T} \to \mathcal{R}_i)}{P(\mathcal{T})\, T(\mathcal{R}_i \to \mathcal{T})}\right]. \tag{11}$$

Because the outcome of a given step belongs to a finite set of possible states and depends only on the state immediately preceding it, the random walk may be identified as a Markov chain[63]. As such, it will converge to a unique distribution as long as it is ergodic and not periodic. Ergodicity implies that there is a finite probability to move from any state to any other state in a finite number of steps. Periodicity is inhibited as long as there is a finite probability for the walk to remain in the state that it is in. The Metropolis walk obeys these conditions and is expected to converge to a unique distribution, i.e., the acceptance function in Eq. (11), leads to the equilibrium distribution P.

To summarize, the Metropolis algorithm generates a sequence of configurations which satisfy the requirements of a Markov chain. It is known that a Markov chain will converge to a unique distribution. In the case of the Metropolis algorithm, that distribution is determined by the details of the acceptance term A. The transition amplitude T, which is arbitrary as long as it is positive for any move, should be chosen so as to move through phase space as efficiently as possible. If the ratio of accepted steps to total steps (acceptance ratio) is high, it means the moves are too small. Generally one desires a walk that minimizes the autocorrelation time for such variables as the local energy. The autocorrelation function of an observable is given by $\langle \mathcal{O}(t)\mathcal{O}(t+\tau)\rangle$ and the autocorrelation time is defined as the time τ, for which there is no correlation between $\mathcal{O}(t)$ and $\mathcal{O}(t+\tau)$ Very often this turns out to result in an acceptance ratio near 0.5. A common choice for T is the uniform move where a move is made anywhere within a "box" of dimension δ^d, where d is the dimension of the move.

Diffusion Monte Carlo (DMC)

Diffusion Monte Carlo[44] is a method that allows one to sample the *exact* ground state of a quantum many-body boson problem. As discussed below, this is because the ground state wavefunction is everywhere positive for bosons and thus can be used as a positive definite probability function. However, for other cases – a system of many Fermions or an excited state of any system – the wavefunction cannot be everywhere positive. This leads to the well-known "sign problem" in Monte Carlo methods.[1] We will first discuss the diffusion method ignoring the sign problem, and then address the complications introduced when treating fermions.

Consider the Schroedinger equation in imaginary time,

$$\frac{\partial |\Phi\rangle}{\partial t} = -\hat{H}|\Phi\rangle, \tag{12}$$

or in a real space representation,

$$\frac{\partial \Phi(\mathrm{R}, t)}{\partial t} = -\frac{\nabla^2}{2m}\Phi(\mathrm{R}, t) + V(\mathrm{R}, t)\Phi(\mathrm{R}, t), \tag{13}$$

where $\vec{\nabla}$ denotes the sum of laplacian operators for all degrees of freedom, $\sum_i^N \vec{\nabla}_i$, and R denotes the coordinates of all the particles. If $|\Phi, t=0\rangle$ is expanded in eigenfunctions of the Hamiltonian $|\phi_i\rangle$ with eigenvalues ϵ_i then $|\Phi\rangle$ evolves in time as

$$|\Phi, t\rangle = c_0 e^{-\epsilon_0 t}|\phi_0\rangle + c_1 e^{-\epsilon_1 t}|\phi_1\rangle + \dots . \quad (14)$$

For large time, t, the dominant component of $|\Phi\rangle$ will be the ground state $|\phi_0\rangle$,

$$|\Phi, \infty\rangle \cong c_0 e^{-\epsilon_0 t}|\phi_0\rangle, \quad (15)$$

since the ground state has the lowest eigenvalue ϵ_0. Further, if we shift the zero of energy by a carefully chosen constant E_T, the decay or growth of the amplitude may be controlled. For example, if

$$\frac{\partial |\Phi\rangle}{\partial t} = (-H + E_T)|\Phi\rangle, \quad (16)$$

then at long time t, the function is given by

$$|\Phi, \infty\rangle \cong c_0 e^{(E_T - \epsilon_0)t}|\phi_0\rangle. \quad (17)$$

The ideal choice for E_T is ϵ_0, the exact ground state energy. However, this is precisely what we will be interested in calculating, so it is unlikely we will know it *a priori*. We will consider the choice of E_T again when the details of the diffusion Monte Carlo algorithm are discussed.

The argument above demonstrates that we are interested in finding a quantum mechanical operator $\hat{G}(t', t)$ which projects an initial state $|\Phi, t\rangle$ forward in imaginary time to $|\Phi, t'\rangle$. This is known as the time propagator or alternatively as the Green's function for Eq. (13) The relationship between the propagator and the wavefunction is expressed below:

$$|\Phi, t'\rangle = \hat{G}(t', t)|\Phi, t\rangle. \quad (18)$$

Putting this into a real space representation and inserting a complete set of states, we have

$$\Phi(R', t') = \int dR\, G(R', t'; R, t)\Phi(R, t). \quad (19)$$

$\hat{G}(t', t)$ is well-known as $\exp[-\hat{H}\tau]$ where $\tau = (t' - t)$ so

$$G(R', R, \tau) = \langle R'|\exp[-\hat{H}\tau]|R\rangle. \quad (20)$$

Here $\hat{H} = \hat{T} + \hat{V}$, a sum of the kinetic and potential energy operators. The Green's function does not have a simple analytic form in the case of interacting particles($\tilde{V} \neq 0$). However, for small values of τ we may use the commutator expansion,

$$e^{-\hat{H}\tau} = e^{-\hat{V}\frac{\tau}{2}} e^{-\hat{T}\tau} e^{-\hat{V}\frac{\tau}{2}} + O(\tau^2), \quad (21)$$

to write an expression for the Green's function which is correct in the small τ limit:

$$G(R', R, \tau) \cong \langle R'|e^{-\hat{V}\frac{\tau}{2}} e^{-\hat{T}\tau} e^{-\hat{V}\frac{\tau}{2}}|R\rangle \quad (22)$$

$$\cong \int dR''' \int dR'' \langle R'|e^{-\hat{V}\frac{\tau}{2}}|R'''\rangle \langle R'''|e^{-\hat{T}\tau}|R''\rangle \langle R''|e^{-\hat{V}\frac{\tau}{2}}|R\rangle \quad (23)$$

$$\cong \int dP\, \langle R'''|P\rangle e^{-DP^2\tau} \langle P|R''\rangle e^{-(V(R)+V(R'))\frac{\tau}{2}} \quad (24)$$

Note that the last form of this equation uses a shorthand notation for clarity, where the operator P is the sum of momentum operators of the individual particles $P = \{p_i\}$ and the integral over the states $|P\rangle$ denotes a sum over the complete set of eigenstates of the momentum operator. The simplicity of this form results from the fact that V is diagonal in real space, i.e., $\langle R|\tilde{V}|R'\rangle = V(R)\delta(R'-R)$. Furthermore, from the identity $\langle R|P\rangle = e^{iP\cdot R}$ we find that the short time Green's function can be written

$$G(R', R, \tau) \cong \int dP\, e^{-iP(R'-R)} e^{-DP^2\tau} e^{-(V(R)+V(R'))\frac{\tau}{2}} \quad (25)$$

$$\cong e^{-\frac{(R'-R)^2}{4D\tau}} e^{-(V(R)+V(R'))\frac{\tau}{2}}, \quad (26)$$

with $D = \frac{\hbar^2}{2m}$. Now, we may rewrite Eq. (19) as

$$\Phi(R', t+\tau) = \int dR\, e^{-\frac{(R'-R)^2}{4D\tau}} e^{-(V(R)+V(R'))\frac{\tau}{2}} \Phi(R, t) + O(\tau^2) \quad (27)$$

We can understand this equation as an iterative form for the evolution of the wavefunction in imaginary time. The kinetic energy term is exactly what one would expect from a pure diffusion equation (i.e. $\hat{V} = 0$) and the potential energy term corresponds to a weighting, or branching. The former term leads to diffusion, i.e., a spreading of a function as imaginary time increases. The latter term is analogous to a rate process which increases the function in regions where the potential energy is low and decreases the function where it is high. The combination of these effects leads to a stable function in the long time limit.

To simulate this equation on a computer, we begin with an ensemble of configurations distributed according to $\Phi(R, t = 0)$. Each configuration is simply a specific set of particle coordinates $\{r_i\}$. A new configuration R' is chosen with the probability $e^{-\frac{(R'-R)^2}{4D\tau}}$ and is then weighted according to $e^{-(V(R)+V(R'))\frac{\tau}{2}}$. As described later, it is more efficient on a computer not to weight configurations, but rather to "kill" or "multiply" configurations according to a random algorithm that will preserve the balance over a long simulation. This is called "branching" and it is this step that leads to more probable inclusion of configurations of low energy and exclusion of those of high energy, so that the algorithm in principle will lead to the *exact* energy. The time associated with the configuration R' is incremented by τ, and after a sufficient number of steps we would approach the large time limit distribution which, according to our previous argument, is the ground state of the operator $\hat{H}$. What we have done, is build up the projection forward in time by using successive jumps in small time steps. Unfortunately, in most cases of interest, the approach described thus far is not sufficient because the fluctuations are too large for these expressions to be used directly. This is because $V(R)$ often has large variations. For example, it diverges when two charged particles approach each other. To get around this difficulty, we reintroduce the concept of importance sampling as discussed in the following section.

Importance Sampling

The ideas of importance sampling can be used at this point to rewrite the evolution equation for the wavefunction in imaginary time in a form in which there is less variation in the integrand, i.e., to remove large variations in the potential energy function in the weighting factors in Eq. (27). Consider multiplying Eq. (19) on the left by a guiding function $\Psi_G(R')$ and inserting $\Psi_G^{-1}(R)\Psi_G(R)$ in the integral. Then we have

$$\Psi_G(R')\Phi(R', t+\tau) = \int dR\, \Psi_G(R')G(R', R, \tau)\Psi_G^{-1}(R)\, \Psi_G(R)\Phi(R, t). \quad (28)$$

With few exceptions[45], the guiding function, $\Psi_G(R)$, is the same as the trial wavefunction, $\Psi_T(R)$, used in variational calculations. We can simplify notation in the following discussion by using $\Psi_G(R) = \Psi_T(R)$ exclusively. This may be interpreted as an evolution equation for $f(R,t) = \Psi_T(R)\Phi(R,t)$ with the Green's Function $\tilde{G}(R',R,\tau) = \Psi_T(R')G(R',R,\tau)\Psi_T(R)^{-1}$. Multiplying Eq. (13) by $\Psi_T(R)$ generates an equation which describes the time development of $f(R,t)$,

$$-\Psi_T(R)\frac{\partial \Phi(R,t)}{\partial t} = -D\Psi_T(R)\nabla^2 \Phi(R,t) + V(R)\Psi_T(R)\Phi(R,t). \tag{29}$$

If we make use of the following identity,

$$\Psi_T \nabla^2 \frac{f}{\Psi_T} = \vec{\nabla} \cdot \left[\vec{\nabla}f - 2f\frac{\vec{\nabla}\Psi_T}{\Psi_T} + f\frac{\nabla^2 \Psi_T}{\Psi_T}\right] \tag{30}$$

we can rewrite Eq. (29) as

$$-\frac{\partial f(R)}{\partial t} = -D\nabla^2 f + D\vec{\nabla} \cdot 2\left(f(R)\frac{\vec{\nabla}\Psi_T(R)}{\Psi_T(R)}\right) + ((E_L(R') + E_L(R))/2 - E_T)f. \tag{31}$$

Here $E_L(R)$ is the local energy $H\Psi/\Psi$ defined previously in Eq. (29) and the energy has been shifted by the constant E_T. This looks like Eq. (13) with $V(R)$ replaced with $E_L(R)$ and an additional term $2D\vec{\nabla} \cdot (f\frac{\vec{\nabla}\Psi_T}{\Psi_T})$. As we will see, this term acts like a drift force on the configuration R. A proper choice of guiding function $\Psi_T(R)$ will make $E_L(R)$ a smooth function with no singularities. To solve for the Green's function $\tilde{G}$, let us consider the operator $\tilde{H}$

$$\tilde{H} = \underbrace{-D\nabla^2 + D\vec{\nabla} \cdot \vec{F}(R)}_{\tilde{T}} + \underbrace{((E_L R') + E_L(R))/2 - E_T)}_{\tilde{V}}, \tag{32}$$

where $\vec{F}(R) = \frac{\vec{\nabla}\Psi(R)}{\Psi(R)}$. Following the previous procedure, we may write an expression for $\tilde{G}$ as follows:

$$\tilde{G}(R',R,\tau) = \langle R'|\exp[-\tilde{H}\tau]|R\rangle \tag{33}$$

To solve for $\tilde{G}$ we assume that the displacement allowed in R for short time steps τ is small enough that $\vec{F}(R)$ is fairly constant and we may move it to the left of the gradient. As before, we write

$$\langle R'|\tilde{G}|R\rangle = \langle R'|e^{-\tilde{V}\frac{\tau}{2}}e^{-\tilde{T}\tau}e^{-\tilde{V}\frac{\tau}{2}}|R\rangle + O[\tau^3] \tag{34}$$

To evaluate $\tilde{T}$, we again insert a complete set of momentum eigenstates $|P\rangle$. $\tilde{V}$ is diagonal in the real space representation and using $\langle R''|R\rangle = \delta(R'' - R)$ and $\langle R|P\rangle = \exp[-iP \cdot R]$ we have,

$$\tilde{G}(R',R,\tau) = \frac{1}{(4\pi D\tau)^{3/2}} e^{-((E_L(R)+E_L(R')+E_T)\tau} e^{-(R'-R-D\tau\vec{F})^2/4D\tau} \tag{35}$$

We have arrived at an analytic expression which approximates the Green's function to Eq. (29) for small τ. In addition, Eq.(28) demonstrates how to generate $f(R', t+\tau)$ given $f(R,t)$.

Now we turn to the question of how this diffusion procedure can be implemented as a practical algorithm. Consider $f(R,t)$ as a probability distribution for the configurations

R. $\tilde{G}(R', R, \tau)$ may be thought of as a transition probability to R' given R. Each time step τ will involve sampling R' from a gaussian centered about the drifted position $R - D\tau \vec{F}(R)$ and a weighting term proportional to $\exp[-(E_L(R') + E_L(R))/2 + E_T]$. The three fundamental steps of the diffusion Monte Carlo algorithm are drift, diffusion, and weighting (or branching). These steps are repeated until $f(R,t)$ has reached its asymptotic state which is

$$f_\infty(R) = f(R, \infty) = \Psi_T(R)\phi_0(R). \tag{36}$$

Note that although it follows from Eq. (17) that $\phi(R)$ is the exact ground state wavefunction, nevertheless the function which has been derived, $f_\infty(R)$, is a mixed function involving the trial function and the exact ground state wavefunction.

Despite the fact that the method just described leads to a mixed function f, it is straightforward to show that it may be used to derive the *exact* value for the most important physical observable, the ground state energy ϵ_0. The proof is quite simple and has been given in various places such as the text by Koonin.[13] The ground state energy may be expressed as

$$\epsilon_0 = \frac{\int dR \phi_0(R) \hat{H} \phi_0(R)}{\int dR \phi_0(R) \phi_0(R)} \tag{37}$$

$$= \frac{\int dR \phi_0(R) \hat{H} \Psi(R)}{\int dR \phi_0(R) \Psi(R)}, \tag{38}$$

where the second equation follows from the fact that the Hamiltonian can act to the left of the exact ground state to give the exact energy. However, this same expression can be used with the Hamiltonian acting to the right. If we also multiply and divide by the trial function, then we arrive at the desired expression:

$$\epsilon_0 = \frac{\int dR \Psi(R) \phi_0(R) [\hat{H}\Psi(R)/\Psi(R)]}{\int \Psi(R) \phi_0(R)} \tag{39}$$

$$= \frac{\int dR f_\infty(R) E_L(R)}{\int dR f_\infty(R)}. \tag{40}$$

Finally, Eq. (39) may be rewritten in the spirit of Eq.(2), as an estimate based upon a Monte Carlo sampling method:

$$E_0^N = \frac{1}{N} \sum_{\substack{i=1 \\ R \in f(R)}}^{N} E_L(R_i). \tag{41}$$

where R_i are chosen from the distribution $f(R, \infty)$. Thus we have derived an expression for the exact energy written in terms of an average over the local energy $E_L(R_i)$ where the points R_i are configurations distributed according to the mixed distribution f. This is a remarkable result since this expression, as well as the algorithm for finding the distribution f, involves only the local energy which is determined *strictly by the trial function*; yet it results in the *exact* energy. For a sampling of N configurations chosen randomly, Eq. (41) gives an estimate of the exact energy and the variance provides an estimate of the accuracy of this estimate. The expectation value of any operator that commutes with the Hamiltonian also may be calculated in this way.

For operators, $\hat{O}$ that do not commute with the Hamiltonian, there is no analogous expression giving an exact expectation value. Nevertheless, it is possible to approximate the expectation value of $\hat{O}$ in the ground state through a combination of the variational and mixed distribution estimate. The mixed distribution estimate is the result of averaging over the mixed distribution $\Psi_T \phi_0$ in the DMC algorithm. If we assume that the trial and guiding wavefunction Ψ_T is equal to the ground state ϕ_0 plus a small orthogonal function one can derive approximate forms for the expectation values. We will not go into such expressions here, but we note that they are needed for such interesting properties as the kinetic and potential energies, radial distribution functions, etc.[15],

Instead of a weighting function that weights each configuration in terms of the local energy as described above, it is more convenient to carry out the simulation with "branching", i.e., eliminating or multiplying confiugurations. The branching probablitity is calculated as described below in the practical algorithm. Each configuration is carried forward independently and by adjusting the trial energy E_T a steady state population of configurations is maintained.

When $\Psi_T \to \phi_0$, the branching term disappears since the local energy $E_L(R)$ would be the constant E_0. In this case, we would like our Green's function to sample the exact ground state regardless of the time step τ. One approach is to include a Metropolis-like step where a move from $R \to R'$ is accepted with probability $A(R \to R') = min[1, W(R', R)]$ where

$$W(R', R) = \frac{|\Psi_T(R')|^2 G(R' \to R, \tau)}{|\Psi_T(R)|^2 G(R \to R', \tau)}. \tag{42}$$

Expanding the exact Green's function from Eq.(33) in eigenfunctions of $\tilde{H}$, we have

$$\tilde{G}(R', R, \tau) = \sum_\alpha \Psi_T(R') \phi_\alpha(R') e^{(E_\alpha - E_T)\tau} \phi_\alpha(R) \Psi_T^{-1}(R) \tag{43}$$

Putting this expression for the exact Green's function into Eq. (42), we find $W(R', R) = 1$, as expected. This just means that when $\tilde{G}$ is exact, all moves are accepted. When $\tilde{G}$ is not exact, the metropolis rejection step ensures that when the trial wavefunction Ψ_T approaches the exact ground state, the detailed balance condition is satisfied and that the stable asymptotic distribution is the ground state $|\phi_0|^2$. We may also note here that without branching, this step will ensure that the trial function $|\Psi_T|^2$ is sampled. This demonstrates a sampling procedure for VMC and makes clear the close relationship between VMC and DMC. As we discuss below, the difference in the algorithms is that DMC includes the branching step whereas VMC does not.

A Practical Algorithm for Diffusion Monte Carlo

We may now list the steps of a practical diffusion Monte Carlo algorithm. These steps have been described in the paper by Reynolds, et.al.,[44] and here we give a closely related description. Recall that Eq.(25) describes the time evolution of the mixed distribution function $f(R, t)$, which we interpret as a probability distribution function for R. The function, $f(R, t)$ is represented on a computer as an ensemble of configurations $\{R\}$. This ensemble typically consists of 100 to 1000 configurations. In principle, any reasonable choice of $f(R, 0)$ is acceptable to begin a simulation. In practice, $|\Psi_T(R)|^2$ is used. An ensemble of configurations $\{R\}$ distributed according to $|\Psi_T(R)|^2$ is easily generated by an equilibrated variational calculation using $\Psi_T(R)$ as the trial wavefunction. The diffusion algorithm for bosons, where $\Psi_T(R)$ is positive everywhere, proceeds as follows:

1. Begin with an ensemble of N_w configurations, or "walkers", distributed according to $|\Psi_T(R)|^2$.

2. Choose one walker R_i from the ensemble.

3. Move an individual particle j of walker R_i from position r_j to r_j' using the expression

$$r_j' = r_j + D\tau F_j(R) + \chi \qquad (44)$$

where F_j is the drift force on particle j defined earlier and χ is a three-dimensional gaussian variable with mean equal to zero and variance equal to $2D\tau$. This represents the drift and diffusion parts of the Green's function Eq.(25).

4. Accept the move with probability $A(R \rightarrow R', \tau)$, as defined by Eq.(42).

5. Repeat the drift and diffusion move for each of the particles ($j = 1, N$) in the walker R_i.

6. Assign τ additional time units to the new configuration R' and calculate quantities of interest such as $E_L(R')$.

7. Calculate the multiplicity of the new configuration R' using the branching term,

$$M = \text{int}\left[\left(\frac{N_w}{N_c}\right)^{\tau_c} \exp\left[-\left(\frac{E_L(R) + E_L(R')}{2} + E_T\right)\tau\right] + \zeta\right], \qquad (45)$$

where ζ is a random variable between 0 and 1, and N_c is the present number of walkers. The factor τ_c determines the sensitivity of the feedback on the population variation. When the population N_c is greater than the desired number of walkers N_w, we wish to lower the multiplicity. The term $(\frac{N_w}{N_c})^{\tau_c}$, can be brought into the exponential as $\tau_c \log(\frac{N_w}{N_c})$, where it is more readily seen as an adjustment to the trial energy. Place M copies of the configuration R' at the bottom of the ensemble list. Weight the calculated physical observables by M.

8. When all the particles in a configuration have been moved, re-estimate E_T by setting it equal to the cumulative average of the local energy $E_L(R)$. Choose the next configuration and go to step(2). Repeat this procedure until a steady state is reached for quantities like $\langle E_L(R) \rangle$. When the asymptotic steady state is reached, the algorithm generates configurations R distributed according to the mixed distribution $\Psi_T(R)\phi_0(R)$.

We may also note here that if one omits the branching step, then the algorithm leads to sampling of the original trial function $|\Psi_T|^2$. So in effect, it demonstrates a sampling procedure for VMC which is in fact used in actual calculations. It also makes clear the close relationship between VMC and DMC. The difference is that DMC includes the branching step and VMC does not. It is this branching step that favors low energy configurations and reduces the weight of high energy configurations that generates the improvements in the DMC sampling over that of the VMC approach. For bosons the improvement leads to the exact ground state energy (and other expectation values), but for Fermions there are additional problems.

The Fixed-Node Method for Fermions

One of the fundamental assumptions of the diffusion Monte Carlo algorithm is that the mixed distribution may be interpreted as a probability distribution. This implies that $\phi_0(R)\Psi_T(R)$ is greater than zero over the space of possible R. For bosons, this is not a problem since the exact bosonic ground state $\phi_0(R)$ has no nodes and any reasonable guiding function, $\Psi_T(R)$, would be constructed without nodes also. (This is trivial to accomplish in practice and was, of course, present from the very beginning in Monte Carlo studies of boson systems.[27]) However, for fermions, where spin statistics demand that the wavefunction be antisymmetric under exchange of identical particles, the ground state wavefunction, $\phi_0(R)$, must have both positive and negative regions. This is the famous "sign problem" which is the bane of all QMC calculations for Fermion systems with more than one dimension.[1]

To continue to interpret the mixed distribution $f(R,t)$ as a probability, the positive regions of $\Phi(R)$ and $\Psi_T(R)$ must overlap. The same is true for negative regions. It is clear that if $\Psi_T(R)$ had the exact nodes of the ground state $\phi_0(R)$ we could proceed with the algorithm as before with no caveats. However, one typically does not know the nodal boundaries of the ground state a-priori.

The fixed-node approximation for treating Fermions was first introduced by Anderson [48] and extensively used in later work. [49] [44] This is an easily implemented technique which assumes that $\phi_0(R)$ has the same nodes as $\Psi_T(R)$. In essence this assumption means that the state $\Phi(R,\infty)$ will be the bosonic ground state (i.e., will have no nodes) in each nodally bound region. This approximation provides an upper bound estimate of the ground state energy. Typically, the error incurred through the use of the fixed-node approximation is about 10% of the difference between variational and diffusion results[64]. To demonstrate this variational property of the fixed-node approximation, consider the bosonic ground state solution for any nodally bounded region V_α. Let $\Phi_\alpha(R)$ be the solution in this region with no nodes, with $\Phi_\alpha(R)$ defined to vanish everywhere outside this region. Using this solution, we may construct a properly antisymmetrized wavefunction φ by permuting identical particles and multiplying by -1 for odd permutations,

$$\varphi(R) = \sum_P (-1)^P \Phi_\alpha(PR) \tag{46}$$

This state is defined throughout all space, and its energy ϵ_α is a variational estimate of the exact ground state energy.

The fixed-node condition is exceedingly simple to implement. In step (2) of the DMC algorithm given above, where R' is chosen, if the sign of $\Phi(R)\Psi_T(R)$ is not equal to $\Phi(R')\Psi_T(R')$, then the move to R' is rejected and the walker is simply returned to the list. This procedure was shown to give the correct statiscal average by Umrigar, et. al. [65]

Beyond the fixed-node approximation, there are other methods such as transient estimate[49] and release node[45] which promise a truly exact Fermion ground state energy. However, these methods are numerically unstable and give expectations values that grow increasingly noisy with successive generations. The error increases as the exponential of the energy difference between the Fermi and Bose ground state energies. Therefore these methods are difficult to apply and have been carried out only in certain cases, such as the homogeneous electron gas[49] or in small systems.[45] Recently Zhang and Kalos[46][47] have proposed a method for treating the sign problem exactly but it is limited to systems of very few fermions.

The other commonly used exact ground state method, GFMC, will not be discussed here. It avoids the short time approximation inherent in DMC by constructing an algorithm for the exact Green's function. Valuable references for this approach include Ceperley[66] and Lee[51].

As we have seen, the Monte Carlo method produces estimates of the ground state properties in the quantum many-body problem. The algorithms are relatively simple to implement on a computer and do not require excessive amounts of memory. Continued advances in computational capability and improvements in algorithmic efficiency will make progress possible on a wide variety of previously intractable problems of physics and chemistry.

THE HOMOGENEOUS ELECTRON GAS IN 2 AND 3 DIMENSIONS

One of the most important applications of quantum Monte Carlo to Fermion systems is the ground state energy of the homogeneous interacting electron gas as a function of density. The first Monte Carlo calculations[35] for 2-dimensional and 3-dimensional gasses were variational calculations based upon a trial function of the Slater-Jastrow type,

$$\Psi(\mathbf{r}_1,\ldots,\mathbf{r}_N) = \exp\left[-\sum_{i\leq j} u(r_{ij})\right] D(\mathbf{r}_1,\ldots,\mathbf{r}_N). \tag{47}$$

Here the Slater determinant (really a product of two determinants of up and down spin electrons) ensures the antisymmetry of the wavefunction, and two-body correlations are included in the symmetric Jastrow factor which involves the correlation functions (or "pseudopotentials") $u(r_{ij})$. In the homogeneous system the determinant is composed of products of plane waves with wave vectors k_i chosen to satisfy the boundary conditions (periodic) and fill the Fermi sea, i.e., occupy k states below the Fermi K_F. Several different forms for the pseudopotentials have been used and optimized variationally; remarkably, those derived simply from the RPA[67] were shown to be quite accurate.[35] The quality of the RPA form presumably comes from the fact that it gives exactly both the small r (cusp conditions) and large r (plasmon) behaviors. The critical aspects of these variational calculations can be found in the original paper[35] and also in more recent papers using the variational approach. In particular, the recent paper of Pickett and Broughton[68] reports a completely independent variational study of the 3-dimensional homogeneous gas. Also many aspects of variational calculations have been discussed by Fahy, et.al.,[38] although these authors considered inhomogeneous cases.

The most exact calculations on homogeneous systems have been reported based upon calculations using the fixed-node DMC method, with a release node approach to establish the final results. For the 3-dimensional case, the calculations of Ceperley and Alder[49] set the standard for the field by determining the correlation energy to great precision. Their energies have been the basis for current density functional calculations, since the forms used have been fit to the Ceperley-Alder results. In addition, this work provided rigorous tests of approximate methods over a large range of densities. This work established the most definitive results thus far for the density at which the homogeneous gas becomes unstable to formation of a Wigner crystal, above $r_s \approx 100$.

In addition, Ceperley and Alder found a small range of densities for which a polarized Fermi fluid was found to be more stable than either the unpolarized fluid or the Wigner crystal. Recently, another Diffusion Monte Carlo study of the 3-dimensional gas has been reported by Ortiz and Ballone.[69][70] This work concentrated upon the density range where the fluid phase is stable (r_s from 1 to 10) and calculated the energies of systems with partial spin polarization. These calculations used different boundary conditions (an fcc supercell) and carefully analyzed the size dependence and the extrapolation to infinite size. Finally, Ortiz and Ballone have reported the most detailed study thus far on the pair correlation functions. Together with the original work of Ceperley and Alder, these independent studies provide an in-depth investigation of the 3-dimensional homogeneous gas with a careful investigation of the possible sources of error.

The 2-dimensional electron gas is of current interest in large part because of experiments on actual systems realized in semiconductors. The unique features include interesting transport, the quantum Hall effect, and possible Wigner crystalization of the electrons at low densities. In addition, there are other 2-dimensional problems such as high temperature superconductivity, which apparently is closely related to the planar nature of the copper oxide superconductors.

Tanatar and Ceperley[54] carried out a detailed study in which they determined the correlation energy and investigated the low density regime where the Wigner transition occurs. Recently, Kwon, Ceperley, and Martin[43] have carried out a new study with an improved form for the variational trial function. These workers used a "backflow" correlation of the form suggested by Feynman and Cohen,[71] and previously used in 3-dimensional He.[72][56] There are two aspects of this work which are important for the present discussion. First, the backflow form of the function changes the nodes of the trial wavefunction. This is accomplished because the positions in the Slater determinant are not the electron positions but modified positions that depend upon the other particles in a surrounding region, with a form that leaves the entire wavefunction antisymmetric.[71][43] This is particularly important for our present considerations because only by changing the nodes is the DMC energy improved. Therefore this work gives some ideas about the accuracy of the typical fixed node results and how much they can be improved. Second, it is found that the variational energy is significantly improved, capturing most of the correlation energy that was found by the best previous DMC calculations. Since it is a primary goal of many-body calculations to determine the most important correlations between the electrons, this is a significant result in itself.

Here we mention selected results from Ref. [43] to show the magnitude of various contributions to the energies in the 2-dimensional gas, as well as the accuracy of different levels of Monte Carlo calculations. The Hartree-Fock energy (in Rydbergs) is given by $E_{HF} = 1/r_s^2 - 1.2004/r_s$. For comparison, the RPA energy which includes HF energy, the summation of the ring diagrams, and the second-order exchange diagram, is given in a high density limit by $E_{RPA} = E_{HF} - 0.38(4) - 0.172 r_s log(r_s) + O(r_s)$. This is valid only for small r_s, giving $E_{RPA} = -0.58(4)$ at $r_s = 1.0$ and no valid result for the larger values of r_s. In Table 1 we give the comparison of Hartree-Fock, Slater-Jastrow, and the improved wavefunction with both backflow and 3-body correlations. As shown in Ref. [43] backflow is much more important than 3-body correlation at high density, whereas they are of about equal importance at the largest r_s studied. As we can see from the variational calculations reported in the table, the 2-body correlation in the Slater-Jastrow function is responsible for the largest improvement over Hartree-Fock.

Table 1. Variational and fixed-node GFMC energies with various trial wave functions for $N = 58$ in Ry per electron. In the last line is shown the most exact DMC result extrapolated to infinite size from several finite size calculations. All results are from Kwon, et.al. The abbreviations denote different trial functions: S.J., the Slater-Jastrow function; 3BD, three-body correlation, and BF, backflow correlation.

Method	ψ_{trial}	$r_s = 1.0$	$r_s = 5.0$	$r_s = 10.0$	$r_s = 20.0$
HF		-0.2004	-0.2001	-0.1100	-0.0575
VMC	S.J.	-0.3879(2)	-0.2936(1)	-0.16837(2)	-0.09164(1)
	S.J.+3BD	-0.3894(5)	-0.2947(1)	-0.16895(2)	-0.09195(2)
	S.J.+BF	-0.4024(5)	-0.2972(2)	-0.16962(2)	-0.09199(2)
	S.J.+3BD+BF	-0.4029(5)	-0.2976(1)	-0.17000(2)	-0.09225(2)
DMC	S.J.	-0.4043(5)	-0.2980(1)	-0.17037(2)	-0.09248(1)
	S.J.+3BD+BF	-0.4087(2)	-0.2991(1)	-0.17086(1)	-0.09265(1)
DMC	S.J.+3BD+BF	-0.4195(6)	-0.2990(2)	-0.17071(4)	-0.09258(2)

Nevertheless, backflow and 3-body effects are important: In fact, they give a variational energy which captures most of the correlation energy beyond the Slater-Jastrow energy, as may be seen by comparing with the DMC energy, given in the table. Therefore, this is a significant improvement in the variational energy which is very important for the goal of finding accurate energies from the variational method. We also see that the changes in the nodes due to the backflow function leads to significant changes in the fixed-node DMC energy. This change in DMC energies, however, is quite small in comparison to the change in variational energies, as expected.

The last row of the table shows the energies extrapolated to infinite size from the values calculated for five different sizes ranging up to 114 particles. The form of the corrections for size dependence is very similar to that discussed later in the section on hydrogen. From the change between the energies for N=58 and the extrapolated value, we see that finite size corrections are quite important, and must be done carefully to have accurate results for the thermodynamic limit.

Finally, the DMC results were compared with energies found previously by different methods.[73, 74, 75, 76] The comparisons are shown here in Fig. 1 taken from Ref. [43]. We see overall agreement but important differences among the methods. The solid line in the figure is a Pade approximant fit to the DMC data that is constrained to have the correct high density limit. It turns out that the extrapolated form at low density is also close to the expected form for the Wigner crystal, even though the calculations were not done at such large r_s. Thus we believe that the analytic fitted form for the correlation energy[43] is quite accurate for a large range of r_s.

THE METALIZATION OF HYDROGEN AT HIGH PRESSURES

It was first pointed out by Wigner and Huntington[77] that at sufficiently high pressures, hydrogen should dissociate from its molecular state to form an atomic metal. Metallic hydrogen has been the focus of much theoretical and experimental work[78] because of its special role as the simplest of all metals and its importance in the evolution and composition of heavy planets. Although the structure of metallic hydrogen plays a crucial role in determining whether it is a metal or an insulator, there is no consensus on any characteristics of the structure. There have been many diverse predictions, with some studies supporting low-coordination anisotropic structures[79][80][79][81][82], others concluding that high-coordination isotropic structures are favored[36][83], and yet others considering the possibility of a quantum liquid state[84]. The main reason for these differences lies in the difficulty in treating the large zero point motion of the protons and the effect of electronic correlation near the atomic-molecular transition.

Ceperley and Alder[36] were the first to carry out quantum Monte Carlo(QMC) calculations on hydrogen under pressure. They included both electrons and protons as quantum particles; however, they only considered high symmetry structures. In more

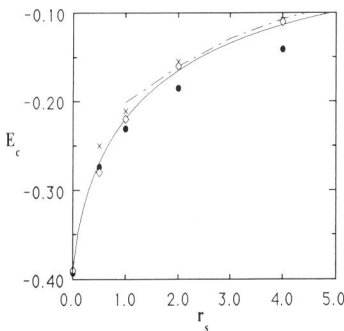

Figure 1. The correlation energy, E_c in units of Ry per electron, as a function of the density parameter r_s. The solid line shows our MC result and the dashed curve is that of Sim, et.al.; crosses represent the STLS calculation of Jonson. Also shown are the results of Freeman by the coupled-cluster summation in the ring approximation (filled circles) and in the ladder approximation (diamonds).

recent work, we have also considered low symmetry structures and have shown that hydrogen is indeed expected to form a sequence of complex structures as a function of pressure.[15, 57] This latter work also involved LDA calculations so that detailed comparisons with LDA results [80, 85] could be made.

The QMC calculations involve a trial function of the Slater-Jastrow form given above with correlation functions between all particles, electron-electron, proton-proton, and electron-proton, and the electrons described by orbitals in the Slater determinant. The protons are treated as distinguishable particles, which is accurate, except for very small energies (which are important for molecular phases[15] that we will not discuss here). The pair correlation functions derived within the RPA were found to work well for metallic hydrogen at high pressure,[36] similar to the previous work on the homogeneous gas.[35]

Ceperley and Alder[36] assumed the single body wavefunctions $\phi_k(r_l)$ were simple

Table 2. The static and dynamic LDA, VMC and DMC energies, for the different crystal structures. Also shown are the zero-point energy and the pressure calculated from the virial theorem.

Structure	E_{LDA}(Ry)	E_{VMC}(Ry)	E_{DMC}(Ry)	E_{ZPE}(Ry)	P(Mbar)
			static		
Bcc	-1.01392	-1.0062(3)	-1.0101(3)		2.53(4)
Sim. Cub.	-1.01898	-1.0144(4)	-1.0185(3)		2.85(3)
Sim. Hex.	-1.02406	-1.0275(4)	-1.032(1)		3.30(2)
Diamond	-1.02413	-1.0204(4)	-1.0235(2)		3.27(4)
β-Sn	-1.02711	-1.022(1)	-1.0256(7)		2.89(4)
			dynamic		
Bcc		-0.9716(1)	-0.9810(7)	0.0291(7)	2.89(3)
Sim. Cub.		-0.9747(1)	-0.9873(3)	0.0312(3)	3.11(2)
Sim. Hex.		-0.9828(2)	-0.988(1)	0.044(1)	3.54(1)
Diamond		-.9821(2)	-0.993(1)	0.0305(1)	3.47(4)
β-Sn		-0.981(1)	-0.992(2)	0.036(2)	3.33(9)

plane waves $e^{ik \cdot r_l}$ with $|k| < k_f$, whereas Natoli, et. al.,[57] used single body wavefunctions from the self-consistent solutions to the Kohn-Sham equations in the LDA, as was done in a previous VMC study of bcc hydrogen[37]. These functions may be represented as,

$$\phi_{k_l}(r_m) = \sum_G c_{Gk_l} e^{i(k_l+G) \cdot r_m} \qquad (48)$$

where G are the reciprocal lattice vectors of the unit cell. Since the LDA wavefunctions include electron-proton correlations, the electron-proton pair potential term in Eq.(1) is modified.[57] In addition, the occupied orbitals are taken to be those given by LDA, which are different from the spherical Fermi surface used by Ceperley and Alder. Improvement in the trial function is important because it allows more rapid convergence of the diffusion calculation and because the nodes of the wavefunction are determined by the Slater determinant and are not improved by the fixed-node DMC method.

The most serious systematic problem in the calculations is the extrapolation to infinite size cells. Since the occupied k-states of the single body wavefunctions change with the size and shape of the periodic cell, it is essential to carefully extrapolate our calculations for systems of finite size to get a result applicable to the infinite system. This is particularly important in a metal because the delocalized electronic wave functions are sensitive to the boundary conditions. For each separate crystal structure Natoli, et. al. fitted finite size calculations (which range from eight to 432 atoms) to a form similar to that used in the homogeneous case:

$$E_N = E_\infty + c_1 \left(E_\infty^{lda} - E_N^{lda} \right) + \frac{c_2}{N} \qquad (49)$$

Good quality fits were obtained in all cases by varying E_∞, c_1, and c_2.

Table 2 shows the results of our total energy calculations on the static and dynamic lattice at one value of the density $r_s = 1.31$, which was chosen because this is the region in which Ceperley and Alder predicted the transition to a metallic state. For the static lattice, VMC gives energy differences between structures in excellent agreement with the DMC results while LDA does not. Since the difference in energy between E_{VMC} and

E_{DMC} is a measure of the quality of the variational wavefunction, it is apparent that the accuracy is comparable in all the static structures. However we see that full DMC calculations are needed to establish the dynamic crystal structure of atomic hydrogen at this density since VMC makes errors as large as 13 mRy/atom. The results for the static lattice indicate that the simple hexagonal lattice is the most stable at this density and that in general, anisotropic structures with lower coordination are preferred.

The zero point motion in this problem is extremely important. Its magnitude is larger than the differences in static energies discussed above, and it has been proposed by Straus and Ashcroft[83] that the dependence on structure will be great enough to ultimately favor isotropic structures. A proper treatment of the lattice degrees of freedom is critical to obtaining accurate results for the real crystal. Since the protons and electrons equilibrate on very different time scales, including protonic degrees of freedom slows the convergence by a factor of about 10. To avoid many costly calculations, one DMC calculation was done on a moderately sized system ($\approx$ 60 atoms) for each structure and the result was extrapolated to the infinite bulk using the VMC fitting parameters. Table I shows the results of our VMC and DMC calculations on the dynamic lattice. We define the zero-point energy(ZPE) to be the difference between the static and dynamic diffusion energies. For these structures, in general, lower coordination leads to higher ZPE as was discussed by Straus and Ashcroft[83] and Ceperley and Alder[36]. Diamond, however, has amongst the lowest ZPE so that its final energy, including all contributions, is lowest of all the structures considered.

The statistical errors for all calculations were found to be <1 mRy/atom. There are systematic errors which arise from the finite time step, which Ceperley and Alder estimated to be on the order of 1 mRy/atom. The fixed node errors were estimated to be approximately 0.5 mRy/atom for the static lattice and 1.5 mRy/atom for the dynamic lattice, based on results of "released node" calculations on the homogeneous electron gas.[64] These estimates of the absolute accuracy sets the scale of confidence for the predictions for hydrogen under pressure. As one can see from Table 2, these estimates suggest that we can confidently conclude that the ordering of energies is correct for these structures; however, of course, we cannot say anything about other possible structures which we have not considered.

In order to determine the stable structure as a function of pressure, one should consider the enthalpy $H = E + PV$ vs pressure P. Since pressure can be calculated from the ground state wavefunction, it follows that this can be done directly from the QMC calculations. In Fig. 2 we show the results reported in Ref. [57][15], which indicate that at pressures where hydrogen should transform from molecules to non-molecular phases, it is predicted to transform gradually from 4- to 6- to higher-coordination structures as the pressure is increased. Only above $\approx$ 10 Mbar, does it finally stabilize in a close packed strucure. Of course, these comparisons are only for the structures considered; we cannot rule out other structures which have not been tested.

Further discussion of the nature of hydrogen at high pressure and calculations on the molecular phases are given by Natoli, et. al.[15, 57]. For our purposes here, the primary conclusion is that it is feasible to carry out QMC calculations with such accuracy - including all statistical and estimated systematic errors - that predictions can be made for complex real materials. As shown above hydrogen turns out to be extremely complex and the requirements are quite severe for theory to be able to distinguish between many structures of nearly the same energy.

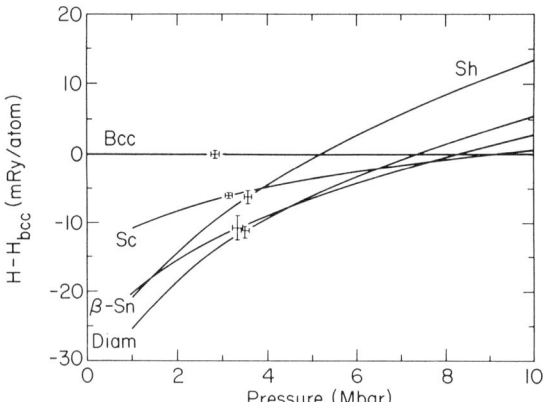

Figure 2. Enthalpy difference from bcc vs. pressure for the five dynamic crystal structures considered. The points with error boxes are derived from the DMC calculations done at $r_s = 1.31$ and the lines indicate extensions to other densities using LDA and a functional form for the zero-point energy as described in the text. Molecular structures not considered here are stable at pressures below approximately 3Mbar.

THE CORE PROBLEM

Although core electrons are relatively inert, their presence has crucial effects upon the active valence electrons. The core states are problematic in sampling methods because their characteristic energy scales are large and time scales small; in DMC this causes an increase in computational time to achieve a given accuracy for the total energy which scales as $Z^{6.5}$, where Z is the atomic number.[86] The reasons for this rapid increase with Z is that the time scales as the cube of total number of electrons Z^3, with additional factors of Z^2 because the step size in the simulation must be decreased and $Z^{1.5}$ because of the increase in total energy. For this reason, direct calculations are not feasible even for heavy atoms, let alone molecules or solids! One approach involves full many-body calculations on the valence electrons only, with effects of the core electrons replaced by a pseudopotential[87] (PP) or pseudohamiltonian[86, 88] (PH). This breaks the problem into two parts: a) finding accurate pseudopotentials and b) simulating valence electrons moving in the non-local potentials which are more complex than ordinary local pseudopotentials. Fahy, Wang, and Louie[38] have carried out non-local pseudopotential calculations using VMC methods, with remarkable agreement with experiment for the cohesive energy of carbon and silicon. Li, et.al.,[59] have carried out DMC calculations with a PH and shown that it is indeed possible to calculate the total valence energy to a precision of order 0.05eV per atom. This is sufficient for many real problems and leads to properties of Si in good agreement with experiment, including improvement of the well-known errors in the cohesive energy found in the local density approximation (LDA).

The second part of our work addresses the question: how can one generate a pseudopotential or pseudohamiltonian that treats the effects of the cores with sufficient accuracy that they are worthy to be used in such accurate valence calculations. Progress in this direction has been made in recent work,[41, 42] which has shown how to estimate the non-local term from a variational function while treating the local terms in the full fixed-node DMC. We also must face the fact that up to now pseudopotentials have been generated only in approximate one-electron methods such as LDA or Hartree Fock.[87] Any errors made in generating the potentials propagate directly into the final answers. To overcome this defect, we have recently developed methods for a more rigorous many-body "core-valence partitioning" that can be used to generate improved "quasiparticle pseudopotentials"[91, 90].

CALCULATIONS ON OTHER SOLIDS

In Monte Carlo calculations on general solids the valence electrons can be described by the same type of variational functions as was used in hydrogen. The first such calculations were done by Fahy, et.al.,[38] who used a slightly generalized Slater-Jastrow form,

$$\Psi(\mathbf{r}_1,\ldots,\mathbf{r}_N) = \exp\left[\sum_{i=1}^{N}\chi(\mathbf{r}_i) - \sum_{i\leq j}u(r_{ij})\right]D(\mathbf{r}_1,\ldots,\mathbf{r}_N), \quad (50)$$

where D is a Slater determinant of single particle states, and the factor $\chi(\mathbf{r}_i)$ is used to modify the density.[34, 38] This factor could be included in the wavefunctions, but it convenient to keep in this factorized form. The one-body term $\chi(\mathbf{r})$ has been used by Fahy, et.al.,[38] to construct an improved VMC charge density, using the fact that the density given by LDA calculations appears to be remarkably close to the true density. LDA calculations also are used to construct good trial orbitals in the Slater determinant. Finally, the dependence upon cell size can be determined from LDA calculations and used in the extrapolate of the QMC results to large cell size.

Calculations have been done independently by two groups for silicon.[38, 59] Fahy, et.al.,[38] carried out VMC calculations using a non-local pseudopotential (PP), whereas Li, et.al.,[59] used a pseudohamiltonian[86] (PH) which can be employed directly in both VMC and DMC methods. We will discuss explicitly the work of Li, et. al.,[59] which was done with a cubic supercell containing 64 atoms with periodic boundary conditions. In the Slater-Jastrow wavefunction, two cutoffs for the LDA orbitals were used, 7 Ry and 15 Ry, to test the influence on the calculated energy. In VMC, the energy obtained with orbitals cutoff at 15 Ry is 0.44 eV lower than the energy with a 7 Ry cutoff, while the two DMC energies are the same within statistical errors (0.04 eV/atom). This test suggests that the error in the nodal locations caused by the truncation of the LDA trial function is small. Results found using the larger cutoff are discussed here. The difference in an LDA calculation between a 64 atom and infinite system is 0.11 eV, and our QMC results are corrected assuming that they have the same size dependence as LDA. The number of walkers in the DMC ensemble is chosen to be 200 and the initial distribution was obtained from VMC. A time step of 0.015 in atomic units was used. A test calculation using half the time step gave identical results, showing that the time step error is less than 0.03eV/atom. A typical run with 3×10^4 steps, took 20 hours of CRAY-XMP time. To achieve the same error bars, the DMC calculation takes only 2.6 as much computer time as VMC but does not require systematic search of trial functions.

Figure 3 shows the energy as a function of lattice constant from the LDA, VMC and DMC calculations.[59] The curves are least square fits to the Murnaghan equation

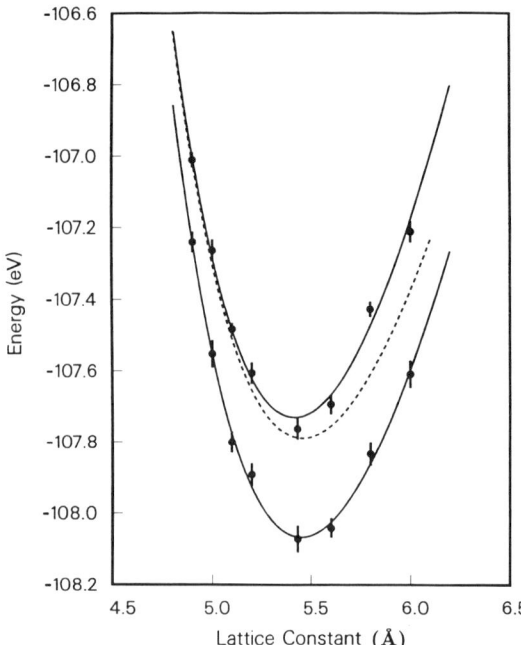

Figure 3. Total energy of silicon versus lattice constant using a pseudohamiltonian (PH) from VMC (upper curve), DMC (lower), and LDA (dashed). The same corrections for finite size effects have been applied to VMC and DMC points and error bars show the estimated statistical errors.

of state. The total energy dropped 0.21(3) eV in the atom and 0.34(3) eV/atom in the solid at zero pressure (with the most accurate variational function) in going from the VMC to DMC. This difference reflects the fact that it is easier to construct a good trial function in the atom than in the solid. In VMC it is important to construct equally good trial functions at all the lattice constants, otherwise there will be a systematic bias in the results.

Also shown in Fig. 3 are the LDA total energies using the same pseudohamiltonian (PH). Since the minimum in the LDA curve is close to the experimental lattice constant, it follows that the QMC results are also in good agreement with experiment. For semiconductors like silicon, LDA is known to work very well, and indeed the total energies from LDA are very close (~ 0.2 eV) to those from QMC. A larger change occurs in the atom where the spin polarized LDA energy is (~ 0.8 eV) higher than DMC. The difference 0.6 eV is the correction to the LDA binding energy, bringing the computed result much closer to experiment. Our DMC cohesive energy, bearing in mind the unknown transferability of the PH, should be between 4.51 (assuming no correction) and 4.73 eV (assuming the LDA gives correctly the difference between the PH and PP), in general agreement with the most quoted experimental value, 4.63(8) eV.

There has also been similar QMC calculations on other systems, such as variational calculations on carbon[38], molecules[44] [90], metal surfaces[58], clusters and other solids [92]; however, we will not describe these works further.

CONCLUSIONS

There are three primary conclusions of the work summarized here. The first is that it is feasible to carry out *exact* calculations for the ground state of many-body Boson systems, whereas for systems of Fermions, the famous "sign problem" has prevented the development thus far of methods which are assymtotically exact. Nevertheless, *nearly exact* results can be found for real molecules and condensed matter using the fixed-node method, in which the nodes on the many-body wavefunction are assumed to be the same as those of a Slater determinant made of appropriate orbitals. It has been found that orbitals from independent-electron methods like LDA and HFA give orbitals which are accurate for these purposes. See, for example, Refs. [38], [59], [57], and [15].

Second, it is feasible to carry out accurate simulations of systems with hundreds of particles. These are sufficient for large finite systems as well as to extrapolate to infinite system size to represent extended states of condensed matter. Extrapolation is a matter of concern for realistic problems where it is difficult to have results for enough different finite sizes, each with sufficiently small statistical errors, that confident extrapolations can be made. At present, extrapolations are done primarily by making the assumption that, for large sizes, the finite size effects are similar to those in independent-particle calculations.

Finally, the largest impediment for calculations on general materials is the way core electrons are treated. If the core is replaced by a pseudopotential or pseudohamiltonian, then valence-only DMC calculations can be carried out with an absolute accuracy of order 0.05 eV per atom in a real solid.[59] This approaches the fundamental limitation of the fixed node approximation, which introduces errors of this magnitude, as has been shown in atomic and molecular calculations. Developments are in progress to have effective methods for fixed node calculations with non-local potentials.[41, 42] The accuracy of the pseudopotential then becomes the paramount issue for accurate calculations. This has been addressed recently using a new approach to *ab initio* many-body "core/valence partitioning." The results have been transformed into "quasiparticle pseudopotentials,"[91, 90] which have been shown to accurately include the effects of the cores (including many-body core-valence correlations) in valence only calculations.

Together these results are promising for future calculations that can provide accurate *ab initio* many-body calculations for general, complex molecules and states of condensed matter.

ACKNOWLEDGMENTS

The ideas described here are in large part due to our collaborator, David M. Ceperley, inventor of many aspects of current quantum Monte Carlo methods, as well as many of our colleagues at (or formerly at) the University of Illinois, especially P. Acioli, J. Grossman, Y. Kwon, X. P. Li, W. Magro, L. Mitas, G. Ortiz, and E. L. Shirley.

This work was supported by the NSF grants DMR-91 17822 and DMR- 89 20538 and the U. S. Department of Energy under grant DE FG02- 91ER4539. Computations were done at the National Center for Supercomputer Applications at the University of Illinois.

References

[1] See, for example, (a) Monte Carlo Methods in Statistical Physics, edited by K. Binder (Springer, Berlin, 1979); and (b) Monte Carlo Methods in Statistical Physics II, edited by K. Binder (Springer, Berlin, 1984).

[2] P. W. Anderson, "Basic Notions of Condensed Matter Physics" (Benjamin/Cummings, Menlo Park, CA, 1984).

[3] P. Fulde, "Electron Correlation in Molecules and Solids" (Springer, Berlin, 1993).

[4] See, for example, W. E. Pickett, Rev. Mod. Phys. **61**, 433 (1989).

[5] See, for example, papers in "Theory of the Inhomogeneous Electron Gas", ed. by S. Lundquist and N. H. March (Plenum Press, New York, 1983), and W. E. Pickett, Comput. Phys. Rep. **9**, 115 (1989).

[6] G. D. Mahan, "Many Particle Physics" (Plenum Press, New York, 1990).

[7] P. Hohenberg and W. Kohn, Phys. Rev. **136**, B864 (1964); and W. Kohn and L. J. Sham, Phys. Rev. **140**, A1133 (1965).

[8] D. M. Ceperley and M. H. Kalos, in Ref. 1a.

[9] K. E. Schmidt and M. H. Kalos, in Ref. 1b.

[10] See, for example, references in Ref. [3].

[11] D. M. Ceperley, and B. Bernu, J. Chem. Phys. **89**, 6318(1988).

[12] Y. Kwon, R. M. Martin, and D. M. Ceperley, to be published.

[13] S. E. Koonin, "Computational Physics" (Benjamin/Cummings, Menlo Park, CA, 1986)

[14] H. Gould and J. Tobochnik, "An Introduction to Computer Simulation Methods - Applications to Physical Systems" (Addison-Wesley, Reading, MA, 1988), especially part II.

[15] V. D. Natoli, thesis, University of Illinois, 1993, and to be published.

[16] R. M. Martin, X. P. Li, E. L. Shirley, L. Mitas, and D. M. Ceperley, "Quantum Monte Carlo Calculations on Materials: Tests on Crystalline Silicon and the Sodium Dimer", Invited paper at Seventh Int. Conf. on Progress in Many Body Theories, 1991; published in "Progress in Many Body Theories, Vol. 3", ed. T. L. Ainsworth, et. al., (Plenum, New York) 1992.

[17] R. M. Martin, Y. Kwon, X. P. Li, V. Natoli, L. Mitas, E. L. Shirley, and D. M. Ceperley, "Quantum Monte Carlo Calculations on Real Materials", Proceedings of the 1992 Taniguchi Symposium; Springer Series in Solid Sate Physics, Vol. 114, ed. K. Terakura and H. Akai, p 191, 1993.

[18] Courant, Friedrichs, and Lewy, Math Ann. **100**, 32 (1928).

[19] N. Metropolis and S. Ulum, J. Am. Stat. Assn **44**, 335 (1949).

[20] M. Donskar and M. Kac, in *Mark Kac: Probability, Number Theory, and Statistical Physics, Selected Papers*, edited by K. Baclawski and M. D. Donsker (MIT Press, Cambridge, Mass., 1979), p. 351.

[21] R. Blankenbecler, D. J. Scalapino, and R. L. Sugar, Phys. Rev. D **24**, 2278 (1981).

[22] J. E. Hirsch, Phys. Rev. B **28**, 4059 (1983).

[23] D. M. Ceperley and E. Pollock, Phys. Rev. B **39**, 2084 (1989).

[24] D. M. Ceperley, Phys. Rev. Lett. **69**, 331 (1992).

[25] D. J. Scalapino, in Proceedings of the Los Alamos Symposium on High Tmpeerature Superconductivity, ed. K. Bedell, et. al., (Addison-Wesley, Redwood City, CA) 1990, p 314.

[26] Some authors distinguish between GFMC and DMC. Our DMC calculations use a short time approximation to the Green's function (see Ref. [44]), and we have tested that the time step is small enough that this is not an essential approximation.

[27] W. L. McMillan, Phys. Rev. A **138**, 442 (1965).

[28] D. Schiff and L. Verlet, Phys. Rev. **160**, 208 (1967).

[29] C. C. Chang and C. E. Campbell, Phys. Rev. B **15**, 4238 (1977).

[30] J. Hansen and D. Levesque, Phys. Rev. **165**, 293 (1968).

[31] J. P. Hansen, D. Levesque, and D. Schiff, Phys. Rev. A **3**, 776 (1971).

[32] M. H. Kalos, D. Levesque, and L. Verlet, Phys. Rev. A **9**, 2178 (1974).

[33] J. P. Hansen and R. Mazighi, Phys. Lett. A **81**, .

[34] D. M. Ceperley, G. V. Chester, and M. H. Kalos, Phys. Rev. B **16**, 3081 (1977).

[35] D. Ceperley, Phys. Rev. B **18**, 3126 (1978).

[36] D. M. Ccpcrlcy, and B. J. Aldcr, Phys. Rcv. B **36**, 2092 (1987).

[37] X. W. Wang, J. Zhu, S. G. Louie, and S. Fahy, Phys. Rev. Lett. **65**, 2414 (1990).

[38] S. Fahy, X. W. Wang and S. G. Louie, Phys. Rev. Lett. **61**, 1631(1988); Phys. Rev. **B42**, 3503(1990).

[39] D. Ceperley, J. Comp. Phys. **51**, 404 (1983).

[40] C. J. Umrigar, K. G. Wilson, J. W. Wilkins, Phys. Rev. Lett. **60**, 1719 (1988).

[41] L. Mitáš, E. L. Shirley and D. M. Ceperley, J. Chem. Phys. **95**, 3467 (1991).

[42] L. Mitáš, in Computer Simulations in Condensed Matter Physics IV, ed. D. P. Landau, Springer (1992); earlier references on approaches to deal with core electrons and non-local potentials in QMC are given in this paper.

[43] Y. Kwon, D. M. Ceperley, and R. M. Martin, Phys. Rev. (1993).

[44] P. J. Reynolds, D. M. Ceperley, B. J. Alder, and W. A. Lester, J. Chem. Phys. **77**, 5593 (1982).

[45] D. M. Ceperley and B. J. Alder, J. Chem. Phys. **81**, 5833 (1984).

[46] S. Zhang and M. H. Kalos, Phys. Rev. Lett. **67**, 3074 (1991).

[47] M. H. Kalos, preprint.

[48] J. B. Anderson, J. Chem. Phys. **63**, 1499 (1975); **65**, 4121 (1976); **73**, 3897 (1980).

[49] D. M. Ceperley and B. J. Alder, Phys. Rev. Lett. **45**, 566 (1980).

[50] J. W. Moskowitz and K. E. Schmidt, J. Chem. Phys. **85**, 2868 (1986).

[51] M. A. Lee and K. E. Schmidt, Computers in Physics **6**, 192 (1992).

[52] M. H. Kalos, Phys. Rev. **128**, 1791 (1962).

[53] M. H. Kalos, J. Comp. Phys. **1**, 127 (1966).

[54] B. Tanatar and D. M. Ceperley, Phys. Rev. B 39, 5005 (1989).

[55] M. Lee, K. E. Schmidt, M. H. Kalos, and G. V. Chester, Phys. Rev. Lett. **46**, 728 (1981).

[56] R. M. Panoff and J. Carlson, Phys. Rev. Lett. **62**, 1130 (1989).

[57] V. Natoli, R. M. Martin, and D. M. Ceperley, Phys. Rev. Letters 70, 1952 (1993).

[58] X. P. Li, R. J. Needs, R. M. Martin, and D. M. Ceperley, Phys. Rev. **45**, 6124 (1992).

[59] X. P. Li, D. M. Ceperley, and R. M. Martin, Phys. Rev. **44**, 10929 (1991).

[60] M. H. Kalos and P. Whitlock, *Monte Carlo Methods* (John Wiley and Sons, New York, 1986).

[61] W. H. Press and S. A. Teukolsky, *Numerical Recipes* (Cambridge University Press, Cambridge, 1992).

[62] N. Metropolis *et al.*, J. Chem. Phys. **21**, 1087 (1953).

[63] M. Allen and D. Tildesley, *Computer simulation of liquids* (Oxford University Press, Oxford, 1989).

[64] D. Ceperley, in *Recent progress in many-body theories*, edited by J. G. Zabolitsky (Springer-Verlag, Berlin, 1981), p. 262.

[65] C. J. Umrigar, K. J. Runge, and M. P. Nightingale, in *Monte Carlo Methods in Theoretical Physics*, edited by S. Caracciolo and A. Fabbrocini (ETS, Pisa, 1990), p. 161.

[66] D. Ceperley, J. Stat. Phys. **43**, 815 (1986).

[67] T. Gaskell, Proc. Roy. Soc. London **77**, 1182 (1981).

[68] W. E. Pickett and J. Q. Broughton, Phys.Rev. **48**, 14859, (1993).

[69] G. Ortiz and P. Ballone, Europhys. Lett. **23**, 7, (1993).

[70] G. Ortiz and P. Ballone, preprint, 1993.

[71] R. P. Feynman and M. Cohen, Phys. Rev. **102**, 1189 (1956).

[72] K. E. Schmidt, M. Lee, and M. H. Kalos, Phys. Rev. Lett. **47**, 807 (1981).

[73] M. Jonson, J. Phys. C **9**, 3055 (1976).

[74] D. L. Freeman, Solid State Commun. **26**, 289 (1978).

[75] D. L. Freeman, J. Phys. C **16**, 711 (1983).

[76] H. K. Sim, R. Tao, and F. Y. Wu, Phys. Rev. B **34**, 7123 (1986).

[77] E. Wigner and H. B. Huntington, J. Chem. Phys. **3**, 764, (1935)

[78] N. W. Ashcroft, Nature **340**, 345 (1989)

[79] E. G. Brovman, Yu Kagan and A. Kholas, Soviet Phys. JETP **34**, 1300, (1972); ibid **35**, 783, (1972)

[80] T. W. Barbee, A. Garcia, J. L. Martins and M. L. Cohen, Phys. Rev. Lett. **62**, 1150 (1989);T. W. Barbee and M. L. Cohen, Phys. Rev. B **44**, 11563 (1991)

[81] K. Ebina and H. Miyagi, Phys. Lett. A, **142**, 237 (1989)

[82] H. Nagara, J. Phys. Soc. Japan **58**, 3861 (1989)

[83] David M. Straus and N. W. Ashcroft, Phys. Rev. Lett. **38**, 415, (1977)

[84] See for example A. H. MacDonald and C. P. Burgess, Phys. Rev. B **26**, 2849 (1982);J. Oliva and N. W. Ashcroft, Phys. Rev. B **23**, 6399, (1981); K. K. Mon, G. V. Chester and N. W. Ashcroft, Phys. Rev. B **21**, 2641 (1980); D.M. Ceperley, *Simple Molecular Systems at Very High Pressure*, Plenum (1988)

[85] E. Kaxiras J. Broughton and R. J. Hemley, Phys. Rev. Lett. **67**, 1138 (1991).

[86] G. B. Bachelet, D. M. Ceperley and M. G. B. Chiochetti, Phys. Rev. Lett. **62**, 2088(1989).

[87] D. R. Hamann, M. Schlüter and C. Chiang, Phys. Rev. Lett. **43**, 1494(1979); G. B. Bachelet, D. R. Hamann and M. Schlüter, Phys. Rev. **B26**, 4199(1982).

[88] W. M. C. Foulkes and M Schlüter, Phys. Rev. **B42**, 11505(1990).

[89] X. W. Wang , S. Fahy, and S. G. Louie, Phys. Rev. Lett. **65**, 2414(1990).

[90] E. L. Shirley, L. Mitáš, and R. M. Martin, Phys. Rev. **44**, 3395 (1991).

[91] E. L. Shirley,thesis, University of Illinois, 1991; E. L. Shirley and R. M. Martin, Phys. Rev. **47**, 15413 (1993).

[92] L. Mitáš, work in progress.

DENSITY FUNCTIONALS, MOLECULAR DYNAMICS, AND MORE

R.O. Jones

Institut für Festkörperforschung
Forschungszentrum Jülich
D-52425 Jülich, Germany

INTRODUCTION

The Third Gordon Godfrey Workshop addresses the role of computational methods in understanding (and developing) new materials. There is no doubt that these methods, particularly the use of computer simulations, will play a very important role in developing our understanding of the structures and properties of molecules, clusters, and bulk materials with complex structures. In the present chapter, I shall focus on methods for calculating the stable structures of such systems, the problems that must be faced, and ways of overcoming them. I shall show that the combination of density functional and molecular dynamics schemes provides a powerful way of calculating structures, although it is by no means the answer to all our problems in this area.

In my view, the geometrical arrangement of the constituent atoms is one of the most important properties of any material. This is very apparent to chemists, as the study of molecules and their interactions implies a knowledge of the atomic positions. Trends in physical properties, such as the boiling points of alkanes C_nH_{2n-2}, can be correlated both with the size of the molecule (n in this case) and the amount of branching in the structure. In the context of clusters or small molecules, for example, we could ask why the tetrahedral forms of P_4 and As_4 are common components of the respective vapours, while tetrahedrane $(CH)_4$, with the same number of valence electrons, has yet to be synthesized.[1] The importance of geometrical structure in molecular biology was quite apparent to Francis Crick, who wrote: "If you want to study function, study structure".[2] In my experience, it is sometimes less obvious to physicists, some of whom are interested in finding universal rules that apply to all systems. In focusing on the geometrical structure and related properties, we address at the outset problems that are *specific* to individual systems. We shall see, however, that the study of sets of such problems (such as families of molecules containing atoms of a particular group of the periodic table) give rise to *patterns* that are quite fascinating.

In principle, the stable geometric arrangements of atoms in any material can be found if we can determine the total energy E of the system of electrons and ions. We adopt a set of coordinates $\{\mathbf{R}_I\}$ for the nuclei, determine E, and then repeat the calculation for all possible configurations, the most stable structure is "simply" that with the lowest energy. There are then two quite distinct problems associated with this procedure: the calculation of E for one given geometry, and the determination of the most stable of the possible structures.

These problems, which are the central issues in the present lectures, are illustrated in Fig. 1, where we show the energy surfaces of the ozone molecule O_3, where the O-O bonds are assumed to be of equal length, for different electronic states.[3] There are several structures for which the energy has a local minimum, two of them with the symmetry $^1A_1$. The most stable state of ozone has a bond angle α of 116.8°, but there is also a minimum with α=60°. There is no obvious reason for favouring the open structure over the equilateral triangle, as can be seen from the near degeneracy of these structures in the case of the valence isoelectronic molecule S_3. Even with as few as three atoms, the most stable isomer can only be found with certainty from an extended set of calculations.

The calculation of the energy E plays a central role in structure calculations as described above. One obvious approach would be to utilize our knowledge of properties of small molecules or bulk systems to parameterize the energy surfaces, and much of the work on cluster simulations has been based on simple force laws. We shall see that this approach also has its limitations, since it is often semi-empirical by nature, and it is difficult to know how much the final results are affected by the choice of parameters.

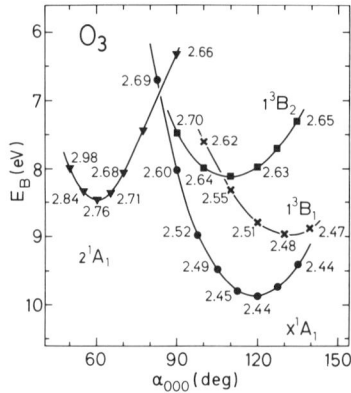

Fig. 1. Energy surfaces of states of the ozone molecule, O_3.

A more satisfactory approach would be to calculate E from the exact wave function Ψ of the system. We could then calculate, in principle, not only the total energy

$$E_{GS} = \langle\Psi|\widehat{\mathcal{H}}|\Psi\rangle/\langle\Psi|\Psi\rangle, \tag{1}$$

where $\widehat{\mathcal{H}}$ is the Hamiltonian of the system, but also many other properties of interest. In practice, however, approximations to Ψ are essential and we shall see that the numerical effort required to calculated accurate energies increases rapidly as the number of electrons increases. The situation was summarized some years ago by Martin et al.,[4] who noted that "an *ab initio* calculation of the total energy of a ten-atom metal cluster, including the effects of correlation, is a state-of-the-art calculation even if it performed for only one point on the energy surface."

The second problem – the determination of the most stable structures amongst the many possible – is at least as difficult, since the number of geometrical configurations grows rapidly as the number of atoms N increases (there are $3N-6$ independent *coordinates*). The determination of the number of possible structures ("isomers") consistent with a given chemical formula is one of the oldest in theoretical chemistry. It was clear to Cayley 120 years ago that the number of isomers increased very rapidly with increasing N, and this has been confirmed by many subsequent studies.[5] An example is the work of Hoare and McInnes[6] on clusters with pairwise forces of the Lennard-Jones:

$$V_2^{LJ}(r) = 4\epsilon_0 \left[\left(\frac{\sigma}{r}\right)^{12} - \left(\frac{\sigma}{r}\right)^6 \right], \tag{2}$$

and Morse forms:

$$V_2^M(r) = D_e \left[\exp-\{2a(r-r_e)\} - 2\exp-\{a(r-r_e)\} \right], \tag{3}$$

with parameters chosen to give a realistic description of isolated clusters.

For clusters with up to 13 atoms, Hoare and McInnes[6] sought to find *all* the structures corresponding to local energy minima, and the number they found is given in Table 1. While the differences between the two potential forms is interesting, it is more significant for the present discussion that the number of minima grows rapidly with increasing N. For clusters of atoms interacting with a pairwise potential, in fact, Wille and Vennik[7] have shown that there is no known algorithm for finding the exact ground state energy (and structure) that has a polynomial time dependence on N. The exponential increase in the number of possible isomers means that the problem belongs to the class NP-hard,[8] and is generally described as "intractable".

Table 1. Number of isomers in clusters with Morse (M) and Lennard-Jones (LJ) potential (Ref.[6]).

N	M	LJ
6	1	2
7	3	4
8	5	8
9	8	18
10	16	57
11	24	145
12	22	366
13	36	988

The above discussion shows that those interested in calculating the structure of a cluster by means of a total energy calculation are faced with two problems: (1) The number of structures corresponding to a local energy minimum increases dramatically with increasing N, even for simplified force laws, (2) the determination of the *exact* energy from a solution of the many-particle wave function for a *single* geometry is itself a problem which tractable only for very small clusters. In such cases, the global minimum can be found, in principle, by locating and enumerating *all* the local minima. However, the search is confined in practice to small parts of configuration space based on intuition, experimental information and/or symmetry restrictions.

In systems where the ground state is unknown or there are many local minima, it is necessary to develop alternative methods for finding solutions that are near to optimal. Kirkpatrick et al.[9] noted the connection between statistical physics and the minimization of a function of many variables, and suggested "simulated annealing" based on a Monte Carlo sampling as a generally applicable way of finding such solutions. Such a method is well known to crystal growers, as all methods for improving the perfection of a crystal rely on raising the temperature. The motion of the atoms means that the system is unlikely to become trapped in an unfavourable local minimum in the energy surface, and slow cooling can result in energetically favourable structures.

We shall start the lectures by discussing the methods for calculating the energy E for a given set of atomic coordinates. We shall show that interatomic force laws based on our knowledge of the energy surfaces of small molecules can not be expected to describe reliably the energy surfaces in larger systems. On the other hand, highly accurate calculations can be performed only for systems with very few electrons, and some of the common approximations lead to unreliable energies. The density functional (DF) formalism,[10] with a local spin density (LSD) approximation for the exchange-correlation energy, provides a tractable method for energy calculations with predictive value in a variety of systems. The strategy of simulated annealing described above can also be implemented by using molecular dynamics (MD), and Car and Parrinello[11] showed that it could be combined with the DF scheme to give a parameter-free method for calculating electronic properties that makes no assumptions about ground state geometries. The use of finite temperatures allows an efficient sampling of the potential energy surface, and the DF method does not require the parameterization of the interatomic forces common in MD schemes.[12]

The MD/DF method has been applied in recent years to a range of molecules and clusters, as well as to simulations of extended systems such as amorphous materials and liquids. The examples I have selected for the present lectures include elements of the main groups IIIa (Al, Ga), Va (P, As) and VIa (S, Se). It is not my aim to be exhaustive, since this can also be exhausting for the reader. I shall focus, however, on the *trends* that can be found in the structural and bonding properties across the periodic table. The observation of such trends raises the hope that the prediction of bonding properties may often be possible without requiring vast calculations.

STRUCTURES FROM TOTAL ENERGY CALCULATIONS

Parameterization of forces

There is an enormous literature on diatomic molecules, and the binding energy curves – or the effective potential between the atoms – could be useful input for describing the energy surfaces of larger systems. We have seen an example previously in the Lennard-Jones potential:

$$V_2^{LJ}(\mathbf{R}_{IJ}) = 4\epsilon_0 \left[\left(\frac{\sigma}{\mathbf{R}_{IJ}}\right)^{12} - \left(\frac{\sigma}{\mathbf{R}_{IJ}}\right)^6 \right], \tag{4}$$

where $\mathbf{R}_{IJ}$ is the distance between atoms I and J. This form has been used very extensively in simulations of rare gas solids and liquids, such as argon. The success of such calculations indicates that pairwise additive potentials can provide a good approximation to the energy surface in some systems.

The situation is more complicated in molecules and clusters where bonds of a directional or "chemical" nature occur. We take as an example the molecule ozone (O_3), whose energy surface we have already seen in Fig. 1. It is well known that O_2 has a bond length of 1.2075 Å and a dissociation energy D_e of 5.12 eV.[13] If we viewed the ozone molecule as simply a superposition of O-O pairs, then we would expect an equilateral triangular structure (three bonds compared with two in an "open" structure) with a dissociation energy of over 10 eV. In fact, we have seen (Fig. 1) that the open structure ($\alpha \sim 120°$) is more stable than the triangular, and the dissociation energy is only 1.1 eV![3] It is easy to find further examples, such as the linear bonds in C_3 and CO_2, that underscore the fact that the strength of a chemical bond is not generally additive.

If we cannot describe the energy surfaces of a molecule or cluster using a superposition of pairwise forces, it is natural to extend the interaction to further neighbours:

$$E(\mathbf{R}_1, ..., \mathbf{R}_N) = \sum_{I<J=1}^{N} V_2(\mathbf{R}_{IJ}) + \sum_{I<J<K=1}^{N} V_3(\mathbf{R}_I, \mathbf{R}_J, \mathbf{R}_K) + ... \tag{5}$$

By the appropriate choice of the parameters it is often possible to reproduce known experimental information about molecules or related extended systems, and then to make predictions about other systems. A good example for such an approach is the work of Stillinger and Weber,[14] who used a formula of this type with

$$V_3^{SW}(\mathbf{R}_I, \mathbf{R}_J, \mathbf{R}_K) = h(\mathbf{R}_{IJ}, \mathbf{R}_{IK}, \Theta_{JIK}) + h(\mathbf{R}_{JI}, \mathbf{R}_{JK}, \Theta_{IJK}) \\ + h(\mathbf{R}_{KI}, \mathbf{R}_{KJ}, \Theta_{IKJ}) \quad (6)$$

in a simulation of the structural properties of liquid sulphur. The calculations provided valuable insight into the dynamics of sulphur at elevated temperatures, but it was also found that the most stable structures predicted for small clusters were generally different from those found in other investigations.[15] While parameterized expressions for the energy are often the only way of obtaining information about large systems or for long simulations, it is clear that the empirical basis of the approach remains a source of uncertainty.

Energy surfaces from the wave function Ψ

The energy of a system is given in terms of its exact wave function Ψ by eq. (1). If we seek instead a reliable estimate of the wave function, it is common to rely on the Rayleigh-Ritz principle, which says that

$$E' = \langle \Phi | \widehat{\mathcal{H}} | \Phi \rangle / \langle \Phi | \Phi \rangle \geq E_{GS}, \quad (7)$$

for an approximate solution Φ, with the equality applying to the exact solution Ψ. In practice, this means that improvements in the wave function are reflected in the energy expression (7), with small decreases in E' implying a wave function that is approaching convergence. Before using this (variational) method to discuss approximations that can be applied to systems in general, we shall show how it can be applied to atoms and molecules with very few electrons.

Variational wave functions for few-electron systems. In the case of the hydrogen atom, we have

$$\widehat{\mathcal{H}} = -\frac{\hbar^2}{2m}\nabla^2 + \frac{e^2}{r}. \quad (8)$$

If we assume an approximate wave functions of the form $\phi(r) = \exp-(cr)$, then we find

$$\nabla^2 \phi = \left(c^2 - \frac{2c}{r}\right)\exp-(cr). \quad (9)$$

The minimum energy is now found for $c = me^2/\hbar^2 = a_0^{-1}$, where a_0 is the Bohr radius, and

$$E_{min} = -\frac{me^4}{2\hbar^2} = -\frac{e^2}{2a_0}. \quad (10)$$

In this case, the choice of the exponentially decaying radial function has led to the exact solution, but we may also test an approximation of the form $\phi(r) = \exp-(cr^2)$. In this case, minimization of the energy expression as a function of c leads to

$$c = \frac{8}{9\pi a_0^2} \quad ; \quad E_{min} = -\frac{4}{3\pi}\frac{e^2}{a_0}. \quad (11)$$

Although we have assumed a form of the radial function that we *know* to be incorrect, we see that the variational principle leads to a total energy that is within $\sim 15\%$ of the exact value.

Let us now study the Hamiltonian of a two-electron system, the He atom with electrons at r_1 and r_2:

$$\widehat{\mathcal{H}} = -\frac{\hbar^2}{2m}\left(\nabla_1^2 + \nabla_2^2\right) - 2e^2\left(\frac{1}{r_1} + \frac{1}{r_2}\right) + \frac{e^2}{r_{12}}. \quad (12)$$

If we neglect the interaction between the electrons, the last term in the Hamiltonian, we can separate the equation by assuming that $\phi = u(r_1)v(r_2)$. In this case, we obtain Schrödinger equations for the functions u and v:

$$\widehat{\mathcal{H}}_1 u(r_1) = E_1 u(r_1)$$

$$\widehat{\mathcal{H}}_2 v(r_2) = E_2 v(r_2),$$

with $E = E_1 + E_2$. The functions $u(r_1)$ and $v(r_2)$ are solutions of the Schrödinger equation for the singly charged He atom:

$$u(r_1) = \left(\frac{Z^3}{\pi a_0^3}\right)^{\frac{1}{2}} \exp-(Zr_1/a_0); \quad v(r_2) = \left(\frac{Z^3}{\pi a_0^3}\right)^{\frac{1}{2}} \exp-(Zr_2/a_0). \tag{13}$$

Evaluation of the energy expression is straightforward,[16] leading to

$$E' = Z\left(2Z - 4(Z-2) - \frac{5}{4}\right)E_H, \tag{14}$$

where E_H is the energy of the hydrogen atom. This expression has a minimum where $Z = 27/16 = 1.6875$, with $E_{min} = 5.695 E_H$. The reduction of the effective charge Z below the value of the singly charged He atom (2.0) is to be expected, since the presence of the second electron tends to screen the attractive potential of the nucleus, and the energy E_{min} is in rather good agreement with the value found from extensive variational calculations, $5.807 E_H$.[16]

The above examples show the power of the variational method in evaluating the total energy of a quantum mechanical system. Even with rather simple approximations to the wave function, we can obtain surprisingly good energies. The example of the two-electron system illustrates a further point: Useful approximations can result if we assume the electrons to be independent (i.e., to satisfy single-electron Schrödinger equations) and then treat their interactions using the variational principle. This approach is the basis of one of the most important methods of calculating the electronic structure of molecules, and we shall now discuss it.

Hartree-Fock approximation. One of the earliest approximations for Ψ is due to Hartree, who viewed the electrons as independent, moving in the average field of the other electrons in the system. In terms of the variational approach, we may consider the wave function as a product of single-particle functions, i.e.,

$$\Psi(\mathbf{r}_1, \mathbf{r}_2, ...) = \psi_1(\mathbf{r}_1)........\psi_n(\mathbf{r}_n) \tag{15}$$

For this form of wave function, the variational principle requires that each of the functions $\psi_i(r_i)$ satisfies a one-electron Schrödinger equation of the form

$$\left(-\frac{\hbar^2}{2m}\nabla^2 + V_{ext} + \Phi_i\right)\psi_i(\mathbf{r}) = \epsilon_i \psi_i(\mathbf{r}), \tag{16}$$

where V_{ext} is the potential due to the nuclei, and the Coulomb potential Φ_i is given by Poisson's equation and arises from the average field of the other electrons. The state of the system is then defined by the single particle functions $\psi_i(\mathbf{r})$, the eigenvalues ϵ_i, and the occupancy of the "orbitals" so defined. Fermi statistics can be incorporated into this picture by replacing the product wave function by a single (Slater) determinantal function, often called a "configuration". The resultant (Hartree-Fock, HF) approximation leads to an additional "exchange" term V_x^{HF} in the Schrödinger equation that is non-local in the sense that:

$$V_x^{HF}\psi(\mathbf{r}) \equiv \int d\mathbf{r}' \, V_x^{HF}(\mathbf{r}, \mathbf{r}')\psi(\mathbf{r}') . \tag{17}$$

Nevertheless, the single-particle picture, with the wave function described in terms of orbitals with particular spins and occupation numbers, is unchanged. The inclusion of

Fermi statistics improves the total energy calculation, and the HF approximation has been an indispensable benchmark in molecular physics since its inception over 60 years ago.

Hartree-Fock (often referred to in the chemical literature as "self-consistent field") calculations, and the picture that results, are very familiar, but it has been known for many years that the total energies so found are not accurate enough for many purposes. The lowest-lying configuration is generally only one of many with comparable energies, and an improved energy results if we take a linear combination. This procedure, where effects beyond the Hartree-Fock approximation ("correlation effects") are included by improving the many-particle wave function, is known as "configuration interaction" (CI). It leads, in principle, to the exact wave function, from which many properties of interest can be calculated, but the numerical effort required increases dramatically with increasing electron number. In practice, numerically exact total energies can be found using CI only for systems with well below 100 electrons.

Density functional (DF) formalism

We have seen that semi-empirical parameterizations of energy surfaces can allow us to perform calculations for large numbers of particles, although we must be cautious about the predictions that result. On the other hand, the reliability of accurate calculations of the wave function of the interacting system is beyond question, but the calculations can only be performed for systems with relatively few electrons. In the remainder of these lectures, I shall focus on a method that seeks the middle ground: We should like to use a method that can lead to reliable predictions of structures and energy differences in molecules, clusters and extended systems, and is free of adjustable parameters. The method is the density functional formalism, and we shall see that it can be combined with molecular dynamics to address the difficulties that arise from the existence of multiple minima in the energy surface. As in the Thomas-Fermi approach, the focus of the DF formalism is on the electron density $n(\mathbf{r})$, rather than the wave function of the system.

The basic theorems of the DF formalism were derived by Hohenberg and Kohn.[17] They showed that:
(1) Ground state (GS) properties of a system of electrons and ions in an external field V_{ext} can be determined from the electron density $n(\mathbf{r})$ alone.
(2) The total energy E is such a functional of the density, and $E[n]$ satisfies the variational principle $E[n] \geq E_{GS}$. The density for which the equality holds is the ground state density, n_{GS}.

The usual implementation of this scheme results from the observation by Kohn and Sham[18] that the minimization of $E[n]$ is simplified if we write (we adopt atomic units with $e = \hbar = m = 1$):

$$E[n] = T_0[n] + \int d\mathbf{r}\, n(\mathbf{r})\left(V_{ext}(\mathbf{r}) + \frac{1}{2}\varphi(\mathbf{r})\right) + E_{xc}[n], \qquad (18)$$

where T_0 is the kinetic energy that a system with density n would have in the absence of electron-electron interactions, $\varphi(\mathbf{r})$ is the Coulomb potential, and E_{xc} defines the exchange-correlation energy.

The variational principle yields

$$\frac{\delta E[n]}{\delta n(\mathbf{r})} = \frac{\delta T_0}{\delta n(\mathbf{r})} + V_{ext}(\mathbf{r}) + \Phi(\mathbf{r}) + \frac{\delta E_{xc}[n]}{\delta n(\mathbf{r})} = \mu, \qquad (19)$$

where μ is the Lagrange multiplier associated with the requirement of constant particle number. If we compare this with the corresponding equation for a system with an effective potential $V(\mathbf{r})$ but *without* electron-electron interactions,

$$\frac{\delta E[n]}{\delta n(\mathbf{r})} = \frac{\delta T_0}{\delta n(\mathbf{r})} + V(\mathbf{r}) = \mu, \qquad (20)$$

we see that the mathematical problems (19, 20) are identical, provided that

$$V(\mathbf{r}) = V_{ext} + \Phi(\mathbf{r}) + \frac{\delta E_{xc}[n]}{\delta n(\mathbf{r})}. \tag{21}$$

The solution of (20) can be found by solving the Schrödinger equation for non-interacting particles,

$$(-\frac{1}{2}\nabla^2 + V(\mathbf{r}))\,\psi_i(\mathbf{r}) = \epsilon_i \psi_i(\mathbf{r}), \tag{22}$$

yielding

$$n(\mathbf{r}) = \sum_{i=1}^{N} |\psi_i(\mathbf{r})|^2 \tag{23}$$

It is necessary to satisfy the condition (20), and this can be achieved in a self-consistent procedure. From the solutions ψ_i we determine the density and the related potential, and then we solve (22) again to determine new values of ψ_i. This cycle is repeated until the input and output solutions are identical.

From the solution of a single-particle equation we can determine the total energy and density of the lowest state of a system of electrons and ions, and all quantities derivable from them. In contrast to the Hartree-Fock potential, the effective potential $V(\mathbf{r})$ is *local*. With the exception of the exchange-correlation energy E_{xc}, all terms in the energy expression (18) are straightforward to evaluate and the equations present, in principle, no more numerical difficulties than the solution of Hartree's equations. Approximations to E_{xc} are currently unavoidable, however, and it is important to discuss this term in some detail.

Exchange-correlation energy, E_{xc}. This is the only contribution to the energy for which approximations are necessary. The most widely used is the local spin density (LSD) approximation

$$E_{xc}^{LSD} = \int d\mathbf{r}\, n(\mathbf{r})\, \varepsilon_{xc}[n_\uparrow(\mathbf{r}), n_\downarrow(\mathbf{r})], \tag{24}$$

where $\varepsilon_{xc}[n_\uparrow, n_\downarrow]$ is the exchange and correlation energy per particle of a homogeneous, spin-polarized electron gas with spin-up and spin-down densities $n_\uparrow$ and $n_\downarrow$, respectively. This approximation is free of adjustable parameters, but its application to atoms, molecules and solids cannot be justified by small departures from homogeneity, as none of these systems satisfy this criterion. In their original work, Kohn and Sham[18] commented that "we do not expect an accurate description of chemical bonding" with the local density approximation, and a decade passed before the first tests of the LD approximation were made in this context. It is remarkable that it was found to reproduce many measurable quantities satisfactorily, as it is by no means obvious that this should be the case.

The crucial simplification in the density functional scheme is the relationship between the interacting system, whose energy and density we seek, and the fictitious, non-interacting system for which we solve Eqs. (21, 22). This can be studied by considering the interaction $\lambda V_{ee} \equiv \lambda/|\mathbf{r} - \mathbf{r}'|$ and varying λ from 0 (non-interacting system) to 1 (physical system). This is done in the presence of an external potential, V_λ,[19] such that the ground state of the Hamiltonian

$$H_\lambda = -\frac{1}{2}\nabla^2 + V_{ext}(\mathbf{r}) + V_\lambda + \lambda V_{ee} \tag{25}$$

has density $n(\mathbf{r})$ for all λ. The exchange-correlation energy of the interacting system can then be expressed in terms of an integral over the coupling constant λ.[20,21,22]

$$E_{xc} = \frac{1}{2}\int d\mathbf{r}\, n(\mathbf{r}) \int d\mathbf{r}'\, \frac{1}{|\mathbf{r}-\mathbf{r}'|} n_{xc}(\mathbf{r}, \mathbf{r}' - \mathbf{r}), \tag{26}$$

with

$$n_{xc}(\mathbf{r}, \mathbf{r}' - \mathbf{r}) \equiv n(\mathbf{r}') \int_0^1 d\lambda \left(g(\mathbf{r}, \mathbf{r}', \lambda) - 1\right). \tag{27}$$

The function $g(\mathbf{r}, \mathbf{r}', \lambda)$ is the pair-correlation function of the system with density $n(\mathbf{r})$ and Coulomb interaction λV_{ee}. The exchange-correlation hole, n_{xc}, describes the effect of the interelectronic repulsions, i.e., the fact that an electron present at the point $\mathbf{r}$ reduces the probability of finding one at $\mathbf{r}'$. The exchange-correlation energy may then be viewed as the energy resulting from the interaction between an electron and its xc-hole.

Three observations should be made here. First, since $g(\mathbf{r}, \mathbf{r}')$ tends to unity as $|\mathbf{r} - \mathbf{r}'| \to \infty$, the above separation into electrostatic and exchange-correlation energies can be viewed as an approximate separation of the consequences of long- and short-range effects, respectively, of the Coulomb interaction. We may then expect that the total interaction energy will be less sensitive to changes in the density, since the long-range part can be calculated exactly. Second, there are important consequences of the isotropic nature of the Coulomb interaction V_{ee}.[21] A variable substitution $\mathbf{R} \equiv \mathbf{r}' - \mathbf{r}$ in (26) yields

$$E_{xc} = \frac{1}{2} \int d\mathbf{r}\, n(\mathbf{r}) \int_0^\infty dR\, R^2 \frac{1}{R} \int d\Omega\, n_{xc}(\mathbf{r}, \mathbf{R}). \tag{28}$$

Eq. (28) shows that the xc-energy depends only on the spherical average of $n_{xc}(\mathbf{r}, \mathbf{R})$, so that approximations for E_{xc} can give an *exact* value, even if the description of the non-spherical parts of n_{xc} is arbitrarily inaccurate.

The final observation follows from the definition of the pair-correlation function, namely that there is a sum-rule that requires that the xc-hole contains one electron, i.e., for all $\mathbf{r}$,

$$\int d\mathbf{r}'\, n_{xc}(\mathbf{r}, \mathbf{r}' - \mathbf{r}) = -1. \tag{29}$$

This means that we can consider $-n_{xc}(\mathbf{r}, \mathbf{r}' - \mathbf{r})$ as a normalized weight factor, and define locally the radius of the xc-hole,

$$\left\langle \frac{1}{R} \right\rangle_{\mathbf{r}} = -\int d\mathbf{r}\, \frac{n_{xc}(\mathbf{r}, \mathbf{R})}{|\mathbf{R}|}. \tag{30}$$

This leads to

$$E_{xc} = -\frac{1}{2} \int d\mathbf{r}\, n(\mathbf{r}) \left\langle \frac{1}{R} \right\rangle_{\mathbf{r}}, \tag{31}$$

showing that, provided the sum-rule (29) is satisfied, the exchange-correlation energy depends only weakly on the details of n_{xc}.[21] In fact, we can say that it is determined by the first moment of a function whose second moment we know exactly.

LSD calculations – sources of error. The above arguments show that the LSD approximation can give a reasonable description of systems where the density is far from homogeneous. As experience with molecular calculations developed, however, it became clear that LSD calculations gave rise to some persistent, almost systematic errors for a variety of systems.[10] In particular, the binding energy of sp-bonded molecules is often overestimated by ~ 1 eV per bond. Although the errors seem to be rather regular, discrepancies of this size are far from acceptable, and it is essential to obtain insight into the origins. To do this we study some simple model systems, focusing on the exchange energy, where the presence of explicit formulas simplifies the discussion greatly.

The HF exchange energy can be expressed in terms of exchange integrals,[23]

$$I_{ij} = e^2 \int d\mathbf{r} \int d\mathbf{r}'\, \frac{\Phi_i^*(\mathbf{r})\Phi_j(\mathbf{r})\Phi_i(\mathbf{r}')\Phi_j^*(\mathbf{r}')}{|\mathbf{r} - \mathbf{r}'|}, \tag{32}$$

where Φ_i and Φ_j are HF orbitals. Such an integral depends strongly on the nodal structure of Φ_i and Φ_j. If Φ_i and Φ_j have different l and m-quantum numbers, for example, the integrand oscillates and I_{ij} is reduced. Since the LSD approximations for exchange (referred to below as the LSDX approximation) and correlation are expressed only in terms of the charge density, we cannot expect them to include effects of the nodal structure in a precise way.

As an example, we consider the sp-transfer in an F atom, where an $s\downarrow$ electron is transferred to a $p\downarrow$ orbital,[24]

$$1s^2 2s(\uparrow\downarrow) 2p(\uparrow\uparrow\uparrow\downarrow\downarrow) \to 1s^2 2s(\uparrow) 2p(\uparrow\uparrow\uparrow\downarrow\downarrow\downarrow) . \tag{33}$$

The change in the exchange energy due to this process is[23]

$$\Delta E_x = -\frac{9}{25} G^2(2p, 2p) + \frac{2}{3} G^1(2p, 2s) , \tag{34}$$

where the Slater integrals G^k are defined by

$$G^k(i,j) = e^2 \int_0^\infty dr\, r^2 \int_0^\infty dr'\, (r')^2 \frac{r_<^k}{r_>^{k+1}} \phi_i(r)\phi_j(r)\phi_i(r')\phi_j(r') . \tag{35}$$

Here $r_<$ ($r_>$) is the smaller (larger) of r and r', and $\phi_i(r)$ is the radial part of $\Phi_i(\mathbf{r})$. The first term in Eq. (34) is the exchange interaction between a p-electron with $m = -1$ and two p-electrons with $m = 1$ and 0. The second term is the interaction between an s-electron and two p-electrons. The integrand of (32) corresponding to the p-p interaction has two nodal planes as a function of $\mathbf{r}$, while there is only a single nodal plane for the s-p interaction. The latter is larger as a result. Using realistic values for the Slater integrals, we find that $\Delta E_x \sim 6$ eV. The LSDX calculation, however, gives similar radial extents for the s- and p-orbitals. If we assume that the radial parts are *identical* and neglect the small nonspherical corrections, the LSDX approximation predicts that the exchange energy is *unchanged* by the sp-transfer, so that it is not surprising that the LSDX predicts that Δ_{sp} differs from the HF result by 6 eV. The LSDX and LSD (both exchange and correlation) approximations show similar trends in this case, although the deviation between the LSD result and experiment is reduced to 2.6 eV, since the large change in the exchange energy is compensated in part by a change in ΔE_c of opposite sign.

A simple model problem. In order to gain insight into these problems, Gunnarsson and Jones[24] performed extensive LSDX and LSD calculations for atoms and compared the results with HF calculations and experiment, respectively. The differences between the LSDX and HF results are reproduced remarkably well by model calculations where we isolate the effects of the l-dependence of the orbitals by assuming that the s, p and d orbitals have the same *radial* dependence. With this assumption G^k only depends on k. For the $3s$ orbital in calcium, the numerical relations

$$G^0 = \frac{G^1}{0.680} = \frac{G^2}{0.516} = \frac{G^3}{0.414} = \frac{G^4}{0.344} \tag{36}$$

and

$$\int d\mathbf{r}\, \varepsilon_x[n(\mathbf{r}), 0] = 0.451 G^0 \tag{37}$$

were found, where $n(\mathbf{r})$ is the charge density due to a Ca $3s$ electron. Similar coefficients were obtained for the orbitals of other elements of the iron series. Eq. (37) shows that the unphysical self-interaction, $-G^0/2$, is cancelled to within about 10%. In Fig. 2(a) we compare the LSDX and HF exchange energies as a function of the number N of spin up electrons. In the HF case, the shells are filled in the order s, p, d and within each shell in the order $m, m-1, .., -m$. In the LSDX case the small nonspherical contributions are neglected. The LSDX and HF results agree remarkably well.

Fig. 2(b) shows results for the interelectronic exchange energy, for which the self-interaction has been subtracted. With the filling order s, p, d, the LSDX and HF results are in rather good agreement, with the magnitude of the LSDX results being somewhat larger. Fig. 2(b) also shows results for occupations where a subshell is left empty. This does not influence the LSDX results in the present model, but it has a pronounced effect on the HF results. For instance, the curve "s-shell empty" is lower than the curve corresponding to the s, p, d filling, since the p-p exchange interaction is smaller than the s-p interaction. It follows from the orthogonality of the HF orbitals that the exchange hole contains one electron, and a similar sum rule is also satisfied by the LSD and LSDX approximations.[21] Aspects of orthogonality and node formation are then included in all these schemes, although Fig. 2(b) shows that the sum rule does not *guarantee* a good description.

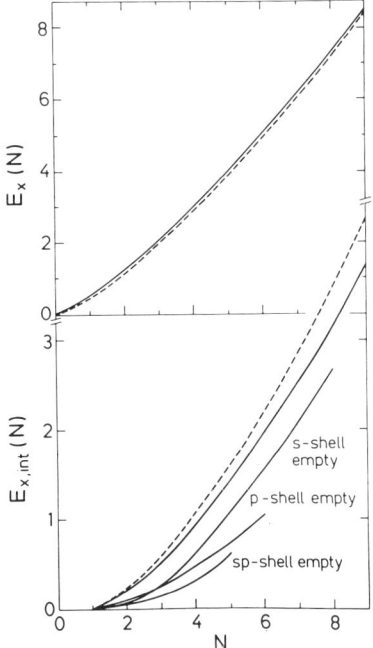

Fig. 2. Exchange energies in model described in text. (a) $E_x^{\rm LSD}$ (dashed curve) and $E_x^{\rm HF}$ (solid curve) as a function of the number of electrons N (in units of $-G^0$); (b) the interelectronic exchange $E_{x,int}$ for these two approximations. In the HF approximation, we show the dependence for different schemes of occupying the orbitals.

This model calculation illustrates two important conclusions: (i) If we occupy the orbitals with the minimum number of nodal planes consistent with the Pauli principle, the trends in the interelectronic exchange energies are reproduced well by the LSDX approximation. The absolute value is overestimated in all the systems considered by Gunnarsson and Jones.[24] (ii) The energy for the transfer from such a state to a state with one additional node is often underestimated substantially in the LSDX approximation.

These results were derived from atomic calculations, but we can extend them to small molecules as follows. In the ground state of H$_2$ and the alkali dimers is a $^1\Sigma_g^+[\sigma_g(\uparrow\downarrow)]$ bond between the valence s orbitals. This is another case where the LSDX approximation reproduces the HF result well, and where the LSD approximation is in good agreement with experiment. The LSDX approximation then gives a satisfactory description of the changes in the self-interaction during the formation of this σ bond, and the LSD approximation accounts for the substantial change in the correlation energy. On the other hand, the $^3\Sigma_u^+[\sigma_g(\uparrow)\sigma_u(\uparrow)]$ state is repulsive, except for a weak minimum at large distances. It may appear that this should be a more favourable case for the LSD approximation than the $^1\Sigma_g^+$ state, since the correlation energy is small and there is no spin flip involved in the bond formation. However, the LSD approximation overestimates the binding energy by 0.5 eV, and the error may be traced to the interelectronic interaction.[24] For small internuclear separations, the σ_u orbital goes over to a p orbital, and the σ_g-σ_u interaction shows similarities with the s-p interaction described above, with a moderate overestimate of the interelectronic exchange energy in both cases. A similar situation occurs in He_2 and for the group IIa dimers.

This discussion has shown that the nodal structure of the wave functions can have a great effect on the accuracy of the LSD approximation, and we have identified classes of problems where the LSD results must be treated with caution. For states with the minimum number of nodal planes consistent with the orthogonality of the orbitals, the LSD approximation usually leads to a moderate overestimate of the exchange-correlation energy. For states with additional nodal planes the exchange-correlation energy is often greatly overestimated. In atoms, the depopulation of s-orbitals can lead to large errors, and similar effects may be expected in bonding situations where sp or sd hybridization reduces the s occupancy.

MD/DF CALCULATIONS - THE METHOD

Theoretical aspects

If we use the DF formalism (with the LSD approximation for the exchange-correlation energy) to describe the energy surfaces of the system in question, then the determination of the most stable structures requires us to address *two* minimization problems. The first is the requirement of the DF variational principle that, for each geometry, we vary the density to minimize the energy, and the second is that we move the ions to find structures with more favourable energies. This may be done by viewing E as a function of two interdependent sets of degrees of freedom: the single particle orbitals $\{\psi_i\}$ that give rise to the density, and the ionic coordinates $\{\mathbf{R}_I\}$,[11]

$$E[\{\psi_i\},\{\mathbf{R}_I\}] = \sum_i \langle \psi_i(\mathbf{r})| -\frac{\nabla^2}{2}|\psi_i(\mathbf{r})\rangle + \int d\mathbf{r}\, n(\mathbf{r})\left(V_{ext}(\mathbf{r}) + \frac{1}{2}\Phi(\mathbf{r})\right) +$$
$$+ E_{xc}[n(\mathbf{r})] + \frac{1}{2}\sum_{I\neq J}\frac{Z_I Z_J}{|\mathbf{R}_I - \mathbf{R}_J|} \;. \quad (38)$$

To minimize this function, we use a dynamical simulated annealing technique to follow the trajectories of $\{\psi_i\}$ and $\{\mathbf{R}_I\}$ given by the Lagrangian

$$\mathcal{L} = \sum_i \mu_i \int_\Omega d\mathbf{r}\, |\dot\psi_i^* \dot\psi_i| + \sum_I \frac{1}{2} M_I \dot{\mathbf{R}}_I^{\,2} - E[\{\psi_i\},\{\mathbf{R}_I\}] + \sum_{ij}\Lambda_{ij}\left(\int_\Omega d\mathbf{r}\,\psi_i\psi_j^* - \delta_{ij}\right) \quad (39)$$

and the corresponding equations of motion

$$\mu \ddot\psi_i(\mathbf{r},t) = -\frac{\delta E}{\delta \psi^*(\mathbf{r},t)} + \sum_k \Lambda_{ik}\psi_k(\mathbf{r},t) \;,$$

$$M_I \ddot{\mathbf{R}}_I = -\nabla_{\mathbf{R}_I} E \;. \quad (40)$$

Here M_I denote the ionic masses and μ_i are fictitious "masses" associated with the electronic degrees of freedom, dots denote time derivatives, and the Lagrangian multipliers Λ_{ij} are introduced to satisfy the orthonormality constraints on the $\psi_i(\mathbf{r},t)$. From these orbitals and the resultant density $n(\mathbf{r},t) = \sum_i |\psi_i(\mathbf{r},t)|^2$ we can use Eq. (38) to evaluate the total energy E, which acts as the classical potential energy in the Lagrangian (39). The artificial Newton's dynamics for the electronic degrees of freedom, together with the assumption $\mu_i \ll M_I$, effectively prevent transfer of energy from the classical to the quantum degrees of freedom over long periods of simulation. The method is well-suited for traditional MD applications as well as to simulated annealing.

Computational details

Density functional calculations have been performed with a great variety of numerical techniques and with numerous basis sets to represent the single particle functions ψ_i. On the other hand, almost all implementations to date of DF calculations with molecular dynamics have used a pseudopotential representation of the electron-ion interaction and a basis set of plane waves. The applications of the MD/DF scheme given below all use the latter approach, and we now give some details.

The interaction between electrons and ions V_{ext} is given by a pseudopotential

$$V_{ext}(\mathbf{r}) = \sum_I v_{ps}(\mathbf{r} - \mathbf{R}_I),$$

$$v_{ps}(\mathbf{r}) = \sum_{l=0}^{\infty} v_l(\mathbf{r}) \hat{P}_l, \qquad (41)$$

where $\hat{P}_l$ is the angular momentum projection operator. There are several prescriptions for determining $v_l(\mathbf{r})$, and we have used parameterizations given by Bachelet et al.[25] and Stumpf et al.[26] Our experience indicates that accurate calculations require components of v_l up to at least $l = 2$. The eigenfunctions of (22) are expanded in the plane wave basis

$$\psi_i(\mathbf{r}) = \psi_{j\mathbf{k}}(\mathbf{r}) = \sum_{n=1}^{M} c_{j\mathbf{G}_n}^{\mathbf{k}} \exp\left(i(\mathbf{k} + \mathbf{G}_n) \cdot \mathbf{r}\right), \qquad (42)$$

leading to the self-consistent eigenvalue problem

$$\sum_n H_{mn}\, c_{j\mathbf{G}_n} = \epsilon_j\, c_{j\mathbf{G}_m},$$

$$H_{mn} = \delta_{mn}|\mathbf{G}_m|^2 + V_{ext}(\mathbf{G}_m - \mathbf{G}_n) + \Phi(\mathbf{G}_m - \mathbf{G}_n) + V_{xc}(\mathbf{G}_m - \mathbf{G}_n), \qquad (43)$$

where the index $\mathbf{k}$ has been dropped.

We have generally used a face-centred-cubic unit cell [lattice constant 15.9 Å] with constant volume [1000 Å^3] and periodic boundary conditions, although larger unit cells have been used to check the calculations. This supercell geometry leads to a large separation between the individual clusters, and the accurate reproduction of the symmetries in S_6, S_8 and S_{12}, for example, shows that the results are indeed insensitive to the choice of boundary conditions. The cut-off energy for the plane wave expansion (42) of the electronic eigenfunctions ψ_i was 10.6 - 14.0 Ry, leading to $\sim$ 4000-6000 plane waves for a single point ($\mathbf{k} = 0$) in the Brillouin zone (BZ). The convergence properties of the equilibrium geometries and total energies with respect to basis set size and number of k-points must be checked thoroughly. It is clear that the shape of the energy surfaces of the clusters is less sensitive to the cut-off than is the absolute value of the energy.

The above considerations could apply for any electronic structure calculation using a plane wave basis. The specifically MD aspects of the procedure are initiated by displacing the atoms randomly from an arbitrary geometry, with velocities $\dot{\psi}_i$ and $\dot{\mathbf{R}}_I$ set equal to zero. For this geometry we then determine those ψ_i that minimize E via an efficient self-consistent iterative diagonalization technique. With the electrons initially in their ground state, the dynamics (40) generate Born-Oppenheimer (BO) trajectories without the need for additional diagonalization/self-consistency cycles for the electrons over several thousand time steps. In typical applications, the "mass" μ_i of the electronic degrees of freedom was 300–1800 a.u., and the MD time step $\Delta t = 1.7 - 3.4 \times 10^{-16}$ s. For such time steps, the equations of motion can be integrated accurately using the Verlet algorithm.[27] If we define a "temperature" T by the mean classical kinetic energy of the atoms, the atomic configurations represent physical molecules at elevated temperatures. The energy surfaces are probed by varying T, and the minima in the potential energy surfaces are found by reducing T slowly to zero. Further details are given in the original papers.

MD/DF CALCULATIONS - SOME APPLICATIONS

To demonstrate the usefulness of the MD/DF approach, particularly the simulated annealing aspects, we now discuss some representative applications. We have chosen ring molecules of sulphur, the oxides of S_7, clusters of phosphorus and arsenic, and molecules containing elements of both groups Va and VIa. Some recent results on clusters of aluminium and gallium are also included. One of the advantages of the DF method is that it can be applied to all elements, even those with large atomic numbers. We shall then focus on the bonding trends that become apparent, and on the reasons for them. The reader is referred to the original articles for full details of the calculations.

Group VIa elements: oxygen, sulphur, selenium

The group VIa elements provide some of the best characterized atomic clusters. The elements S and Se, in particular, are unique in that many allotropes are molecular crystals comprising regular arrays of well-separated rings of two-fold coordinated atoms. X-ray structure analyses have been performed for S_n, $n = 6\text{–}8, 10\text{–}13, 18, 20$ and $Se_n, n = 6, 8$.[28,29] There are five crystalline modifications of selenium: four comprise Se_8 and Se_6 rings, and trigonal Se consists of parallel helical chains with high molecular weight. Several mixed crystals of the form Se_nS_m[30] and a range of sulphur oxides (S_nO, n=5–10; S_7O_2, $S_{12}O_2$)[31] and ions are also known. The preparation of these clusters has been reviewed by Steudel,[29] and the presence of so many molecules with well established structures provides us with ideal tests of our method. We note also that many studies of the disordered amorphous and liquid systems have indicated that the clusters show structural similarities to the condensed phases.

Clusters of sulphur and selenium. MD/DF calculations[15] performed on clusters up to S_{13} showed that it is possible to determine low-lying energy minima even if the initial geometry is far from the correct structure. Starting from almost linear chains [S_{3-6}] or from nearly planar rings [S_{7-13}], we found structures that agreed well with experiment in all cases where X-ray data were available. As an example we show in Fig. 3(b) the structure of S_{12}. The D_{3d} symmetry is reproduced very well, and the structural parameters (bond length $d = 3.97$ a.u., bond angle $\alpha = 106°$, dihedral angle $\gamma = 88°$) agree well with measured values (3.88 a.u., 106.2°, 87.2°, respectively). Perhaps more interesting is the fact that we can obtain plausible predictions in cases (e.g., S_5, S_9) where single crystal specimens have not yet been prepared. In S_5 we found both the "envelope" (C_s) structure and a C_2 structure with almost the same energy, and the predictions of the structures of S_4, S_5 and S_9 have been confirmed by subsequent correlated wave function calculations.[32]

The results for the S_9 molecule proved to be particularly interesting. While this molecule can be prepared in microcrystalline form,[29,33] the absence of single crystals has ruled out an X-ray structure determination. However, information can be obtained from the Raman spectra, and Steudel *et al.*[33] concluded that the constituent molecules have nearly identical structures, with S-S bonds between 3.84 a.u. and 3.95 a.u., i.e., neither unusually long nor short. From structural trends found in other sulphur ring molecules, they concluded also that the dihedral angles lie in the range $70° \leq \gamma \leq 130°$, so that we can exclude ring geometries with four consecutive atoms that are nearly coplanar. This eliminates the possibility of C_s symmetry, and allows only C_1 and C_2. Our calculations led to a ground state structure, shown in Fig. 3(a), that fulfils all of the above criteria. It has C_2 symmetry, approximately equal bond lengths, and dihedral angles between 66° and 114°.

Some interesting structural trends in sulphur ring molecules have been noted by Steudel and coworkers.[29,34,35] The first is that the bond length d is inversely related to the mean length of the two neighbouring bonds, and this correlation also applies to the calculated structures. The second relationship is between the bond length d and the dihedral angle γ, with d being shortest for $\gamma \sim 90°$. The preference for dihedral angles around 100° that had been noted much earlier by Pauling,[36] who referred to this as the "normal" value for γ in S-S bonds. The dihedral angle is defined by the coordinates of four successive atoms in a chain, so that the substantial variation of bond length with γ is an indication that four-body forces may be important in sulphur ring molecules.

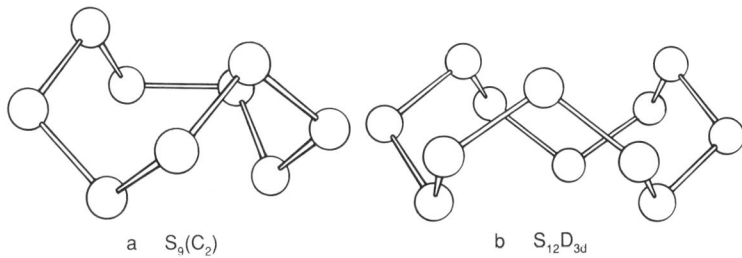

a $S_9(C_2)$ b $S_{12}D_{3d}$

Fig. 3. Structures of (a) S_9 and (b) S_{12}.

Perhaps the most interesting feature of the structural trends in this family of molecules is, however, the difficulty of interpolating between or extrapolating from known structures. In S_9, for example, the knowledge of the structures of several members of the family on either side (n=6-8,10-13), does not allow us to predict unambiguously either the structure or even the pattern of dihedral angles ["motif"][29,37] of the most stable isomer. Isomers with quite different structures can have very similar energies, and the assignment of the ground state depends then on subtle effects that require quantitative measurements or calculations rather than qualitative arguments.

We have focused here not only on comparisons between theory and experiment, but on bonding *patterns*. Selenium clusters up to Se_8 have very similar structures to the corresponding S_n-clusters,[38] the main difference being that the bonds are 12–15% longer. We now show that mixed molecules containing S and Se provide a fascinating area to study.

Isomers of Se_2S_6, Se_6S_2. We have noted above that several ring molecules of the type Se_nS_m are known, and we have performed MD/DF calculations on seven-

and eight-membered rings of this type.[39] The cyclic structures possible in Se_2S_6 and Se_6S_2 are shown in Fig. 4 and the calculated structures (and energies) show several remarkable features. In all eight structures, the bond lengths (d_{S-S}, d_{S-Se}, d_{Se-Se}) are identical to within 0.01 Å (2.08 Å, 2.23 Å, 2.35 Å) *and* the same as found in the earlier calculations for the ring structures of S_8 and Se_8. Structures that maximize the number of like neighbours ("1,2-structures") are the most stable isomers in both Se_2S_6 and Se_6S_2, and the energies of the other structures are *degenerate* to within ±3 meV, lying 0.08 eV above the ground state in each case. These effects can be understood very simply, as we now show.

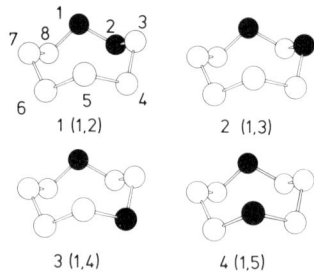

Fig. 4. Structures of Se_2S_6 (and Se_6S_2).

The eight-membered rings are characterized by small deviations from the crown-shaped $[D_{4d}]$ structures familiar from S_8 and Se_8. The well-known correlation between bond length and bond strength[40] suggests that we associate with each bond length (e.g., d_{S-S}) a corresponding contribution to the energy (e.g., E_{S-S}). In the (1,2) structure in Se_2S_6, this leads to

$$E_{12} = E_{Se-Se} + 2E_{S-Se} + 5E_{S-S} . \tag{44}$$

For *all three* remaining structures,

$$E_{13,14,15} = 4E_{S-Se} + 4E_{S-S} . \tag{45}$$

Within the framework of this simple model, the energies of the last three structures are equal, and the energy of the (1,2) differs from them by an amount

$$\Delta E = E_{Se-Se} + E_{S-S} - 2E_{S-Se} . \tag{46}$$

Eq. (44, 45) also apply to the Se_6S_2 structures, with S and Se interchanged. Again we find the (1,2) structure separated from the other three by the energy ΔE in Eq. (46). This argument explains why the energy orderings in Se_2S_6 and Se_6S_2 are the same, and why the most stable (1,2) isomers are separated in energy from three almost degenerate structures. The energy difference ΔE is in reasonable agreement with the measured heats of reaction in the gas and liquid phases.

The results of this model agree with experimental trends found in a range of mixed S-Se systems, particularly the relative abundance of structures with adjacent minority atoms. The symmetry between the energy ordering of the isomers of Se-rich and S-rich

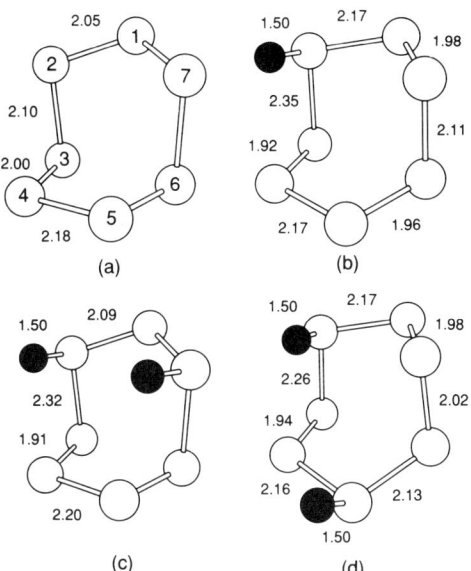

Fig. 5. Structures of (a) S_7 "chair", (b) S_7O, and (c,d) the two most stable isomers found in S_7O_2. Distances are in Å. The white atoms are sulphur, the black atoms oxygen.

molecules is perturbed by differences between the elements such as the atomic radius, but the essential features are unchanged. In this model, the "segregation" of minority components, which could also occur in the disordered liquid and amorphous states, is a consequence of the sign of ΔE in Eq. (46). Systems with a small value of ΔE or with one of opposite sign would behave differently.

Sulphur ring oxides. Sulphur oxides have received much study, and the oxides of S_7 are of particular interest, as this is the only case presently known where a sulphur cluster has both a stable monoxide and dioxide. The most stable isomer of S_7 is the "chair" form shown in Fig. 5(a), and it is remarkable achievement that single crystal samples have been prepared of the monoxide S_7O. In this case, we showed that MD/DF calculations[39] could simulate the change in structure from an eight-membered ring – analogous to S_8 – to the experimentally found form, an O-atom bonded axially to a single member of the S_7 chair structure [Fig. 5(b)]. The dioxide S_7O_2 can be obtained by the oxidation of S_7, S_8, S_7O, or S_8O with CF_3CO_3H. It decomposes readily at 60–62°C with vigorous evolution of SO_2,[41] and it has not yet been possible to perform X-ray diffraction studies. The Raman spectra in CS_2 solution exhibits two infrared bands at 1127 and 1138 cm^{-1}.[41] There must then be at least one S atom between the SO groups, since strong vibrational coupling would lead otherwise to a larger separation between the two SO stretching modes. Even with this information, however, the many candidates for the stable isomers of S_7O_2 include seven- and eight-membered rings, singlet and triplet states, and the possibility of the SO groups being aligned equatorially/axially or *cis-/trans-* relative to the rings.

MD/DF calculations were carried out for a large number (>20) of starting geometries.[42] Several were far from a stable local minimum in the energy surface and annealing led to substantial distortions. In Fig. 5(c-d) we show the two most stable isomers found for S_7O_2. The energy of the C_s form (c) is less than 0.02 eV below (d) (C_1 symmetry), so that both are candidates for the most stable isomer. In analyzing the results, it is very useful to note that there is a definite relationship between the measured Raman frequencies and bond lengths in cumulated sulphur rings.[41] This is shown in Fig. 6 and indicates that the measured Raman frequencies favour the C_1 structure, which was one of the two suggested by Steudel[43] for the most stable isomer.

A study of all the isomer structures we obtained[42] indicates a preference for structures related to the "chair" form of S_7 rather than the "boat" form of similar energy, but the distortion of several structures is so great that the use of the terms "chair" and "boat" is questionable.[42] In agreement with the conclusions of Steudel and Sandow,[41] structures with O-atoms attached to adjacent members of S_7-rings were unstable to annealing at 300K. In the four most stable structures, oxygen is bonded in axial position to atom 2, as in S_7O [Fig. 5(b)]. The general trend for axial SO-bonds to be energetically more favourable than equatorial SO-bonds is well known in ring structures.[44] Eight-membered rings are also possible in the dioxide, and calculations for two representative structures led to energies over 0.6 eV above the isomer 5(c). The lengths of the S-O bonds are very similar to that found in the eight-membered ring of the

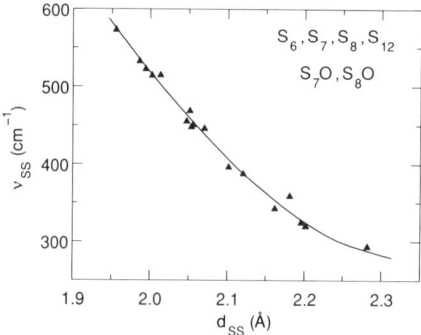

Fig. 6. Relationship between the S-S stretching vibration frequencies ν_{SS} of a bond in homocyclic sulphur rings and the corresponding bond length d_{SS}. The data were taken for the compounds shown.

monoxide (1.75 Å).[39] All the structures discussed above are singlets, i.e., the highest occupied molecular orbital is doubly occupied. Triplet structures were less stable in all cases we have checked. I expect that similar trends should occur in the structures of other oxides of sulphur.

Group Va elements: phosphorus and arsenic

In elemental form, phosphorus shows a structural variety exceeded only by sulphur and possibly boron, and there have been many studies of the crystalline forms.[28,45] The microscopic structures of the amorphous modifications (red, black, grey vitreous) are still under debate. Gas phase clusters have been of interest for many years, and Martin[46] recently detected mass spectroscopically P_n^+ clusters up to $n = 24$. Nevertheless, little is known about the structure of clusters with $n > 4$, and this is also true for arsenic clusters, As_n.

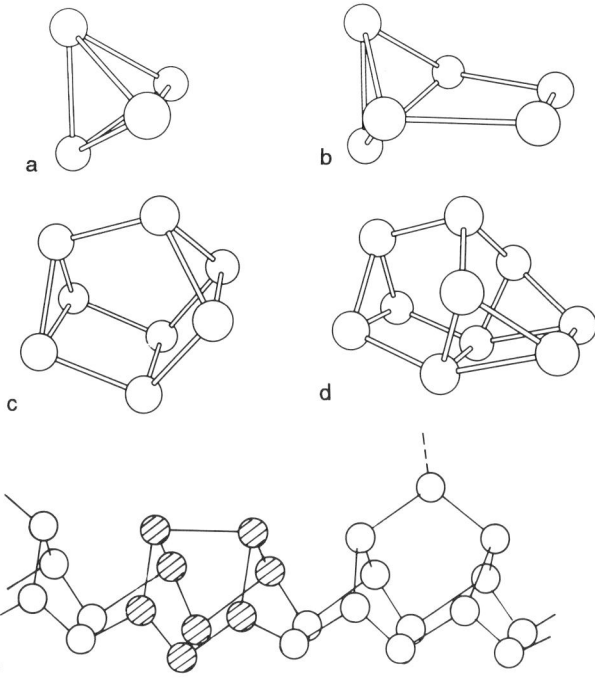

Fig. 7. Structures of the (calculated) most stable isomers of phosphorus clusters: (a) P$_4$, (b) P$_6$, (c) P$_8$, (d) P$_{10}$. The tube that occurs as a component of the Hittorf form of crystalline phosphorus (violet, monoclinic) is shown in (e).

Clusters of phosphorus and arsenic. We have performed MD/DF calculations for neutral and charged clusters of phosphorus and arsenic (with up to 11 atoms).[47,48] The geometries and vibration frequencies in P$_2$ and P$_4$ agree well with experimental data, so we may expect reliable predictions for structures that have not yet been established experimentally. In the larger clusters we obtained some of the most interesting and unexpected structures of all our MD/DF work:
- Although the tetrahedral structure is energetically favoured in P$_4$, there is a large "basin of attraction" for a D_{2d} "roof" structure, i.e., this structure is the closest minimum for a large region of configuration space.
- The "roof" structure is a prominent feature in amorphous phosphorus and in the low-lying isomers of P$_5$ to P$_8$. The calculated ground states in P$_5$, P$_6$, and P$_7$ have a P$_4$-roof with an additional one, two and three atoms, respectively. The structures of P$_9$ to P$_{11}$ can similarly be derived from the most stable isomer of P$_8$.

In Fig. 7(a-d) we show calculated structures for the most stable isomers in P$_4$ to P$_{10}$. It has been known for many years that the tetrahedron is the most stable form of P$_4$, but the remaining structures (and the structures for clusters with an odd number of atoms) had not been found prior to the MD/DF work. In P$_8$, for example, the much-studied cubic (O_h) form corresponds to a shallow local minimum in the energy surface, but simulated annealing led to the C_{2v} structure [Fig. 7(c)], which is much (ca. 1.7 eV) more stable. This "wedge" or "cradle" structure, which may be viewed as a (distorted) cube with one bond rotated through 90°, is a structural unit in violet (monoclinic, Hittorf) phosphorus[49] [see Fig. 7(e)] and forms the basis of the important As_4S_4 (realgar) class of molecules. There is a striking analogy between the structures of the P_8-isomers and those of the isoelectronic hydrocarbons $(CH)_8$. The cubic form of the latter (cubane) has been prepared by Eaton and Cole,[50] and can be converted catalytically to the wedge-shaped form (cuneane).[51]

We have noted above that the structures of P_9 to P_{11} can be derived from the cuneane structure [Fig. 7(c)]. The most stable form of P_9 has a cage structure that is also found as a structural unit in monoclinic phosphorus [Fig. 7(e)] and in many P_nS_m molecules. The tendency of the structures to propagate in a preferred direction is reflected in the crystal structure of violet (monoclinic, Hittorf) phosphorus, which comprises the long tubes shown in Fig. 7(e).

The structures of P_8 and the isomer of P_6 shown in Fig. 7(b) were unexpected, but they have been confirmed by subsequent calculations using correlated wave functions.[52,53] One interesting question that could not be answered definitively by calculations using the LSD approximation is the stability of P_8 relative to two P_4 tetrahedra, as we have seen above that binding energies can sometimes be overestimated substantially by the LSD approximation. This effect is already apparent in the calculated atomization energies in clusters up to P_4, where experimental results are available. The LSD calculations indicated that the C_{2v} isomer of P_8 was slightly more stable than two tetrahedra, while the opposite result was found in calculations using correlated wave functions.[52,53] To examine these issues in more detail, Ballone and Jones[48] have performed calculations on phosphorus and arsenic clusters with up to 11 atoms using a non-local extension of the LSD approximation:[54]

$$E_{xc} = E_{xc}^{LSD} - b \sum_\sigma \int d\mathbf{r}\, n_\sigma^{4/3} \frac{x_\sigma^2}{1 + 6bx_\sigma \sinh^{-1} x_\sigma}. \quad (47)$$

where

$$x_\sigma = \frac{|\nabla n_\sigma|}{n_\sigma^{4/3}}; \quad b = 0.0042 \text{ a.u.} \quad (48)$$

The parameter b was adjusted to reproduce the exchange energies of some closed shell atoms, and it has been found that this modification to the LSD approximation gives significant improvements to calculated atomization energies in numerous small molecules.[54]

The geometries of P_n and As_n clusters are virtually unchanged by the use of the non-local approximation (47, 48) for E_{xc}, and changes in the ordering of the isomer energies for a given cluster are small and restricted to clusters with more than nine atoms. The dissociation energies of P_2-P_4, however, now agree much better with the experimental values, and the wedge-shaped P_8 isomer is slightly less stable than two phosphorus tetrahedra.

The calculations on arsenic clusters[48] gave structures and energy orderings that are very similar to those found in phosphorus clusters. The bond lengths are, however, systematically longer than those in the corresponding phosphorus clusters by between 8% and 10%. This difference can be related to the "size" of the atoms, or the radial extent of the atomic orbitals, and we return to this point below.

Molecules containing elements of groups Va and VIa. In view of the bonding trends found within individual groups of the periodic table, it is natural to ask whether molecules containing elements of more than one main group show similar behaviour. An excellent testing ground is provided by molecules containing elements of groups Va and VIa, since these have been studied intensively for many years. Among the many sulphides of P and As,[55] Jones and Seifert[56] studied the geometries and relative stabilities of isomers of A_4X_3 molecules (A=P,As; X=S,Se,Te), as well as mixed systems such as $P_nAs_{4-n}X_3$, $A_4S_nSe_{3-n}$, $P_4S_nTe_{3-n}$, and P_2X_5. We now discuss some of the results.

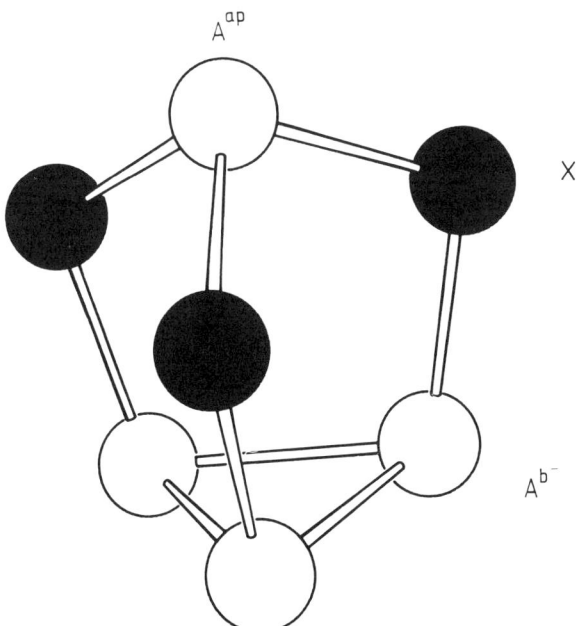

Fig. 8. Structure of A_4X_3 molecules.

The most stable isomers of the A_4X_3 compounds have a C_{3v} cage structure [Fig. 8]. The A-X bond lengths are very close to the sums of the covalent radii determined from calculations on homonuclear molecules, increasing as X goes from S→Se→Te. The bond angles at X are similar to those found in the S_n and Se_n rings, increasing as A goes from P→As as the size of the A_3 base increases. If we assume that P-P and As-As bond energies can be transferred from calculations for the A_4 tetrahedra (see above), we can determine from the calculated binding energies (E_B, Table 2) the bond energies E_{AX}, which are also given in Table 2. If we now turn to the most stable isomers of the compounds $A_4S_nX_{3-n}$ (A=P,As; X=Se,Te), we can compare calculations of the binding energies with the results of a simple model where we use these values of E_{AX}. The excellent agreement found (Table 3) shows how closely the bond energies are related in two distinct families of molecules, and similar results hold for the isomers of the family $P_4S_nTe_{3-n}$ (n=0-3).[56] The transferability of bond energies found in these cases is very encouraging.

The ordering of the bond energies in the A_4X_3 molecules (Table 2):

$$\text{AsS} < \text{PS} < \text{AsTe} < \text{AsSe} < \text{PTe} < \text{PSe}. \tag{49}$$

can be understood from a comparison of the radial functions for the s- and p-orbitals (ℓ=0,1) in atoms of groups Va and VIa [Fig. 9]. The orbital functions in P and As are noticeably more similar than those of S and Se, as we expect from the bond lengths found in the cluster calculations [$\sim$ 10% difference in the former, 12-15% in the latter]. The $3p$ orbital of P and the $4p$ function of Se match better than the $3p$ functions of P and S, although the latter pair belong to the same row of the periodic table. The contraction of the orbital function with increasing nuclear charge (P to S) is compensated if the principal quantum number is increased (P to Se), and the PSe bond is stronger than the PS bond in these systems. The relationship between bond strength and orbital overlap is well established in chemistry,[57] and was introduced in the DF context by Harris and Jones.[40] The results described here show that it has quantitative value.

Table 2. Calculated binding energies (E_B) and bond energies (E_{AX}) between the atoms A and X for the most stable isomers of A_4X_3 molecules (after Ref.[56]).

	$E_B[eV]$	$E_{AX}[eV]$
P_4S_3	-18.17	-1.73
P_4Se_3	-20.68	-2.15
P_4Te_3	-20.44	-2.11
As_4S_3	-15.87	-1.52
As_4Se_3	-18.80	-2.01
As_4Te_3	-18.67	-1.99

Table 3. Binding energies (E_B) for the most stable isomers of $A_4S_nX_{3-n}$, $n = 1, 2$ (X=Se,Te) molecules, and binding energies estimated from bond energies (E_B^{AX}) (after Ref.[56]).

	$E_B[eV]$	$E_B^{AX}[eV]$[a]
P_4S_2Se	-19.01	-19.02
P_4SSe_2	-19.85	-19.86
P_4S_2Te	-18.89	-18.94
P_4STe_2	-19.64	-19.70
As_4S_2Se	-16.84	-16.85
As_4SSe_2	-17.82	-17.83

(a) Bond energies for AX from Table 2, AA from A_4 molecules.[47,48]

Group IIIa elements: aluminium and gallium

The elemental clusters discussed above, as well as the molecules containing group Va and VIa atoms, are typically covalently bonded systems. The bulk systems are generally semiconductors or insulators, and there is a substantial energy gap between the highest occupied and lowest unoccupied molecular orbitals. We now turn to the group IIIa elements aluminium and gallium, which are metallic in the bulk and whose clusters have been studied widely.

Work on Al_n has included magnetic properties,[58] ionization thresholds and reactivities,[59] and the static polarizabilities.[60] There have also been measurements of collision induced dissociation of Al_n^+,[61] and the photoelectron spectroscopy of Al_n^-[62,63] and Ga_n^-,[63] where transitions between states of the anions and states of the neutral clusters can be observed. Gallium clusters with up to more than ten atoms have been detected following laser vaporization of gallium arsenide,[64] and are of are of particular interest in intermetallic compounds with alkali metals, where Ga_8-dodecahedra, Ga_{12}-icosahedra, and Ga_{15} clusters[65] have all been found. The structure of bulk (α)-gallium has been interpreted by von Schnering and Nesper[66] as icosahedra that have been dissected and condensed via edge-sharing.

The dimer is the best studied of the aluminium clusters, although the nature of the ground state has only recently been established. The two candidates for the ground state are the $^3\Pi_u$ ($\sigma_g\pi_u$) and $^3\Sigma_g^-$ (π_u^2) states, and the ease of transfer between σ- and π-electrons is reflected in the fact that each has been favoured at different times. Recent experimental work[70] supports theoretical predictions[67,68] that the $^3\Pi_u$ state is slightly (less than 0.025 eV) more stable. Experimental and theoretical spectroscopic parameters for some low-lying states of Al$_2$ are shown in Table 4, and energy curves for Al$_2$ and Al$_2^-$ in Fig. 10. The MD/DF calculations[65] agree well with available data for Al$_2$,[13,70] although the $^3\Pi_u$ state is slightly (0.08 eV) less stable than the $^3\Sigma_g^-$ state. The equilibrium separations r_e and vibration frequencies ω_e are in excellent agreement with experiment for both states. The well depth [2.03 eV compared with the experimental value 1.5 eV][13] shows an overestimate similar to those found in other sp-bonded systems. The $^5\Sigma_u^-$ state has not been studied before in Al$_2$, but it lies within 1300 cm^{-1} of the $X^3\Sigma_g^-$ state in (valence) isoelectronic B$_2$.[71]

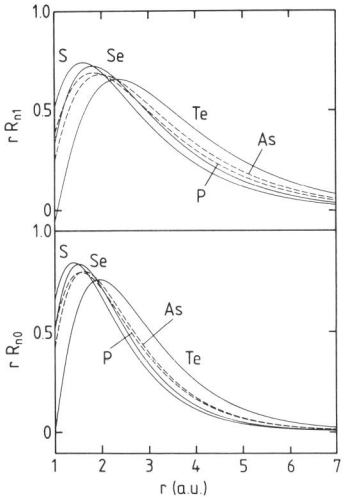

Fig. 9. Radial orbital functions (rR_{nl}) for valence electrons in P and As (group Va, dashed) and S, Se, and Te (group VIa, full curves). The s- and p-functions are shown in the lower and upper frames, respectively (see Ref.[56]).

The most significant difference between the results for the aluminum and gallium dimers is that the bond lengths in Ga$_2$ are 3–7% *shorter* than those in the lighter Al$_2$, a feature evident in Fig. 10. It is unusual to find bonds that are shorter than those between lighter atoms in the same main group, but it is a general feature of these clusters (see below). The vertical excitation energies from states of Al$_2^-$ → Al$_2$ can be observed in photoelectron detachment spectroscopy of negative ions, a technique that has been applied recently to Al$_n$ and Ga$_n$ clusters up to n=15.[63] For Al$_2^-$, there is a peak at ~4.2 eV that shows a vibrational structure with ω_e = 450 ± 40cm^{-1}. This agrees well with both the excitation energy from the anion [Fig. 10] and the vibration frequency of the $^5\Sigma_u^-$ state the neutral dimer [Table 4]. In fact, the overall agreement of

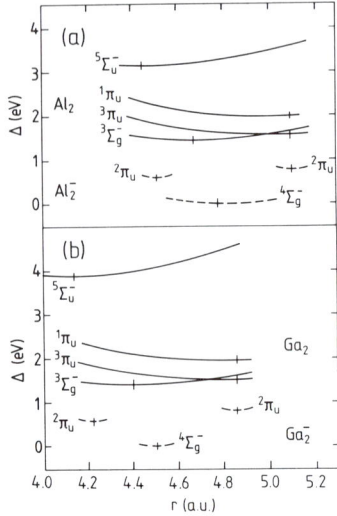

Fig. 10. Energy curves for low-lying states of (a) Al_2^- and Al_2, (b) Ga_2^- and Ga_2.

the MD/DF results with available data on Al_2 and its ions is very encouraging for the application to larger clusters, for which there is much less spectroscopic information.

The results of the MD/DF calculations for the larger clusters show interesting patterns:

- There is a richer variety of structures than we found in clusters of groups Va and VIa elements, and it is easier to transfer electrons between π-orbitals (which dominate in the bonding in planar structures) and σ-orbitals. Planar structures are the most stable for $n < 5$, three dimensional structures for $n > 5$. It is essential to include electron spin in the calculations, as there is a transition at $n = 6$ to ground states with minimum spin degeneracy. The structural variety is consistent with the "metallic" nature of the elements: The valence sp-shells in the atoms are less than half-filled, and there are usually unoccupied bonding orbitals near the highest occupied orbital.

- In both the Al_n and Ga_n families, the stable forms are found by capping smaller clusters, as we show in Fig. 11 for Al_5 to Al_{10}. This prescription alone does not allow us to predict the most stable structures, since the number of possible capped structures increases rapidly with increasing n. The structures comprise triangles packed with particular patterns of dihedral angles. Similar patterns are also found in bulk aluminium and in α-gallium, and the tendency to favour triangular units is found in MD simulations of liquid Al.[72]

- Although the structures of Al_n and Ga_n gallium clusters show many similarities, there are significant differences in (a) the bond lengths [$\sim 5\%$ *shorter* in clusters of the heavier element Ga] and (b) the bond angles, which tend to be closer to $90°$ in gallium clusters.

While it is unusual that clusters of a heavier element (Ga) are more compact than those of an element of the same group with lower atomic number (Al), we note here that the measured sp-promotion energy Δ_{sp} $[^2P(ns^2np) \rightarrow {}^4P(ns^1np^2)]$,[73] is smaller in Al

Table 4. Molecular parameters r_e [atomic units], ω_e [cm^{-1}] for low-lying states of Al$_2$, together with energies (ΔE) relative to the ground state.

		r_e	ω_e	ΔE
$^3\Pi_u(\sigma_g\pi_u)$:				
	(a)	5.135	284.97	
	(b)	5.150	277	
	(c)	5.19	284	
	(d)	5.095	290	
Expt	(e)	5.10	284.2	
$^3\Sigma_g^-(\pi_u^2)$:				
	(a)	4.687	355.15	+0.06
	(b)	4.711	343	+0.02
	(c)	4.78	340	+0.02
	(d)	4.672	340	−0.08
Expt	(e,f)	4.660	350.01	> 0
$^5\Sigma_u^-(\sigma_g^2\sigma_u\pi_u^2\sigma_g)$:				
	(d)	4.444	435	+1.59

(a) Ref.[67]. Coupled-cluster doubles + ST(CCD)
(b) Ref.[68]. Complete active space SCF/ second order CI CASSCF/SOCI.
(c) Ref.[69]. Multireference configuration interaction (MRD-CI).
(d) Ref.[65]. MD/DF.
(e) Ref.[70].
(f) Ref.[13].

than in Ga, and the atomic dipole polarizability of Ga is *smaller* than that of Al.[65] Both effects are consistent with weaker *sp*-hybridization in Ga, and with the observation that Ga clusters have bond angles closer to 90°, a favoured value for unhybridized *p*-orbitals.

In Fig. 12 we show the atomic valence orbitals in group IIIa elements. The *s*-orbital, in particular, is more compact in Ga than in Al. The compact nature of the Ga atom is reflected in the structures of the bulk elements. The nearest neighbour separation is 5.411 a.u. in face-centred-cubic Al, and a weighted average over the seven near neighbours in α-Ga (5.107 a.u.) is 5.3% less. In α-Ga, the angle between the short interlayer bond (4.65 a.u.) and the intralayer bonds are, with one exception, close to 90° (99.5° - 115.4°, with one bond angle of 140.0°).[28]

In addition to the structural differences, we also observed different behaviours of the two elements during the simulated annealing process. The minima in the energy surfaces for the gallium clusters, especially the larger ones, were less pronounced. The barriers between the structures were smaller and it was more difficult to locate the minima of the potential energy surface. The low structural rearrangement energy that we find in Ga$_n$ clusters may be related to the very low melting point of gallium (29.78°C).

DISCUSSION AND CONCLUDING REMARKS

The study of the geometrical structure of molecules and clusters is of central im-

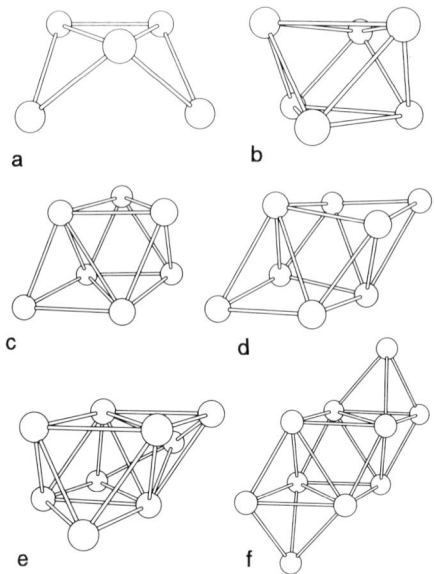

Fig. 11. The most stable isomers of Al_5 to Al_{10}.

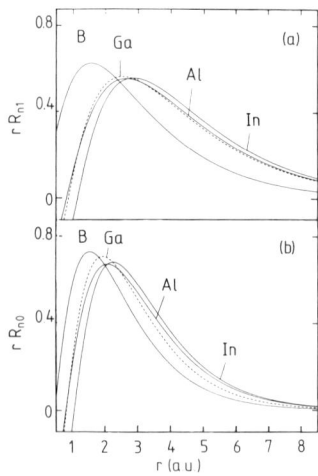

Fig. 12. Radial functions for (a) p- and (b) s-valence orbitals in group IIIa atoms.

portance in chemistry. If one does not already know the answer to a structural problem, then a search of conformational space is unavoidable. We have seen that this involves two distinct aspects: the calculation of the total energy of the system of electrons and ions for a given geometry, and the determination of the most stable of the structures associated with the many minima in the energy surface. Some of the established methods of theoretical chemistry have emphasized the first aspect (accurate calculations of the total energy from the many-electron wave function), others the second (semi-empirical, but necessarily less reliable energy calculations for large regions of configuration space).

The approach we have discussed here addresses both problems with comparable emphasis. The density functional formalism, with the LSD approximation for the exchange-correlation energy, provides us with an approximate method of calculating energy surfaces that has predictive value in a wide range of contexts. It is the most widely used method – without adjustable parameters – for calculating the electronic and cohesive properties of bulk materials, and DF calculations have the advantage that they can be carried out with approximately equal ease for elements throughout the periodic table. We have seen in these lectures that it is then possible to identify and discuss bonding trends. The DF approach has gained acceptance as a method for calculations on clusters and small molecules, and this has been enhanced by the improvements that have resulted from the use of non-local approximations for E_{xc}. When coupled with MD at elevated temperatures (simulated annealing), we have been able to study cases where the most stable isomers are unknown, or where the energy surfaces have many local minima. Examples of both are found in small sulphur and selenium clusters, and we can describe small energy differences between isomers of the heterocyclic molecules Se_nS_m very well. The structures predicted in S_4, S_5 and S_9 are supported by recent calculations using correlated wave functions. In phosphorus and arsenic clusters we predict structures with plausible trends and an interesting analogy to isoelectronic hydrocarbons.

Our present focus, however, is not on the quality of the results in specific cases, but on the structural patterns that are found. We have seen that careful calculations of structures and energy differences in a family of molecules can lead to reliable predictions of these quantities in related compounds. This should also be true in other cases. The Se_nS_m isomers, which show regular deviations from the crown structures of the homocyclic molecules, provide one of many examples of recurring structural patterns, such as coordination numbers and structural parameters. We find bond lengths that transfer remarkably well from one molecule to another, and are consistent with atomic radii in Se that are $\sim 15\%$ larger than in S. The radial orbital functions of the elements show a corresponding behaviour.

The radial functions of the elements and the transferability of the corresponding "atomic radii" between similar structures have also been discussed in the context of phosphorus and arsenic (group Va) where the bond lengths in the former are $\sim 10\%$ shorter. A continuation of this trend to group IVa (C, Si, Ge, ...) would indicate that the bond lengths and radial valence functions in Si and Ge should be very similar. The nearest neighbour separation in bulk Si is, in fact, only 3.7% shorter than that in Ge, and the similarity of the radial functions was noted some years ago by Harris and Jones.[40] The valence radial functions for B, Al, Ga, and In [Fig. 12] show that this pattern continues to group IIIa elements.

The explanation for this trend is not difficult to find. The third-row elements (Ga, Ge, As, Se, Br) differ from other elements in that this is the first row where d-electrons occur in the core. The presence of a full, but relatively weakly bound d-shell means that the 10 d-electrons in the core do not shield completely the extra 10 positive charges on the nucleus. The effective nuclear charge is larger than in the second-row elements (Al, Si, P, S, Cl), leading to a contraction of the valence orbitals. A similar effect arises in the actinides, where f-electrons enter the core for the first time, but it is less pronounced, since the $4f$-core electrons are more compact. Properties related to the nature of the valence orbitals then show an irregular behaviour with increasing atomic number, examples being the orbital functions and the related eigenvalues ε_i.[65] The contraction of the third-row orbitals is even more pronounced if relativistic effects are included,[74] and increases as we move left in the row. Here the d-orbitals are the most extended, the effect of the "d-block contraction" is largest, and the valence orbitals in the heavier element Ga are more compact than in Al.

In these lectures I have covered a lot of ground and have studied a large variety of systems. I have tried to show that *trends* in structures and bond energies can be a fascinating area of research. The use of temperature to "melt" and "anneal" samples has led to surprising and interesting results in many cases, and I hope that I have been able to impart some of the sense of excitement that has accompanied this work.

Acknowledgment

Much of the work described here has been performed in collaboration with, or has been influenced by discussions with many colleagues. In particular, I thank P. Ballone, R. Car, C.-Y. Cha, G. Ganteför, O. Gunnarsson, J. Harris, D. Hohl, M. Parrinello, and G. Seifert. The calculations were performed on Cray computers of the Forschungszentrum Jülich and the German Supercomputer Centre (HLRZ).

REFERENCES

1. R. Hoffmann, *Scientific American*, February 1993, p. 40.
2. F. Crick, in: *What mad pursuit*, Penguin, London (1988), p. 150.
3. R.O. Jones, *J. Chem. Phys.* **82**: 325 (1985).
4. T.P. Martin, T. Bergmann, and B. Wassermann: in *Microclusters*, Proceedings of the First NEC Symposium, Tokyo, 1986, S. Sugano, Y. Nishina and S. Ohnishi, eds., Springer, Berlin (1987), p. 152.
5. A. Cayley, *Phil. Mag. (4)* **47**: 444 (1874); A.C. Lunn and J.K. Senior, *J. Phys. Chem.* **33**: 1027 (1929); G. Polyá, *Acta Math.* **68**: 145 (1937).
6. M.R. Hoare and J.A. McInnes, *Adv. Phys.* **32**: 791 (1983).
7. L.T. Wille and J. Vennik, *J. Phys. A* **18**: L419, L1113 (1985).
8. M.R. Garey and D.S. Johnson, *Computers and Intractability: A Guide to the Theory of NP-Completeness*, Freeman, San Francisco (1979).
9. S. Kirkpatrick, C.D. Gelatt, and M.P. Vecchi, *Science* **220**: 671 (1983).
10. R.O. Jones and O. Gunnarsson, *Rev. Mod. Phys.* **61**: 689 (1989).
11. R. Car and M. Parrinello, *Phys. Rev. Lett.* **55**: 2471 (1985).
12. F. Stillinger, T.A. Weber, and R.A. LaViolette, *J. Chem. Phys.* **85**: 6460 (1986).
13. K.P. Huber and G. Herzberg, *Molecular Spectra and Molecular Structure. IV. Constants of Diatomic Molecules*, Van Nostrand Reinhold, New York (1979).
14. F.H. Stillinger and T.A. Weber, *J. Phys. Chem.* **91**: 4899 (1987).
15. D. Hohl, R.O. Jones, R. Car, and M. Parrinello, *J. Chem. Phys.* **89**: 6823 (1988).
16. See, for example, H. Margenau and G.M. Murphy, *Mathematics of Physics and Chemistry*, Van Nostrand, New York (1955).
17. P. Hohenberg and W. Kohn, *Phys. Rev.* **136**: B864 (1964).
18. W. Kohn and L.J. Sham, *Phys. Rev.* **140**: A1133 (1965).
19. J. Harris and R.O. Jones, *J. Phys. F* **4**: 1170 (1974).
20. D.C. Langreth and J.P. Perdew, *Solid State Commun.* **17**: 1425 (1975).
21. O. Gunnarsson and B.I. Lundqvist, *Phys. Rev. B* **13**: 4274 (1976).
22. J. Harris, *Phys. Rev. A* **29**: 1648 (1984).
23. J.C. Slater, *Quantum Theory of Atomic Structure*, Vol. II, McGraw-Hill, New York (1960), Appendix 21.
24. O. Gunnarsson and R.O. Jones, *Phys. Rev. B* **31**: 7588 (1985).
25. G.B. Bachelet, D.R. Hamann, and M. Schlüter, *Phys. Rev. B* **26**: 4199 (1982).
26. R. Stumpf, X. Gonze, and M. Scheffler, Research Report, Fritz-Haber-Institut, Berlin (April, 1990), unpublished.

27. L. Verlet, *Phys. Rev.* **159**: 2471 (1967).
28. J. Donohue, *The Structures of the Elements*, Wiley, New York (1974), Chapters 8 [group Va] and 9 [group VIa].
29. R. Steudel, in: *Studies in Inorganic Chemistry, Vol. 5*, A. Müller and B. Krebs, eds., Elsevier, Amsterdam (1984).
30. R. Steudel and E.M. Strauss, in: *The Chemistry of Inorganic Homo- and Heterocycles, Vol. 2*, Academic, London (1987), p. 769.
31. H. Bitterer, ed., *Schwefel: Gmelin Handbuch der Anorganischen Chemie*, 8. Aufl., Ergänzungsband 3, Springer, Berlin (1980), p. 8.
32. K. Raghavachari, C.M. Rohlfing, and J.S. Binkley, *J. Chem. Phys.* **93**: 5862 (1990).
33. R. Steudel, T. Sandow, and J. Steidel, *Z. Naturforsch. Teil B* **40**: 594 (1985).
34. R. Steudel, *Angew. Chem.* **87**: 683 (1975) [*Angew. Chem. Int. Edit. Engl.* **14**: 655 (1975)]; *Z. Naturforsch. Teil B* **38**: 543 (1983).
35. R. Steudel, T. Sandow, and R. Reinhardt, *Angew. Chem.* **89**: 757 (1983) [*Angew. Chem. Int. Edit. Engl.* **16**: 716 (1983)].
36. L. Pauling, *Proc. Nat. Acad. Sci. USA* **35**: 495 (1949).
37. F. Tuinstra, *Structural Aspects of the Allotropy of Sulphur and Other Divalent Elements*, Delft (1967).
38. D. Hohl, R.O. Jones, R. Car, and M. Parrinello, *Chem. Phys. Lett.* **139**: 540 (1987).
39. R.O. Jones and D. Hohl, *J. Am. Chem. Soc.* **112**: 2590 (1990).
40. J. Harris and R.O. Jones, *Phys. Rev. A* **19**: 1813 (1979).
41. R. Steudel and T. Sandow, *Angew. Chem.* **90**: 644 (1978); *Angew. Chem. Int. Ed. Engl.* **17**: 611 (1978).
42. R.O. Jones, *Inorg. Chem.*, to be published.
43. R. Steudel, *Phosphorus and Sulphur* **23**: 44 (1985).
44. See, for example, C. Romers, C. Altona, H.R. Buys, and E. Havinga, *Top. Stereochem.* **4**: 39 (1969).
45. D.E.C. Corbridge, *Phosphorus. An Outline of its Chemistry, Biochemistry and Technology*, Elsevier, Amsterdam (1985).
46. T.P. Martin, *Z. Phys. D* **3**: 221 (1986).
47. R.O. Jones and D. Hohl, *J. Chem. Phys.* **92**: 6710 (1990); R.O. Jones and G. Seifert, *J. Chem. Phys.* **96**: 7564 (1992).
48. P. Ballone and R.O. Jones, to be published.
49. H. Thurn and H. Krebs, *Acta Cryst. B* **25**: 125 (1969).
50. P.E. Eaton and T.W. Cole, Jr., *J. Am. Chem. Soc.* **86**: 962, 3157 (1964).
51. L. Cassar, P.E. Eaton, and J. Halpern, *J. Am. Chem. Soc.* **92**: 6366 (1970).
52. R. Janoschek, *Chem. Ber.* **125**: 2687 (1992).
53. M. Häser, U. Schncider, and R. Ahlrichs, *J. Am. Chem. Soc.* **114**: 9551 (1992).
54. See, for example, A.D. Becke, *J. Chem. Phys.* **96**: 2155 (1992); B.G. Johnson, P.M.W. Gill, and J.A. Pople, *J. Chem. Phys.* **97**: 7846 (1992).
55. D.E.C. Corbridge, *The Structural Chemistry of Phosphorus*, Elsevier, Amsterdam (1974).
56. R.O. Jones and G. Seifert, *J. Chem. Phys.* **96**: 2942 (1992).
57. See, for example, R.S. Mulliken, *J. Phys. Chem.* **56**: 295 (1952) and references therein.
58. D.M. Cox, D.J. Trevor, R.L. Whetten, E.A. Rohlfing, and A. Kaldor, *J. Chem. Phys.* **84**: 4651 (1986) [$n = 2 - 25$].
59. D.M. Cox, D.J. Trevor, R.L. Whetten, and A. Kaldor, *J. Phys. Chem.* **92**: 421 (1988) [$n = 2 - 13$].
60. W.A. de Heer, P. Milani, and A. Châtelain, *Phys. Rev. Lett.* **63**: 2834 (1989) [up to $n = 61$].
61. M.F. Jarrold, J.E. Bower, and J.S. Kraus, *J. Chem. Phys.* **86**: 3876 (1987) [$n = 3 - 26$]; L. Hanley, S.A. Ruatta, and S.L. Anderson, *J. Chem. Phys.* **87**: 260 (1987) [$n = 2 - 7$].

62. G. Ganteför, M. Gausa, K.H. Meiwes-Broer, and H.O. Lutz, *Z. Phys. D* **9**: 253 (1988) [$n = 3 - 14$]; K.J. Taylor, C.L. Pettiette, M.J. Craycraft, O. Chesnovsky, and R.E. Smalley, *Chem. Phys. Lett.* **152**: 347 (1988) [$n = 3 - 32$].
63. C.Y. Cha, G. Ganteför, and W. Eberhardt, *J. Chem. Phys.* **100** (1994), in press.
64. S.C. O'Brien, Y. Liu, Q. Zhang, J.R. Heath, F.K. Tittel, R.F. Curl, and R.E. Smalley, *J. Chem. Phys.* **84**: 4074 (1986).
65. R.O. Jones, *Phys. Rev. Lett.* **67**: 224 (1991); *J. Chem. Phys.* **99**: 1194 (1993).
66. H.G. von Schnering and R. Nesper, *Acta Chem. Scand.* **45**: 870 (1991).
67. K.K. Sunil and K.D. Jordan, *J. Phys. Chem.* **92**: 2774 (1988).
68. C.W. Bauschlicher, Jr., H. Partridge, S.R. Langhoff, P.R. Taylor, and S.P. Walch, *J. Chem. Phys.* **86**: 7007 (1987).
69. U. Meier, S.D. Peyerimhoff, and F. Grein, *Z. Phys. D* **17**: 209 (1990).
70. M.F. Cai, T.P. Djugan, and V.E. Bondybey, *Chem. Phys. Lett.* **155**: 430 (1989).
71. M. Dupuis and B. Liu, *J. Chem. Phys.* **68**: 2902 (1978).
72. V.A. Polukhin and M.M. Dzugotov, *Phys. Met. Metall.* **51**: 50 (1981); J. Hafner, *J. Non-Crystalline Solids* **117/118**: 18 (1990).
73. The multiplet averaged values are B 3.57 eV; Al 3.47 eV; Ga 4.71 eV; In 4.35 eV; Tl 5.64 eV. See C.E. Moore, *Atomic Energy Levels*, National Bureau of Standards Circular 467, USGPO, Washington. Vol. I (1949), Vol. II (1952), Vol. III (1958).
74. J.P. Desclaux, *At. Data Nucl. Data Tables* **12**: 311 (1973).

LARGE-SCALE ELECTRONIC STRUCTURE CALCULATIONS IN SOLIDS

Paolo Giannozzi

Scuola Normale Superiore, Piazza dei Cavalieri 7
I- 56126 PISA, Italy

1 Introduction

Electronic-structure calculations in solids have considerably evolved from early approaches (*band structure* calculations in periodic model potentials, aimed at reproducing simple crystals) into very sophisticated and powerful techniques. These techniques usually require no or very little experimental input beyond the basic information on atomic composition and some structural data. This is the origin of the (perhaps too ambitious) definitions of *ab-initio*, or *first-principles*, or (perhaps more appropriately) *parameter-free*, which usually label these techniques. In conjunction with the enormous increase in computer power (and the decrease in computer prices), ab-initio methods now allow us to accurately reproduce and even to *predict* electronic and structural properties of *real materials*, and not just the simplest ones. This *predictive power* makes a strong case in favour of ab-initio methods, whenever they are applicable, with respect to *empirical* or *semiempirical* methods. These are far less computationally demanding but also less reliable.

This paper will introduce ab-initio techniques based on Density Functional Theory, in particular those using plane waves and pseudopotentials. Such an approach has proven to be surprising successful and specially well suited to the study of weakly correlated, s-p bonded materials, like for example semiconductors. Relevant aspects for computer implementation will be examined in some detail. In particular, I will examine problems that arise when dealing with complicated systems, that is those described by unit cells containing many atoms. With present algorithms, both computer time and memory requirements scale unfavourably with the size of the unit cell. As a consequence the study of many interesting physical systems such as alloys, disordered or amorphous materials, defects, surfaces and interfaces, is still difficult or impossible. Possible ways to improve the situation will be pointed out.

2 Theoretical Approach

Solving the many-body Schrödinger equation for electrons in a real material is by no means a trivial task even in the presence of simplifying assumptions (such as perfect periodicity for crystals). For completeness, I list here the main possibilities:

i) 'Exact' solutions: Quantum Monte Carlo calculations, Fermionic simulations with Hubbard-Stratonovich transformations. These very recent developments can yield extremely accurate results, but they are also extremely time-consuming so that only very simple problems can be treated.

ii) 'Green's Function approach'. This is based on a perturbative expansion of the self-energy. The most often used approach is the so-called 'GW approximation'. This method is also very accurate (although not undisputed), but it is also computationally cumbersome and it is still limited to simple crystals.

iii) Hartree-Fock-based methods. Hartree-Fock alone is known not to be very accurate, so that some form of correction (Configuration Interaction or perturbative) has to be added to the Hartree-Fock results in order to get better accuracy. This set of techniques, generally used in Quantum Chemistry, has traditionally been applied to molecules rather than to extended system.

iv) Density Functional Theory (DFT), mainly in the Local Density Approximation (LDA). In contrast to the previous approaches, DFT is a *ground-state* theory in which the emphasis is on the *charge density* as the relevant physical quantity. DFT in the LDA has proved to be highly successful in describing structural and electronic properties in a vast class of materials. Furthermore LDA is computationally very simple. For these reasons LDA has become a common tool in first-principles calculations aimed at describing – or even predicting – properties of complex condensed matter systems.

Many good books and papers have been written on DFT and LDA[1], so that I will recall here only the basic facts. DFT looks deceivingly simple, but it hides a number of subtle points. The interested reader should consult the more specialized literature.

We can start from the obvious statement that an external potential $V(\mathbf{r})$ acting on a system of N interacting electrons will determine the charge density $n(\mathbf{r})$ of the ground state. The opposite statement is far less obvious. However this is exactly what has been demonstrated by Hohenberg and Kohn in 1964[2]: there is only one external potential $V(\mathbf{r})$ which yields a given ground-state charge density $n(\mathbf{r})$. The demonstration is elementary and uses a *reductio ad absurdum* argument.

DFT arises from the Hohenberg and Kohn theorem: the ground state energy E is also uniquely determined by the ground-state charge density. In mathematical terms E is a *functional* [1] $E[n(\mathbf{r})]$ of $n(\mathbf{r})$. We can write

$$E[n(\mathbf{r})] = F[n(\mathbf{r})] + \int n(\mathbf{r})V(\mathbf{r})d\mathbf{r} \qquad (1)$$

where $F[n(\mathbf{r})]$ is a *universal* functional of the charge density $n(\mathbf{r})$ (and not of V). The variational principle implies that the ground-state energy is *minimised* by the ground-state charge density. In this way, DFT reduces the N-body problem exactly to the determination of a 3-dimensional function $n(\mathbf{r})$ which minimises a functional $E[n(\mathbf{r})]$. Unfortunately this is of little utility as $F[n(\mathbf{r})]$ is not known.

One year later, Kohn and Sham (KS) reformulated the problem[3] and opened the way to practical applications of DFT. First, the system of interacting electrons is mapped on to a fictitious system of non-interacting electrons having the same ground state charge density $n(\mathbf{r})$. This is performed by introducing KS orbitals $\psi_i(\mathbf{r})$ for N electrons, such that

$$n(\mathbf{r}) = 2\sum_{i=1}^{N/2} |\psi_i(\mathbf{r})|^2 \qquad (2)$$

---

[1] A functional is a generalisation of the concept of a function: a function associates a value with another value while a functional associates a value with a given function.

assuming double occupancy of all states. Charge conservation requires that the KS orbitals obey orthonormality constraints:

$$\int \psi_i^*(\mathbf{r})\psi_j(\mathbf{r})d\mathbf{r} = \delta_{ij}. \tag{3}$$

The energy functional is rewritten:

$$E = T_0[n(\mathbf{r})] + \frac{e^2}{2}\int \frac{n(\mathbf{r})n(\mathbf{r}')}{|\mathbf{r}-\mathbf{r}'|}d\mathbf{r}d\mathbf{r}' + E_{xc}[n(\mathbf{r})] + \int n(\mathbf{r})V(\mathbf{r})d\mathbf{r}. \tag{4}$$

The first term is the kinetic energy of non-interacting electrons:

$$T_0[n(\mathbf{r})] = -\frac{\hbar^2}{2m}2\sum_{i=1}^{N/2}\int \psi_i^*(\mathbf{r})\nabla^2\psi_i(\mathbf{r})d\mathbf{r}, \tag{5}$$

where m is the electron mass. The second term (called the Hartree energy) has the electrostatic interactions between clouds of charge. The third term, called the *exchange-correlation energy*, contains everything also and all our ignorance of the density functional.

Minimization of Eq. (4) under the constraints $\int n(\mathbf{r})d\mathbf{r} = N$ yields the KS equations:

$$(H - \epsilon_i)\psi_i(\mathbf{r}) = 0 \tag{6}$$

where

$$\begin{aligned} H &= -\frac{\hbar^2}{2m}\nabla^2 + e^2\int \frac{n(\mathbf{r}')}{|\mathbf{r}-\mathbf{r}'|}d\mathbf{r}' + \frac{\delta E_{xc}}{\delta n(\mathbf{r})} + V(\mathbf{r}) \\ &\equiv -\frac{\hbar^2}{2m}\nabla^2 + V_{scf}(\mathbf{r}) \end{aligned} \tag{7}$$

is the so-called KS Hamiltonian.[2] In this way the N-electron problem is remapped on to a 1-electron problem with a self-consistent potential V_{scf} which is given by the sum of the external (ionic) potential and a screening potential:

$$V_{scf}(\mathbf{r}) = V(\mathbf{r}) + e^2\int \frac{n(\mathbf{r}')}{|\mathbf{r}-\mathbf{r}'|}d\mathbf{r}' + \frac{\delta E_{xc}}{\delta n(\mathbf{r})} \tag{8}$$

One still needs a reasonable estimate for the exchange-correlation energy $E_{xc}[n(\mathbf{r})]$. Kohn and Sham[3] introduced the Local Density Approximation, or LDA: they approximated the functional with a *function* of the local density $n(\mathbf{r})$:

$$E_{xc}[n(\mathbf{r})] = \int \epsilon(n(\mathbf{r}))n(\mathbf{r})d\mathbf{r}, \qquad \frac{\delta E_{xc}}{\delta n(\mathbf{r})} \equiv \mu_{xc}(n(\mathbf{r})) = \left(\epsilon(n) - n\frac{d\epsilon(n)}{dn}\right)_{n=n(\mathbf{r})} \tag{9}$$

and for $\epsilon(n(\mathbf{r}))$ used the same dependence on the density as for the homogeneous electron gas (or 'jellium') for which the $n(\mathbf{r})$ is constant.

Even in such simple case the exact form of $\epsilon(n)$ is unknown. However, approximate forms have been known for a long time, going back to Wigner[4]. Numerical results from exact Monte-Carlo calculations in jellium by Ceperley and Alder[5] have been parameterized by Perdew and Zunger[6] with a simple analytical form:

$$\begin{aligned} \epsilon_{xc}(n) &= -0.4582/r_s - 0.1423/(1 + 1.0529\sqrt{r_s} + 0.3334 r_s) & ,r_s \geq 1 \\ &= -0.4582/r_s - 0.0480 + 0.0311\ln r_s - 0.0116 r_s + 0.0020 r_s \ln r_s & ,r_s \leq 1 \end{aligned} \tag{10}$$

---

[2] Functional derivatives $\delta F/\delta n(\mathbf{r})$ are defined implicitly through $\delta F = \int (\delta F/\delta n(\mathbf{r}))\delta n(\mathbf{r})d\mathbf{r}$.

where r_s is the usual parameter in the theory of metals: $r_s = (3/4\pi n)^{1/3}$, and the energy is expressed in Hartree. This form – the 'real' LDA – is often used. More accurate approximations have been recently proposed in Ref.[7]. Usually all forms yield very similar results in condensed-matter calculations (which is not surprising as all the parameterizations are very similar in the range of r_s applicable for solid-state phenomena).

LDA has turned out to be much more successful than expected[8]. Although it is very simple it yields a description of the chemical bond that is superior to that obtained by Hartree-Fock, and it compares well to much weightier Quantum Chemistry methods. For weakly correlated materials such as semiconductors structural and vibrational properties are accurately described: the correct structure is usually found to have the lowest energy, bond lengths, bulk moduli and phonon frequencies are accurate within a few percent or so on[8, 9, 10]. LDA also has some well-known drawbacks. In finite systems the incorrect cancellation of the self-interaction is reflected in an incorrect long-range behaviour[6]. LDA tends to badly overestimate ($\sim 20\%$) cohesive energies and to underestimate to an even worse degree ($\sim 50\%$) the band gaps in insulators. More generally DFT is a ground state theory and KS eigenvalues and eigenvector do not have a clear physical meaning. Moreover LDA is an uncontrolled approximation: it is not clear at all what to do in order to go beyond LDA. There have been several attempts in this direction: recently the *gradient correction*[11] has attracted a lot of interest, but the effectiveness of such approaches is still a controversial issue. Nevertheless LDA is a valuable tool in investigation of material properties.

2.1 Forces in DFT

An important consequence of the variational character of DFT is the possibility of calculating Hellmann-Feynman forces acting on atoms[12]. Forces are the derivative of the total energy with respect to atomic positions $\mathbf{R}_i$:

$$\mathbf{F}_i = -\nabla_{\mathbf{R}_i} E = -\int n(\mathbf{r}) \nabla_{\mathbf{R}_i} V(\mathbf{r}) d\mathbf{r} - \nabla_{\mathbf{R}_i} E_{II}(\mathbf{R}) \qquad (11)$$
$$- \int \left[\frac{\delta F}{\delta n(\mathbf{r})} + V(\mathbf{r}) \right] \nabla_{\mathbf{R}_i} n(\mathbf{r}) d\mathbf{r},$$

where E_{II} is the ion-ion (classical) interaction energy. The electronic part contains an *explicit* dependence on atomic positions through the potential $V(\mathbf{r}) \equiv V_{\{\mathbf{R}\}}(\mathbf{r})$, and an *implicit* dependence through the ground-state charge density $n(\mathbf{r})$. The explicit derivative and the ion-ion term are easy to calculate. The implicit derivative is not. In fact we have to know how the charge density changes when we move the atoms. Fortunately, the variational character of DFT helps us. In fact the last term in Eq. (11) contains the first-order variation of the energy functional around the ground-state energy, which vanishes if the ground state is a minimum. In more mathematical terms, the minimization of the energy under the constraint of constant total charge implies that the variation of the following function:

$$\min \left\{ F[n(\mathbf{r})] + \int n(\mathbf{r}) V(\mathbf{r}) d\mathbf{r} - \lambda \left(\int n(\mathbf{r}) d\mathbf{r} - N \right) \right\}, \qquad (12)$$

with respect to an arbitrary variation $\delta n(\mathbf{r})$, must vanish. λ is a Lagrange multiplier and N the total number of electrons. Using elementary variation calculus, this yields

$$\int \left[\frac{\delta F}{\delta n(\mathbf{r})} + V(\mathbf{r}) - \lambda \right] \delta n(\mathbf{r}) d\mathbf{r} = 0, \qquad (13)$$

that is,
$$\left[\frac{\delta F}{\delta n(\mathbf{r})} + V(\mathbf{r})\right] = \lambda. \tag{14}$$

As a consequence, forces are simply the matrix element of the ground state of the gradient of the external potential plus an ion-ion term:

$$\mathbf{F}_i = -\int n(\mathbf{r})\nabla_{\mathbf{R}_i}V(\mathbf{r})d\mathbf{r} - \nabla_{\mathbf{R}_i}E_{II}(\mathbf{R}). \tag{15}$$

3 Routes to the ground-state

A major goal of electronic structure calculations is to find ground-state ionic positions $\{\mathbf{R}\}$ and the corresponding ground-state electronic states $\{\psi_{gs}\}$ by minimising the total energy $E(\mathbf{R}, \psi) = E_{DFT}(\mathbf{R}, \psi) + E_{II}(\mathbf{R})$, where the second term is the ion-ion classical Coulomb interaction. We assume the validity of the adiabatic (Born-Oppenheimer) approximation. This goal can be achieved through different strategies.

The most straightforward – and historically the first – approach is to do things one at the time using nested iterations. An *outer* iteration on atomic positions minimises $f(\mathbf{R}) = E(\mathbf{R}, \psi_{gs}(\mathbf{R}))$ where the $\psi_{gs}(\mathbf{R})$ are the electronic ground-state KS orbitals for a given atomic configuration $\{\mathbf{R}\}$ (in more formal terms: the $\psi_{gs}(\mathbf{R})$ are 'on the Born-Oppenheimer surface'). The task is much easier if the forces $\mathbf{F}_i = -\nabla_{\mathbf{R}_i} f(\mathbf{R})$ can be calculated.

An *inner* iteration on the potential (or equivalently on the charge density) allows us to find $\psi_{gs}(\mathbf{R})$ and $E(\mathbf{R}, \psi_{gs}(\mathbf{R}))$. Starting from some initial guess for the self-consistent potential (for instance, the bare ionic potential) the KS Hamiltonian is diagonalised and solved:

$$\left(-\frac{\hbar^2}{2m}\nabla^2 + V_{in}(\mathbf{r})\right)\psi_i(\mathbf{r}) = \epsilon_i\psi_i(\mathbf{r}). \tag{16}$$

With the resulting charge density:

$$n_{out}(\mathbf{r}) = 2\sum_{i=1}^{N/2} |\psi_i(\mathbf{r})|^2 \tag{17}$$

a new LDA potential is obtained:

$$V_{out}(\mathbf{r}) = V(\mathbf{r}) + e^2 \int \frac{n_{out}(\mathbf{r}')}{|\mathbf{r}-\mathbf{r}'|}d\mathbf{r}' + \mu_{xc}(n_{out}(\mathbf{r})). \tag{18}$$

This will be equal to the input potential only when self-consistency is attained. The output potential cannot be simply reinserted in the cycle: this will not in general bring convergence. Instead a new V_{in} is generated by some suitable algorithm which takes into account the V_{in}'s and V_{out}'s of preceding interactions. When self-consistency is achieved one can calculate the total energy $E = E(\mathbf{R})$ as the sum of the electronic and ionic terms.[3] We denote this type of iteration of the potential as 'SCF' ('self-consistent field').

This kind of approach, often referred to as the *wheel in the wheel* algorithm,[4] has been widely and successfully used in a variety of solid-state systems[10]. However this is not the only way to achieve our goal, nor necessarily the best. In particular the

---
[3]The general properties of the density functional ensure that *for fixed ions* there is only one minimum, corresponding to the ground state.
[4]Another 'wheel' is generally present in the solution of the KS Hamiltonian, see Sec.5.2.

relaxation of atoms to their equilibrium positions requires an accurate calculation of forces. It may happen that the relaxation brings into a local minimum and never achieves the minimum-energy structure. The SCF iterations of the potential can be quite slow in some cases (e.g. elongated unit cells, low or no symmetry).

An alternative way to find the electronic ground state for fixed ions consists in minimizing directly the density functional, without solving explicitly the KS equations[13]. In practical calculations the density functional, Eq. (4), is a function of the coefficients of the KS orbitals ψ_i in some basis set. Minimization of such a function under the orthonormality constraints of Eq. (3) can be achieved through the use of well-known conjugate-gradient[14] or similar algorithms. This kind of approach – which we will refer to as 'Direct Minimization' – is becoming increasingly common[15]. Along the same line of thought one can minimize directly the density functional very efficiently using the Direct Inversion in Iterative Subspace (DIIS) method[16] borrowed from Quantum Chemistry.

A more radical point of view has been proposed by Car and Parrinello in 1985[17]: do everything at the same time and on the same footing. This is achieved by merging DFT with classical Molecular Dynamics methods into a new and powerful tool – a Quantum Molecular Dynamics – for the study of real materials. Car and Parrinello introduced a fictitious Lagrangian

$$\mathcal{L} = \frac{\mu}{2}\sum_i \int d\mathbf{r}|\dot{\psi}_i(\mathbf{r},t)|^2 + \frac{1}{2}\sum_I M_I \dot{\mathbf{R}}_I^2 - E(\mathbf{R},\psi) \qquad (19)$$

which generates the following set of equations of motion:

$$\mu\ddot{\psi}_i = \left(-\frac{\hbar^2}{2m}\nabla^2 + V_{scf}(\psi)\right)\psi_i - \Lambda_{ij}\psi_j, \quad M_I\ddot{\mathbf{R}}_I = -\nabla_{\mathbf{R}_I}E. \qquad (20)$$

The 'masses' μ associated to the ψ_i's are fictitious: they are used only to generate the dynamics. The Λ_{ij} enforce the orthonormality constraints. The resulting equations of motion can be treated with classical Molecular Dynamics technology. In this way one can simultaneously achieve electronic and ionic relaxation, and even more: one can obtain information on the dynamical behaviour. A good introduction to the Car-Parrinello method and its successes is given in Ref.[18].

The following sections give an overview of technical aspects and of the practical implementation of the SCF approach using plane waves basis sets and pseudopotentials to reproduce the electron-ion interactions. Most of the technical points are relevant for the Car-Parrinello and Direct Minimization approaches as well.

4 Tools

4.1 Supercells

In perfect crystals the natural framework to solve the KS equations takes advantage of periodicity: the crystal is described by a periodically repeated *unit cell* containing one or more atoms (the *basis*), by a crystal lattice and a reciprocal lattice. Using Bloch's theorem, states are classified by a wavevector **k** in the Brillouin Zone (BZ) and form energy bands. In some cases, unit cells can be quite large and contain many atoms. This is the case for example with superlattices, artificial layered materials in which the periodicity of the original lattice – the one of the component materials – is replaced by a 'superperiodicity', with a *supercell* which is larger than the original cell.[5] Another example is given by solid C_{60}, both pure and doped with alkali atoms (such as the superconductor K_3C_{60}).

Many interesting physical systems do not exhibit perfect periodicity. Good examples are disordered superlattices (superlattices with intermixing at the interfaces), quantum wells, substitutional alloys, point defects, and surfaces. In all such cases it is convenient to simulate an aperiodic (or 'almost-periodic') system with a periodically repeated fictitious supercell. The form and the size of the supercell depend on the physical system being examined. For example, the study of point defects requires a defect 'not to see' its periodic replica in order to accurately simulate a truly isolated defect. This approach is actually used also for the extreme case of molecules and clusters. The use of supercells for such demonstrably nonperiodic objects may seem odd but there are important computational advantages in such an approach.[6]

The size of the unit cell – the number of atoms and the volume – is very important. Together with the type of atoms it determines the difficulty of the calculation: large unit cells mean large calculations. Unfortunately many interesting physical systems are described – exactly or approximately – by large unit cells.

4.2 Plane Waves

KS wavefunctions must be expanded in some suitable basis set. For periodic systems, the basis set is formed by Bloch states of wavevector **k**. A large number of different basis sets have been proposed and used. Most fall into the 'localised basis set' category. Some of the most frequently used are Bloch sums of Linear Combinations of Atomic Orbitals (LCAO), Gaussian-type Orbitals (GTO) and Linearised Muffin-Tin Orbitals (LMTO)[9]. These atomic-like functions are tailored for fast convergence, so that only a few (some tens at most) functions per atom are needed. However they are quite delicate to use. In particular it is difficult to check systematically for convergence (a well-known problem of Quantum Chemistry). Another serious drawback is the difficulty of calculating derivatives of the energy (e.g. forces on atoms)[19]. Forces are very important in determining the structure of a complex system. They are crucial quantities in Car-Parrinello and other Quantum Molecular Dynamics applications. Only recently have important advances been made in this field[20].

An opposite approach is to choose completeley extended, atomic-independent Plane Waves (PW) as basis set. For a given reciprocal lattice $\{\mathbf{G}\}$ and for a given vector **k** in the BZ, a PW basis set is defined as

$$|\mathbf{k}+\mathbf{G}\rangle = \frac{1}{V}e^{i(\mathbf{k}+\mathbf{G})\cdot\mathbf{r}}, \quad \frac{\hbar^2}{2m}|\mathbf{k}+\mathbf{G}|^2 \leq E_{cut}, \tag{21}$$

where V is the crystal volume, and E_{cut} is called the 'kinetic energy cutoff', or simply the cutoff. PW's have many attractive features: they are simple, orthonormal by construction, unbiased and it is very simple to check for convergence (by increasing the cutoff).

A distinct computational advantage of PW's is the existence of very fast algorithms (known as the Fast Fourier-Transform, FFT) to perform the discrete Fourier transforms.

---

[5] Strictly speaking, there is nothing 'super' in both the cell and the lattice of a perfect superlattice. One just wants to stress that the unit cell includes several unit cells of the component materials.

[6] The opposite point of view, that is, simulating extended systems with larger and larger clusters, has been traditionally used in Quantum Chemistry. However, it turns out that clusters converge only very slowly in solid state behaviour.

This allows simple and fast transformation from reciprocal to real space and vice versa. The basic one-dimensional FFT executes the following transformation:

$$f_i = \sum_{j=0}^{N-1} g_j e^{2\pi ij/N}, \quad i = 0, ..., N-1,\tag{22}$$

and its inverse

$$g_i = \frac{1}{N}\sum_{j=0}^{N-1} f_j e^{-2\pi ij/N}.\tag{23}$$

The tranformation is usually performed 'in place', that is the result is overwritten on the input vector. This takes $\mathcal{O}(N \log N)$ operations. In three dimensions the discrete Fourier transform maps a function $\tilde{f}(\mathbf{g}_i)$ in reciprocal space into a function $f(\mathbf{r}_i)$ in the unit cell (and vice versa):

$$\mathbf{g}_i = i_1\mathbf{G}_1 + i_2\mathbf{G}_2 + i_3\mathbf{G}_3, \quad \mathbf{r}_i = \frac{j_1}{N_1}\mathbf{R}_1 + \frac{j_2}{N_2}\mathbf{R}_2 + \frac{j_3}{N_3}\mathbf{R}_3 \tag{24}$$

where $\mathbf{R}_1, \mathbf{R}_2, \mathbf{R}_3$ ($\mathbf{G}_1, \mathbf{G}_2, \mathbf{G}_3$) are the three fundamental translations that generate the real-space (reciprocal) lattice, $i_1 = -N_1/2, ..., N_1/2$, and so on. N_1, N_2, N_3 must be sufficiently large to include all available Fourier components; the more Fourier components, the larger the grid in G-space and the finer the grid in R-space. It is easily verified that this 3-d FT can be done in a very fast way by performing 3 inter-nested 1-d FFT. Computers usually have some highly optimised library routines performing FFT.[7]

PW's look like the ideal basis set for solids. Unfortunately their extended character makes it very difficult to accurately reproduce localized functions such as the charge density around a nucleus or even worse, the orthogonalization wiggles of inner ('core') states. In order to describe features which vary on a lengthscale δ, one needs PW's with a cutoff as high as $\sim (2\pi/\delta)^2$, that is one needs $\sim 4\pi(2\pi/\delta)^3/3\Omega$ PW's (where Ω is the dimension of the BZ). A simple estimate for diamond is instructive. The 1s wavefunction of the carbon atom has its maximum around 0.3 a.u., so $\delta \simeq 0.1$ a.u. is a reasonable value. Diamond has an fcc lattice ($\Omega = (2\pi)^3/(a_0^3/4)$) with lattice parameter $a_0 = 6.74$ a.u., thus yielding $\sim 250,000$ PW's. This is clearly too much for practical use.

4.3 Pseudopotentials

Core states prevent the use of PW's. However they do not contribute in a significant manner to chemical bonding and to solid-state properties, only outer ('valence') electrons do.[8] Core states remain there 'frozen' in their atomic configuration making calculations more difficult. The idea of replacing the full atom with a much simpler 'pseudoatom' with valence electrons only arises naturally (apparently for the first time, in a 1934 paper by Fermi[21]). 'Pseudopotentials' (PP's) have been widely used in solid state physics starting from the 1960's. In earlier approaches PP's were devised to reproduce some known experimental solid-state or atomic properties such as energy gaps

---

[7]The 'true' original FFT works only when N_1, N_2, N_3 are powers of 2, but algorithms for more general values do exist. While it is a good practise to keep factors in N as small as possible, it is not always true that powers of 2 are the best choice: it depends on the algorithms used and on the computer architecture.

[8]The border between core and valence electron is often evident on physical grounds but sometimes it is not. For instance, in Cs the 5s and 5p states can sometimes be considered safely as core states but more often they cannot.

or ionization potentials. Already in this 'empirical' phase they proved to be a valuable tool[22].

A major breakthrough occurred in 1979 with the introduction of Norm-Conserving PP's by Hamann, Schlüter, and Chiang[23]. These are *atomic* potentials which must obey the following conditions. Given a reference atomic configuration
1) all-electron and pseudo-wavefunctions must have the same energy, and
2) they must be the same beyond a given 'core radius' r_c, which is usually located around the outermost maximum of the atomic wavefunction;
3) the pseudo-charge and the true charge contained in the region $r < r_c$ must be the same.[9]

The above conditions ensure *transferability*, the ability to yield correct results for configurations different from the reference one, or in different environments (such as in molecules or in condensed matter). In fact Norm-Conserving PP's are able to accurately mimic core scattering properties in an energy region which is not too far from the energy of the reference state. Norm-Conserving PP do not have singularities. They are relatively smooth functions, whose long-range tail goes like $-Z_v e^2/r$ where Z_v is the number of valence electrons. They are *nonlocal* because it is usually impossible to mimic the effect of orthogonalization to core states on different angular momenta l with a single function. There is a PP for every l:

$$\hat{V}^{ps} = V_{loc}(r) + \sum_l V_l(r) \hat{P}_l \qquad (25)$$

where $V_{loc}(r) \simeq -Z_v e^2/r$ for large r and $\hat{P}_l = |l\rangle\langle l|$ is the projection operator on states of angular momentum l. We can recast such a potential in the form

$$\hat{V}^{ps} = V_{loc}(r) + \sum_{lm} Y_{lm}(\mathbf{r}) V_l(r) \delta(r-r') Y_{lm}^*(\mathbf{r}'), \qquad (26)$$

for which the nonlocal character is more evident. Although nonlocality is a drawback, because it makes calculations more difficult, it is not a big one [10] in practical applications.

Tables of PP's for all elements have been published by Bachelet, Hamann, and Schlüter in a a much-quoted 1982 paper[24]. Experience has shown that PP's are practically equivalent to the frozen core approximation[25]: PP and all-electron calculations on the same systems yield almost indistinguishable results (except for those cases in which core states are not sufficiently 'frozen').

4.4 Brillouin-Zone Sampling

In a periodic system, or in a system described by a suitable supercell, the states are classified by a wavevector k in the BZ and by a band index. For a given k, the KS orbitals are expanded into PW's of appropriate periodicity

$$|\psi_{\mathbf{k},i}\rangle = \sum_{\mathbf{G}} \Psi_i(\mathbf{k}+\mathbf{G}) |\mathbf{k}+\mathbf{G}\rangle \qquad (27)$$

up to a given energy cutoff E_{cut}. In order to calculate the charge density $n(\mathbf{r})$ in a periodic system one has to sum over an infinite number of k-points:

$$n(\mathbf{r}) = \sum_{\mathbf{k}} \sum_i |\psi_{\mathbf{k},i}(\mathbf{r})|^2. \qquad (28)$$

---

[9]This last condition explains the name 'Norm-Conserving'. There is an historical reason: some older pseudopotentials had an 'orthogonalization hole' problem which caused violation of condition 3
[10]However, strictly speaking, DFT is valid only for *local* external potentials!

In fact, assuming periodic (Born-Von Kàrmàn) boundary conditions

$$\psi(\mathbf{r} + L_1\mathbf{R}_1) = \psi(\mathbf{r} + L_2\mathbf{R}_2) = \psi(\mathbf{r} + L_3\mathbf{R}_3) = \psi(\mathbf{r}), \tag{29}$$

a crystal has $L_1 L_2 L_3$ allowed k-points. In the limit of an infinite crystal, $L_1 L_2 L_3 \to \infty$ and the discrete sum becomes an integral over the BZ. It is not obvious at all that this integral can be approximated by a discrete sum over an affordable number of k-points. However experience shows that this is actually possible, at least in crystals with completely filled or completely empty bands.

Symmetry – when present – can be used to reduce the number of calculations to be performed. Only one k-point is left to represent each 'star' – the set of k-points that are equivalent by symmetry – with a 'weight' w_i which is proportional to the number of k-points in the star. The infinite sum over the BZ is replaced by a discrete sum over a set of points $\{\mathbf{k}_i\}$ and 'weights' w_i:

$$\frac{1}{V} \sum_{\mathbf{k}} f_{\mathbf{k}}(\mathbf{r}) \longrightarrow \frac{\Omega}{(2\pi)^3} \sum_i w_i f_{\mathbf{k}_i}(\mathbf{r}). \tag{30}$$

The resulting sum is then symmetrized to get the charge density. Other quantities (such as the total energy) which contain sums over the BZ can be dealt with in a similar way.

Suitable sets for BZ sampling in insulators and semiconductors are called 'special points'[26, 27]. Let us consider the case of an fcc crystal having cubic symmetry. The smallest special point grid is formed by the 'mean-value' or Baldereschi point[26]: $\mathbf{k} = (0.6225, 0.2953, 0.0)$ (in units of $2\pi/a_0$). Better accuracy is obtained with the much-used 'two Chadi-Cohen points'[27]: $\mathbf{k}_1 = (1/4, 1/4, 1/4)$, $w_1 = 1$ and $\mathbf{k}_2 = (1/4, 1/4, 3/4)$, $w_2 = 3$. Then there are 6-, 10-point grids and so on, yielding increasing accuracy.[11]

In metals things are more difficult because one needs an accurate sampling of the Fermi surface. The 'Gaussian broadening' and the 'tetrahedron' techniques, or variations of the above[28], are generally used.

When the unit cell gets larger, the BZ become smaller and the need for accurate sampling becomes less stringent. For very large supercells, the sampling obtained with only the Γ point ($\mathbf{k} = 0$) is usually good enough. Of course when supercells are used to simulate aperiodic systems such as clusters and molecules the Γ point is the good choice. There is no point in trying to reproduce the effect of a fictitious periodicity.

5 Algorithmic Aspects

5.1 Potential mixing

The calculation of the self-consistent potential requires a way to mix the input and output potentials to yield a new input potential. The simplest approach ('simple mixing') is the following:

$$V_{in}^{(n+1)} = \alpha V_{in}^{(n)} + (1-\alpha) V_{out}^{(n)}, \tag{31}$$

where the superscripts indicate the iteration number and the value of α must be chosen empirically in order to get fast convergence. As a rule relatively big values ($\alpha = 0.3-0.5$) can be chosen for small cells while smaller values are needed for bigger or elongated

---

[11] Actually the name 'special points' is somewhat misleading in this case. In fact those sets just form uniform grids in the BZ.

cells (like for a superlattice). Simple mixing is not very effective especially for big cells, or even worse for surface calculations. Better results are obtained with more sophisticated algorithms, like for example the family of Anderson algorithms[29] and modified Broyden algorithms[30]. The latter seems to be quite effective in situations of difficult convergence.

Eventually, as the size of the cell increases convergence becomes slower and slower. This adverse behaviour can be traced back to charge oscillations ('charge sloshing') that take place in large cells. Preconditioned Conjugate Gradient algorithms are claimed to be less sensitive to this problem[15]. The phenomenon of convergence-slowing-down is well-known in many different fields of computational physics. The ultimate solution could come from application of *multigrid* concepts[31]. Work along such lines is in progress[32].

5.2 Diagonalization of the Hamiltonian

By far the most time-consuming step in an SCF calculation is the solution of the KS equations, Eq. (7). When the eigenfunctions are expanded on a finite basis set the solution takes the form of a secular equation:

$$\sum_{G'} H(k+G, k+G')\psi_{k,i}(G') = \epsilon_{k,i}\psi_{k,i}(G), \qquad (32)$$

where the matrix elements of the Hamiltonian have the form

$$H(k+G, k+G') = \frac{\hbar^2}{2m}(k+G)^2 \delta_{G,G'} + V_{loc}(G-G') + V_{NL}(k+G, k+G'). \qquad (33)$$

The local contribution V_{loc} includes both the local term in the PP's and the screening potential. The nonlocal contribution V_{NL} comes from the PP's. The problem is reduced in this way to the well-known problem of finding the lowest M eigenvalues and eigenvectors (only the valence states for insulators, a few more for metals) of an $N \times N$ Hermitian matrix (or a real symmetric matrix when inversion symmetry is present). This task can be performed with well-known bisection-tridiagonalization algorithms, for which very good public-domain computer packages (EISPACK or the more recent LAPACK)[33] are available. Unfortunately this straightforward procedure has serious limitations. In fact:

– the computer time required to diagonalise a $N \times N$ matrix grows as N^3;
– the matrix must be stored in memory, requiring $\mathcal{O}(N^2)$ memory.

As a consequence a calculation requiring more than a few hundred PW's becomes exceedingly time- and memory-consuming. As the number of PW's increases linearly with the size of the unit cell it is very hard to study systems containing more than a few (say 5-10) atoms.

Both limitations have been overcome with the introduction of *iterative* techniques[34]. These techniques can be used whenever

i) the number of states to calculate M is much smaller than the dimension of the basis set N, and

ii) a reasonable and economical estimate of the inverse operator H^{-1} is available.

Both conditions are satisfied in an SCF calculation in a PW basis set. In fact $M \ll N$ is always true, and the Hamiltonian matrix is dominated by the kinetic energy at large G (that is, the Hamiltonian is *diagonally dominant*).

Iterative methods are based on a repeated refinement of a trial solution, which is stopped when satisfactory convergence is achieved. The number of iterative steps cannot be predicted in advance. It depends heavily on the structure of the matrix, on the type of refinement used, and on the starting point. A well-known and widely used algorithm is due to Davidson[34]. In this method, a set of correction vectors $|\delta\psi_i\rangle$ to the M trial eigenvectors $|\psi_i\rangle$ are generated as follows:

$$|\delta\psi_i\rangle = \frac{1}{D - \epsilon_i}(H - \epsilon_i)|\psi_i\rangle \qquad (34)$$

where the $\epsilon_i = \langle\psi_i|H|\psi_i\rangle$ are the trial eigenvalues. The $|\delta\psi_i\rangle$'s are orthogonalised and the Hamiltonian is diagonalised (with conventional techniques) in the subspace spanned by the trial and correction vectors. A new set of trial eigenvectors is obtained and the procedure is iterated until convergence is achieved. A good set of starting trial vectors is supplied by the eigenvectors at the preceding iteration of the potential.

An important point is the following. The Hamiltonian matrix is never explicitly required excepted for its diagonal part. Only $H\psi_i$ products are required, which can be calculated in a very convenient way by applying the *dual-space technique*[35]. In fact the kinetic term appearing in Eq. (33) is diagonal in G-space, whereas the local potential term is diagonal in real space. Using FFT's one can go quickly back and forth from real to reciprocal space and perform the products where it is more convenient. There is still a nonlocal term which appears to require the storage of the matrix. The trick is to write V_{NL} in a 'separable' form:

$$V_{NL}(\mathbf{k} + \mathbf{G}, \mathbf{k} + \mathbf{G}') = \sum_{\mu=1}^{N_{at}} \sum_{j=1}^{n} f_j^\mu(\mathbf{k} + \mathbf{G}) g_j^\mu(\mathbf{k} + \mathbf{G}'), \qquad (35)$$

where n is a small number and N_{at} is the number of atoms in the unit cell. This allows us to perform the products by storing only the f and g vectors. The reduction to separable form is exact and straightforward when the Kleinman-Bylander projection (see Appendix) is used. It is more involved but still possible when conventional PP's are used. The trick can also be used with V_{NL} in real space thus taking advantage of the short-range nature of V_{NL} [36].

5.3 Computational workload

When iterative techniques are used to diagonalize the Hamiltonian the time-consuming step is the calculation of the products $H\psi$ followed by orthogonalization. For a system of N_{at} atoms in the unit cell having M occupied states expanded into N PW's, the calculation of the products $H\psi$ for all occupied states require:

- $\mathcal{O}(MN^2)$ operations if the Hamiltonian is stored as a matrix;
- $\mathcal{O}(MN\log N) + \mathcal{O}(N_{at}MN)$ operations if the dual-space technique is used;
- $\mathcal{O}(MN\log N) + \mathcal{O}(N_{at}M)$ operations if the dual-space technique is used with V_{NL} in real space.

Orthogonalization of each trial eigenfunction to all others requires $\mathcal{O}(M^2N)$ operations. Similar considerations apply to the Car-Parrinello and Direct Minimization approaches.

Let us consider how the computational workload scales with the number of atoms in the unit cell. [12] The number of states M is proportional to N_{at}. The number of PW's is proportional to the volume of the unit cell. For this reason $N \propto N_{at}$ as

---

[12] We do not address here the problem of convergence slowing-down with increased size of the cell. We only consider the time needed to execute a basic steps, not how many basic steps are needed in a complete calculation.

well. In summary the $H\psi$ products can be performed in $\mathcal{O}(N_{at}^2)$ operations in the best case. Orthogonalization requires $\mathcal{O}(N_{at}^3)$ operations. For very big cells this will be the dominant part whatever we do to speed up the calculation of the $H\psi$ products.

This unfavourable scaling with the size of the cell has a simple physical root. The Hamiltonian eigenstates of an extended system are usually extended. As a first consequence the computational workload needed to calculate them scales at least as the number of orbitals times their size (the dimension of the basis set), i.e. as N_{at}^2. As a second consequence orbital orthogonalization requires additional work proportional to the square of the number of orbitals times their size, i.e. N_{at}^3.

6 Linear-scaling Algorithms

General physical considerations indicate that the energy and related properties of a system of non-interacting electrons (such as the one appearing in the Kohn-Sham formulation of DFT) should be calculable with a workload proportional to the size of the system. In fact the local nature of the Schrödinger equation and the condition of local charge neutrality imply that the local density of states at a given point – in contrast with the individual eigenfunctions – does not depend on the details of the system far from that point[37].

The search for better-scaling algorithms has become a very active research field in the last few years and many different proposals have appeared in the literature[38, 39, 40, 41, 42].

One possibility is in some way to force the eigenfunctions to be localized in space and to express the electron density in terms of the latter. This idea is being explored following different methods by several authors. In Yang's approach[38] the system is divided into small *independent* subsystems within which the numbers of electrons are determined by a common chemical potential. The electron density is then determined by conventional diagonalization techniques within each subsystem. Another way to achieve the same goal is to use 'confining potentials'. This idea is currently under active development in a Car-Parrinello framework and looks quite promising[39].

Other authors have focussed on the direct solution of the *density matrix*[40] suitably truncated in real space. Another idea is to use a statistical approach, based on the *maximum entropy* principle[41], to extract the relevant information on the density of states.

In the following, I will briefly describe a different approach[42] in which the use of KS orbitals is completely avoided. The charge and energy densities are obtained directly from the one-electron Green's function without going through any Schrödinger-like equations. This method is in principle exact,[13] it displays the right linear scaling with size, and it is still rather simple to implement. The approach is based on a finite-difference real-space discretization of the Hamiltonian and on the Recursion Method[43] to calculate the electron and energy densities from selected elements of the one-electron Green's function.

The starting point is a well-known identity relating the density matrix of a system to its one-electron Green's function,

$$\rho(\mathbf{r},\mathbf{r}') = 2\frac{1}{2\pi i}\oint_{C_F} G(\mathbf{r},\mathbf{r}';z)dz, \qquad (36)$$

---

[13]It is also possible to get rid of the KS orbitals by using Thomas-Fermi like approximations for the density functional. Such approximations are unfortunately quite inaccurate.

where C_F is a contour in the complex energy plane enclosing just the occupied-state eigenvalues. The factor 2 accounts for spin degeneracy and the Green's function $G(\mathbf{r}, \mathbf{r}'; z)$ is the real-space representation of the resolvent of the Hamiltonian,

$$G(\mathbf{r}, \mathbf{r}'; \epsilon) \equiv \langle \mathbf{r} | \hat{G}(\epsilon) | \mathbf{r}' \rangle = \langle \mathbf{r} | \frac{1}{\epsilon - H} | \mathbf{r}' \rangle. \tag{37}$$

According to Eq. (36), the electron density, $n(\mathbf{r}) = \rho(\mathbf{r}, \mathbf{r})$, can easily be calculated from knowledge of the *diagonal* elements of the Green's function. The number of points necessary to numerically evaluate the integral (36) to a given accuracy depends on the valence band width and on the minimum distance of the contour from the energy eigenvalues. Both these properties do not depend on the size of the system, at least for insulators. Therefore if the work needed to calculate Eq. (36) is independent of the size, then the work necessary to obtain the full charge density will scale linearly.

We next discretize the problem by using finite differences on a uniform grid $\{\mathbf{r}_i\}$. The finite-difference representation of the Hamiltonian is sparse. If second-order discretization on a cubic uniform grid is used for the Laplacian the only non vanishing matrix elements $H_{ij} \equiv \langle \mathbf{r}_i | H | \mathbf{r}_j \rangle$ are on the diagonal and between neighbouring points,

$$H_{ij} = \begin{cases} \frac{1}{12h^2} + V_{scf}(\mathbf{r}_i) & \text{if } i = j \\ -\frac{1}{2h^2} & \text{if } |\mathbf{r}_i - \mathbf{r}_j| = h \\ 0 & \text{otherwise,} \end{cases} \tag{38}$$

where h is the spacing between the grid points and a local potential is assumed.[14]

Calculating Eq. (38) by standard factorization techniques would again result in a workload proportional to N_{at}^3. Iterative algorithms to solve elliptic partial differential equations could in principle be used to calculate the inverse of $(\epsilon - H)$ with a cost proportional to N_{at}^2. Of course this price is optimal for the calculation of the *full* inverse matrix. It is too high if only the *diagonal* of the inverse is required.

A convenient way to compute a *single* diagonal element of the Green's function is provided by the Recursion Method of Haydock, Heine, and Kelly[43]. In the Recursion Method, diagonal elements of the Green's function $\langle \phi_0 | \hat{G} | \phi_0 \rangle$ are expressed in terms of a continuous fraction whose coefficients are calculated from a chain of orthonormal states recursively generated from $|\phi_0\rangle$,

$$\begin{aligned} b_1 |\phi_1\rangle &= H|\phi_0\rangle - a_0 |\phi_0\rangle \\ b_2 |\phi_2\rangle &= H|\phi_1\rangle - a_1 |\phi_1\rangle - b_1 |\phi_0\rangle \\ &\cdots \\ b_n |\phi_n\rangle &= H|\phi_{n-1}\rangle - a_{n-1} |\phi_{n-1}\rangle - b_{n-1} |\phi_{n-2}\rangle, \end{aligned} \tag{39}$$

where

$$a_n = \langle \phi_n | H | \phi_n \rangle, \qquad b_n = \langle \phi_n | H | \phi_{n-1} \rangle. \tag{40}$$

The relevant diagonal element of the Green's function is then given by the continuous-fraction expansion,

$$\langle \phi_0 | \hat{G}(\epsilon) | \phi_0 \rangle = \cfrac{1}{\epsilon - a_0 - \cfrac{b_1^2}{\epsilon - a_1 - \cfrac{b_2^2}{\epsilon - a_2 - \cdots}}}. \tag{41}$$

---

[14] Nonlocal PP can also be used exploiting the short-range nature of the nonlocal terms.

The part of the continuous fraction which is not calculated when truncating the chain after n steps (indicated in Eq. (41) by dots '...' in the case $n = 2$) is called the *terminator*, $t_n(\epsilon)$. The terminator describes the influence on $\langle\phi_0|\hat{G}(\epsilon)|\phi_0\rangle$ of that portion of the system which is not spanned by the finite chain. In actual calculations, the continuous fraction (41) is either *truncated* after a finite number n of steps ($t_n(\epsilon) = 0$), or is closed by an approximate terminator. In this case the initial state is localized, $|\phi_0\rangle = |r_i\rangle$, as are the subsequent terms of the chain by virtue of the sparseness of the Hamiltonian (38). Therefore the evaluation of each given step of the chain (39) requires a workload independent of the size of the system. At each step the newly generated state in the chain of states explores further points. The continued fraction can be truncated when the relevant region around a point – the local environment which mostly determines the charge density at that point – is explored. This happens after a number of steps which is independent of the size[37], at least for sufficiently large systems. We conclude that the application of Recursion Method ideas allows us to calculate the electron density – and hence to implement DFT – with a workload which scales linearly with N_{at}. Moreover the procedure is highly suitable for parallel computing as the charge density can be calculated independently at each point of the grid.

The off-diagonal elements of the density matrix necessary to calculate the kinetic energy could be computed along similar lines. The kinetic energy, however, is more conveniently calculated from the relation,

$$E_{kin} = 2\sum_i \epsilon_i - \int n(\mathbf{r})V_{scf}(\mathbf{r})d\mathbf{r}. \tag{42}$$

The sum over occupied-state eigenvalues can again be expressed in terms of diagonal elements of the Green's function:

$$\sum_i \epsilon_i = \frac{1}{2\pi i}\sum_j \oint_{C_F} zG(\mathbf{r}_j,\mathbf{r}_j;z)dz. \tag{43}$$

A first test of this approach has been done in Ref.[42]. It was shown that the number of steps necessary to obtain a satisfactory accuracy was rather large if the chain was simply truncated. This is a consequence of the well-known difficulty of reproducing the properties of an extended system by truncating it to a finite cluster. In fact a truncated chain cannot distinguish between a cluster spanned by the chain functions and the extended system under consideration. Later it has been found [44] that much faster convergence could be achieved when the terminator for free electrons was used. Tests done on large silicon clusters (up to $\sim$ 2000 atoms) have demonstrated the applicability of this approach to realistic cases. A major problem still to be solved concerns accurate force calculation, which still requires an exceedingly high number of chain steps to be performed.

7 Conclusions

Application of iterative techniques, both in SCF and in Car-Parrinello or Direct Minimization approaches, has greatly enhanced the scope of first-principles investigations of real materials[10, 15, 18]. It is now possible to study systems with $\sim$ 100 atoms in the unit cell, including e.g. superlattices, some alloys and disordered materials, simple surfaces, small clusters and fullerene systems.

Many interesting physical systems are still beyond the reach of ab-initio methods based on PW's and PP's. In many cases this is due to the presence of atoms, such as O, F, the rare earths and the transition metals, whose PP's are very hard. A first answer to this limitation is coming from 'ultrasoft' PP's[45]. This approach is computationally quite cumbersome but it is becoming increasingly common[18]. Another idea is to look for new basis sets, hopefully enjoying the advantages of both PW's and localized basis sets but without the respective disadvantages. Two proposals have recently appeared in the literature, 'adaptive grids'[46] and 'wavelets'[47]. Both basis sets have nice properties but their merits in real calculations have to be demonstrated 'on the battlefield'.

Many other interesting systems (e.g. amorphous or disordered materials, alloys, extended defects, carbon microtubules, organic molecules) are described by supercells that are too big ($\sim$ 1000 atoms) for present computers. The development of parallel computers and of parallelized algorithms can bring some of those systems within reach. A recent example of the potential of parallel computing has been the study of the 7×7 reconstruction of the Si (111) surface[48]. To the best of my knowledge this is the largest system (400 atoms) ever studied ab-initio. Eventually the limiting factor is the unfavourable $\mathcal{O}(N_{at}^3)$ scaling of the computer effort with the number N_{at} of atoms. Overcoming such a bottleneck requires new radically different ideas such as those outlined in Sec. 6.

A Appendix

A.1 Generation of Pseudopotentials

The best way to understand what a pseudopotential is is to follow the steps needed in its generation. The basic tool is an atomic LDA program. This usually assumes that the charge density is spherical (a very good approximation even for open-shell atoms) thus allowing separation of variables into the radial and the angular ones. The states are classified in exactly the same way as in an introductory course on quantum mechanics. There is a main quantum number n, an angular momentum $l = 0, .., n-1$, and $m = -l, ..., l$. The atomic configuration is given in terms of occupation numbers (traditionally, $1s^2 2s^2 2p^6$....). The radial KS equations are integrated numerically on a logarithmic grid with one of the many well-known methods (this can be done even on a personal computer in a few seconds). A reasonable configuration is chosen (usually the ground-state) and all-electron self-consistent LDA radial wavefunctions $\phi_l(r)$ are obtained.

At this point there are several possible ways to proceed. A very simple and clear approach is due to Kerker[49]. For each l in the valence shell a nodeless [15] pseudo-wavefunction is constructed in the following way:

$$\begin{aligned} \phi_l^{ps}(r) &= \phi_l(r) &, r \geq r_c \\ &= r^{l+1} e^{p(r)} &, r \leq r_c \end{aligned} \qquad (44)$$

where $p(r)$ is a polynomial $p(r) = a + br^2 + cr^3 + dr^4$ whose coefficients are determined by imposing continuity of $\phi_l^{ps}(r)$ and its first and second derivatives at the matching point r_c, plus the norm-conservation requirement 3) (see Sec. 4.3). The r_c can safely be taken at the outermost maximum. A 'dressed' PP (containing the screening LDA potential as well) is now obtained by inverting the radial KS equation at the all-electron eigenvalue ϵ_l. The final PP is obtained by 'unscreening', that is, removal of the LDA potential generated by valence electrons only.

---

[15]This ensures that there are no lower states with the same l. The inner part of pseudo-wavefunctions is unphysical: orthogonalization to the core states yields meaningless objects.

PP's are usually obtained in numerical form on a grid, sometimes fitted to an analytical form like the one used in ref.[24]. However it is also possible to get PP's directly in numerical form[50]. Assuming a simple analytic form for the PP's which depends on a few parameters $\{\lambda_i\}$, one directly minimizes the following function:

$$f(\{\lambda_i\}) = \sum_l |\epsilon_l^{ps} - \epsilon_l|^2 + \sum_l \int_{r>r_c} r^2 \left(\phi_l^{ps}(r) - \phi_l(r)\right)^2 dr, \tag{45}$$

or some equivalent form, using one of the standard minimisation methods[14]. The author has used this procedure for many simple atoms.

A.2 Towards better Pseudopotentials

After the original Hamann, Schlüter, and Chiang paper, many (real or presumed) improvements and extensions have been proposed. They can be classified according to their main aim.

1) The first goal is to improve the reliability and accuracy of the PP's. One should not forget that a PP is an approximation, albeit a good one, to the true atom (in fact even worse: an approximation to the frozen-core approximation). In some cases, e.g. for alkali atoms with one valence electron, the loss of accuracy due to the neglect of non-linearity in exchange-correlation which is implicit in the unscreening procedure can be intolerable. In such cases the simple 'core correction'[51] is very useful. 'Generalized'[52] and 'extended'[53] PP's have also been proposed to improve transferability. The former allow the use of unbound atomic states as reference states; the latter have scattering properties that are correctly reproduced beyond first order in energy differences.

More recent work concerns extension of PP's to gradient-corrected density functionals [54] (PP's should be generated within the same approximation used for subsequent calculations. Using LDA PP's in gradient-corrected calculations is inconsistent), and PP's for GW and Quantum Monte Carlo calculations[55].

2) The second goal is to improve the computational efficiency of PP's. An important step in this direction is due to Kleinman and Bylander[56]. They proposed projecting the PP on to the reference pseudo-wavefunctions ϕ_l^{ps} in the following way,

$$\hat{V}^{ps} = V_{loc} + V_L + \sum_l \frac{|\delta V_l \phi_l^{ps}\rangle\langle\delta V_l \phi_l^{ps}|}{\langle\phi_l^{ps}|\delta V_l|\phi_l^{ps}\rangle}, \quad \delta V_l(r) = V_l(r) - V_L(r) \tag{46}$$

where $V_L(r)$ is an arbitrary function. By construction the original PP and the projected $\hat{V}^{ps}$ have the same eigenvalues and eigenvectors on the reference states ϕ_l^{ps}. This justifies the hope that the original and projected PP's will yield very similar results on other configurations as well. The Kleinman-Bylander form is much more convenient than the conventional form (for reasons that will be explained in Sec. 5.2). Unfortunately it can happen that spurious states ('ghosts') appear at energies which are lower than or comparable to the reference energy. In such a case the Kleinman-Bylander projection badly fails. Some recipes have been devised to avoid the 'ghost' problem[57].

Another desirable goal is the reduction of the number of PW's needed for calculations. This depends on the type of atoms involved. In typical semiconductors (e.g. Si, Ge, GaAs, AlAs) 100-150 PW's per atom are sufficient for most applications. However, many atoms – transition metals, first-row elements like F, O, and to a lesser extent N and C – are described by strong PP's, requiring impractically large PW basis sets. One can try to exploit the many 'degrees of freedom' which are present in PP generation to get softer PP's. For instance, a rule of thumb states that the larger the matching point r_c is the smoother and less accurate the resulting PP. One can strike a compromise between accuracy and computer budget by pushing r_c outwards. However this will not give dramatic improvements. If an atom has localised d- or p-states, then in any event it will require a lot of PW's. Several recipes have been proposed to get an 'optimally smooth' PP (for example by acting on the form of pseudowavefunctions in the inner, unphysical region). A very simple and effective recipe is described in Ref.[58].

A more radical solution has been proposed by Vanderbilt[45]. His PP's are quite different from traditional PP's: they are definitely much softer, but also much less straightforward to use. The first interesting application to systems containing Oxygen and Copper have already appeared[59].

References

[1] See e.g. *Theory of the Inhomogeneous Electron Gas*, edited by S. Lundqvist and N. H. March (Plenum, New York, 1983); *Density Functional Theory of Atoms and Molecules*, R.G. Parr and W. Yang (Oxford University Press, New York, 1989); R.M. Dreizler and E.K.U. Gross, *Density Functional Theory*, Springer-Verlag, Berlin (1990).

[2] P. Hohenberg and W. Kohn, Phys. Rev. **136**, B864 (1964).

[3] W. Kohn and L.J. Sham, Phys. Rev. **140**, A1133 (1965).

[4] E.P. Wigner, Trans. Faraday Soc. **34**, 678 (1938).

[5] D.M. Ceperley and B.J. Alder, Phys. Rev. Lett. **45**, 566 (1980).

[6] J. Perdew and A. Zunger, Phys. Rev. B **23**, 5048 (1981).

[7] G. Ortiz and P. Ballone, Europhys. Lett. **23**, 7 (1993); G. Ortiz and P. Ballone, to be published (1994).

[8] R.O. Jones and O. Gunnarson, Rev. Mod. Phys. **61**, 689 (1989).

[9] O.K. Andersen, O. Jepsen, and M. Sob, in: *Electronic Band Structure and Its Applications*, edited by M. Yussouf (Springer, Berlin 1987), p. 1.

[10] W.E. Pickett, Computer Phys. Reports **9**, 115 (1989).

[11] A.D. Becke, Phys. Rev. A **38**, 3098 (1988).

[12] H. Hellmann, *Einführung in die Quantenchemie* (Deuticke, Leipzig, 1937); R.P. Feynman, Phys. Rev. **56**, 340 (1939).

[13] I. Štich, R. Car, M. Parrinello, and S. Baroni, Phys. Rev. B **39**, 4997 (1989).

[14] W.H. Press, B.P. Flannery, S.A. Teukolsky, W.T. Vetterling, *Numerical Recipes*, 2nd ed., Cambridge University Press (1991)

[15] M.C. Payne, M.P. Teter, D.C. Allen, T.A. Arias, and J.D. Joannopoulos, Rev. Mod. Phys. **64**, 1045 (1992).

[16] J. Hutter, H.P. Lüthi, and M. Parrinello, Comput. Mat. Sci., in press (1994).

[17] R. Car and M. Parrinello, Phys. Rev. Lett. **55**, 2471 (1985).

[18] G. Galli and A. Pasquarello, in *Computer Simulation in Chemical Physics*, edited by M.P. Allen and D.J. Tildesley (Kluwer, Amsterdam, 1993), p. 261.

[19] P. Pulay, Mol. Phys. **17**, 197 (1969).

[20] R. Yu, D. Singh, and H. Krakauer, Phys. Rev. B **43**, 6411, (1991); M. Methfessel and M. van Schilfgaarde, Phys. Rev. B **48**, 4937 (1993).

[21] E. Fermi, Nuovo Cimento **11**, 157 (1934).

[22] See the papers in *Solid State Physics*, edited by H.E. Ehrenreich, F. Seitz, and D. Turnbull, vol.24 (Academic Press, New York, 1970).

[23] D.R. Hamann, M. Schlüter, and C. Chiang, Phys. Rev. Lett. **43**, 1494 (1979).

[24] G.B. Bachelet, D.R. Hamann and M. Schlüter, Phys. Rev. B **26**, 4199 (1982).

[25] A subtle point about the validity of frozen-core approximation is discussed in U. von Barth and C.G. Gelatt, Phys. Rev. B **21**, 2222 (1980).

[26] A. Baldereschi, Phys. Rev. B **7**, 5212 (1973).

[27] D.J. Chadi and M.L. Cohen, Phys. Rev. B **8**, 5747 (1973); H.J. Monkhorst and J.D. Pack, Phys. Rev. B **13**, 5188 (1976).

[28] Two recent references on this subjects: M. Methfessel and A.T. Paxton, Phys. Rev. B **40**, 3616 (1989); J. Hama, M. Watanabe, and T. Kato, J. Phys.: Condens. Matter **2**, 7445 (1990).

[29] D.G. Anderson, J. Assoc. Comput. Mach. **12**, 547 (1965).

[30] D.D. Johnson, Phys. Rev. B **38**, 12807 (1988).

[31] For an introduction: W.L. Briggs, *A Multigrid Tutorial*, SIAM, Philadelphia (1987).

[32] S. Baroni and M. Buongiorno Nardelli, unpublished.

[33] E. Anderson et al, *LAPACK Users' Guide*, SIAM (Philadelphia, 1992).

[34] For a recent review, see E.R. Davidson, Computer Phys. Commun. **53**, 49 (1989).

[35] N. Troullier and J.L. Martins, Phys. Rev. B **43**, 8861 (1991).

[36] R.D. King-Smith, M.C. Payne, and J.S. Lin, Phys. Rev. B **44**, 13063 (1991).

[37] J. Friedel, Adv. Phys. **3**, 446 (1954); F.J. Dyson, unpublished, as quoted in: C. Kittel, *Quantum theory of Solids* (Wiley, New York, 1963), p. 339.

[38] W. Yang, Phys. Rev. Lett. **66**, 1438 (1991); W. Yang, Phys. Rev. A **44**, 7823 (1991).

[39] G.Galli and M. Parrinello, Phys. Rev. Lett **69**, 3547 (1992); F. Mauri, G. Galli, and R. Car, Phys. Rev. B **47**, 9973 (1993); F. Mauri and G. Galli, Phys. Rev. B, in press (1994); P. Ordejón, D.A. Drabold, M.P. Grumbach, and R.M. Martin, Phys. Rev. B **48**, 14646 (1993).

[40] X.-P. Li, R.W. Nunes, and D. Vanderbilt, Phys. Rev. B **47**, 10891 (1993); S. Goedecker, preprint; M. Daw (unpublished).

[41] D.A. Drabold and O.F. Sankey, Phys. Rev. Lett. **70**, 3631 (1993).

[42] S. Baroni and P. Giannozzi, Europhys. Lett. **17**, 547 (1992).

[43] For a review, see: *Solid State Physics*, edited by H. Ehrenreich, F. Seitz, and D. Turnbull (Academic, New York, 1980), Vol. 35.

[44] A. Franceschetti and S. Baroni, unpublished; A. Franceschetti, Ph.D. Thesis, SISSA-Trieste, 1993 (unpublished).

[45] D. Vanderbilt, Phys. Rev B **41**, 7892 (1990).

[46] F. Gygi, Europhys. Lett. **19**, 617 (1992).

[47] K. Cho, A. Arias, J.D. Joannopoulos, and P.K. Lam, Phys. Rev. Lett. **71**, 1808 (1993).

[48] I. Štich, M.C. Payne, R.D. King-Smith, J.-S. Lin, and L.J. Clarke, Phys. Rev. Lett. **68**, 1351 (1992); K. Brommer, M. Needels, B. Larson, and J.D. Joannopoulos, Phys. Rev. Lett. **68**, 1355 (1992).

[49] G. Kerker, J. Phys. C **13**, L189 (1980).

[50] U. von Barth and R. Car, unpublished.

[51] S.G. Louie, S. Froyen, and M.L. Cohen, Phys. Rev. B **26**, 1738 (1982).

[52] D.R. Hamann, Phys. Rev. B **40**, 2980 (1989).

[53] E.L. Shirley, D.C. Allan, R.M. Martin, and J.D. Joannopoulos, Phys. Rev. B **40**, 3652 (1989).

[54] G. Ortiz and P. Ballone, Phys. Rev. B **43**, 6376 (1991).

[55] E.L. Shirley and R.M. Martin, Phys. Rev. B **47**, 15413 (1993).

[56] L. Kleinman and D.M. Bylander, Phys. Rev. Lett. 48, 1425 (1982).

[57] X. Gonze, P. Kaeckell, and M. Scheffler, Phys. Rev. B **41**, 12264 (1990); X. Gonze, R. Stumpf, and M. Scheffler, Phys. Rev. B **44**, 8503 (1991).

[58] N. Troullier and J.L. Martins, Phys. Rev. B **43**, 1993 (1991).

[59] K. Laasonen, A. Pasquarello, R. Car, C. Lee, and D. Vanderbilt, Phys. Rev. B **47**, 10142 (1993).

COMPUTER SIMULATION OF MATERIALS USING PARALLEL ARCHITECTURES

Priya Vashishta, Rajiv K. Kalia, Aiichiro Nakano, Wei Jin, and Jin Yu

Concurrent Computing Laboratory for Materials Simulations
Department of Physics & Astronomy and Department of Computer Science
Louisiana State University, Baton Rouge, LA 70803-4001, USA

ABSTRACT

Algorithms are designed to implement molecular dynamics (MD) and quantum molecular dynamics (QMD) simulations on emerging concurrent architectures. A highly efficient multiresolution algorithm is designed to carry out large-scale MD simulations for systems with long-range Coulomb and three-body covalent interactions on distributed-memory MIMD (multiple instruction multiple data) machines. The performances of these algorithms are tested on the Intel Touchstone Delta and IBM SP-1 systems. The computational complexities of these algorithms are O(N) and parallel efficiencies close to 0.9. The core computational kernel of the QMD approach consists of solutions of parabolic partial differential equations (PDE) such as the time-dependent Schrödinger equation or time-dependent Kohn-Sham equation. This problem is coupled with another computationally intensive problem, i.e., solution of elliptic PDEs (the Poisson equation) for the long-range electron-electron interaction. We have designed parallel algorithms for both problems on SIMD (single instruction multiple data) machines.

In the past three years, we have used the parallel computer architectures in our *Concurrent Computing Laboratory for Materials Simulations* (CCLMS) to carry out MD and QMD simulations on network glasses, ceramic composites, nanophase materials, solid C_{60} and graphitic tubules, and quantum transport in nanoscale devices.

Structural transformation, intermediate-range order and dynamical behavior of SiO_2 glass at high pressures are investigated with molecular dynamics. At high densities, the height of the first sharp diffraction peak is considerably diminished, its position changes from 1.6 to 2.2 Å^{-1}, and a new peak appears at 2.85 Å^{-1}. At twice the normal density, Si-O bond length increases, Si-O coordination changes from 4 to 6, and O-Si-O bond-angle changes from 109° to 90°. This is a tetrahedral to octahedral transformation, which was reported recently by Meade, Hemley, and Mao.

Molecular dynamics simulations of porous silica, in the density range 2.2 - 0.1 g/cm^3, are carried out on a 41,472-particle systems using a MIMD computer. The internal surface area, pore surface-to-volume ratio, pore-size distribution, fractal dimension, correlation length, and mean-particle size are determined as a function of the density. Structural transition between a condensed amorphous phase and a low-density porous phase is characterized by these quantities. Various dissimilar porous structures with different fractal dimensions are obtained by controlling the preparation schedule and temperature.

Pore interface growth and the roughness of fracture surfaces in silica glasses are investigated by MD simulations with 1.12-million particles. During uniform dilation, the pores coalesce and grow in size. When the mass density is reduced to 1.4 g/cm^3, the pores grow catastrophically to cause fracture. The roughness exponent for fracture surfaces, $\alpha = 0.87 \pm 0.02$, supports experimental claims about the universality of α.

Lattice dynamics of solid C_{60} is studied using a unified interaction model which consists of a tight-binding potential for the intra-molecular interaction and a Lennard-Jones and bond charge model for the inter-molecular interaction. Phonon dispersion and density of states of solid C_{60} are calculated in the energy range from 0 to 210 meV. The inter-molecular phonon density of states shows peaks around 2.3 meV and 3.7 meV, and extends to 7.6 meV. The calculated phonon spectrum agrees well with inelastic neutron scattering experiments. The effects of orientational disordering and pressure on the inter- and intra-molecular phonons of solid C_{60} are investigated.

Recently a new form of carbon -- graphitic tubule -- has been discovered. It is the fourth member of the carbon family with dimension of one (diamond in 3D, graphite in 2D, and fullerene in 0D). Using the tight binding molecular dynamics method (TBMD), the structural and dynamical properties of graphitic tubules are studied. The phonon dispersion and density of states of graphitic tubules with various helicities and diameters are calculated. Compared with graphite, phonon modes in tubules are softened by the curvature. Unique features of the graphitic tubule, with special emphasis on low-frequency modes, are discussed. The symmetry of phonon modes is analyzed, and infrared and Raman active modes are identified. Sound velocities in graphitic tubules are also calculated as functions of tubule helicity and diameter.

INTRODUCTION

Despite significant recent developments in materials-simulation techniques,[1-30] the goal of reliably predicting the properties of new materials in advance of fabrication and measurement has not yet been achieved. The primary reason for this lack of success is the inability of sequential machines to handle large-scale simulations. For example, molecular dynamics (MD) simulations for long-range interactions scale as N^2 where N is the number of particles in the system. In many physical systems, the desired system sizes are in the range of 10^6 particles. These are beyond the compute power of most sequential machines. However, the MD technique has considerable inherent parallelism. By exploiting this parallelism on emerging parallel architectures, it is possible to perform large-scale simulations for complex materials.[4-7,10-12]

Over the years, SiO_2 has been the focus of many investigations in solid state physics, microelectronics, geosciences, materials science and engineering.[31-61] In the past decade, numerous attempts have been made to investigate the structure and dynamics of crystalline and glassy states of SiO_2 at high pressures.[41-53] Brillouin,[41] Raman,[41-43] infrared,[42,44] neutron,[45] and x-ray[46] scattering techniques have been used to investigate pressure-induced effects in SiO_2 glass. At high pressures, irreversible changes, indicating permanent densification, have been observed in the Brillouin and Raman spectra[41] of recovered SiO_2 glass samples. Infrared absorption measurements,[44] however, indicate reversible changes in the SiO_2 glass at 20 GPa.

There has been a growing interest in porous materials because of their many technologically important applications. Much of the recent work has focused on aerogel silica, a form of porous SiO_2 which is prepared by hypercritical drying of an alcoholic silica gel.[55-61] It is an environmentally safe material with a large thermal resistance which makes it a suitable alternative to chlorofluorocarbon (CFC)-foamed plastic in thermal insulation of commercial and household refrigerators.[55] The application of porous glasses results from their unique selective separation capabilities, molecular transport, thermal resistance, and mechanical properties. All of these characteristics depend crucially on structural correlations such as the pore size, internal surface area, surface-to-volume ratio, and interface texture.

In recent years, a great deal of progress has been made in understanding the morphology of surfaces and interfaces. Scale-invariant surface fluctuations related to different growth processes have been observed in a wide variety of systems:[62,63] vapor deposition; fluid flow in porous media; sedimentation of granular materials; and thin-film growth. The root mean square surface fluctuations averaged over a distance l obey the scaling relation,[62,63] $W \sim l^\alpha$. Recent experiments on a wide variety of materials reveal that fracture surfaces exhibit the scaling properties with the roughness exponent $\alpha \approx 0.8$.[64,65] This led them to suggest that the roughness exponent for fracture surfaces has a universal value. However the universality of the roughness exponent on the nanometer scale is still an unresolved issue.[66]

We investigate the structural correlation of densified[54] and porous[61] SiO_2 glasses, and the fracture of SiO_2 glasses[67] by MD simulations. Interparticle potentials used in our MD simulations include two-body (V_2) and three-body (V_3) terms,[39]

$$V = \sum_{i,j} v_2(r_{ij}) + \sum_{i,j,k} v_3(\mathbf{r}_{ij}, \mathbf{r}_{ik}) \ . \tag{1}$$

Since the breakthrough in the synthesis of C_{60}[68,69] and the discovery of superconductivity in K_3C_{60} and Rb_3C_{60},[70] tremendous effort has been made to understand the structure, orientational order, rotational dynamics, and inter- and intra-molecular interactions in fullerite and their compounds.[71-90] It is now well established that upon crystallization, the centers of C_{60} clusters form a face-centered-cubic (*fcc*) crystal with fullerene molecules rotating almost freely at room temperature. The *fcc* phase has the symmetry of the space group $Fm\bar{3}m$ with a lattice constant of $a_0 = 14.2$ Å. On cooling below 250 K, solid C_{60} undergoes a first order structural phase transition from an orientationally disordered *fcc* phase to an orientationally ordered simple cubic (*sc*) structure with four molecules per unit cell located on four different cubic sublattices having different orientations. The low temperature *sc* phase belongs to the space group $Pa\bar{3}$ with a lattice constant of $a_0 = 14.05$ Å.

Several theoretical approaches[91-105] have been used to investigate the physical properties of solid C_{60}. A force-field calculation[94] and classical molecular dynamics simulations[96] based on a simple atom-atom type inter-molecular van der Waals interaction have been unable to explain the stability of the $Pa\bar{3}$ ground state of solid C_{60}. Attempts have been made to improve the van der Waals model by considering the non-spherical character of the charge distribution around carbon atoms. An inter-molecular bond-bond charge interaction model[97] and an inter-molecular atom-bond interaction model[98] have been proposed. Meanwhile, by fitting to first-principles calculations of electronic band structure and volume-dependent binding energies of various crystalline carbon phases, Xu, Wang, Chan, and Ho (XWCH) have proposed a transferable tight-binding potential for carbon.[99] It is therefore natural to combine the inter-molecular (van der Waals and bond charge) and the intra-molecular (tight-binding) interaction models to investigate the structural and dynamical properties of solid C_{60}. This unified approach permits the calculation of the inter-molecular phonon dispersion and density of states, and it also predicts modifications of the intra-molecular vibrational modes and their dispersion when the molecules form a solid and when the solid is under pressure.[104,105]

Using the unified approach, we have calculated the inter-molecular and intra-molecular phonon density of states and dispersion for crystalline C_{60}. Effects of orientational disorder on the lattice dynamics of solid C_{60} have been studied. The calculated phonon spectrum agrees well with inelastic neutron-scattering experiments.[83,84] The specific heat as a function of temperature has been calculated, and the result is also in good agreement with experiments.[85] The effect of pressure on the inter- and intra-molecular phonon spectra of solid C_{60} has also been studied. In addition, the pressure dependence of sound velocities, elastic constants, bulk modulus and specific heat have been calculated.

Recently a new form of carbon -- graphitic tubule was discovered in carbon rods under arc discharge by high-resolution transmission-electron microscopy.[106] It is the fourth member of the carbon family with dimension of 1 (diamond in 3D, graphite in 2D, and fullerene in 0D). A microtubule has the form of a rolled graphite sheet with a diameter of a few nanometers. The carbon-atom hexagons on the tubule are usually arranged in a helical fashion about the tubule axis.

Due to crystalline perfection, various possible helical structures, and the dimensionality as well as the high efficiency of production,[107] graphitic tubules may possess unusual mechanical, electronic, and optical properties which may find considerable technological applications (e.g. nanoscale devices, light-weight and high-strength composite materials, etc.). Recent theoretical studies[108,109] have shown that the electronic properties of a graphitic tubule strongly depend on its helical structure. It can be metallic or semiconducting, which implies that the electronic property of graphitic tubule can be tuned by changing its geometric parameters.

Intensive studies of the structure and electronic properties of graphitic tubules and related compounds have been carried out,[106-121] and the relationship between the structure symmetry and lattice dynamics of graphitic tubule have been discussed.[111] However, to the best of our knowledge, there are not detailed theoretical results for the phonon dispersion or density of states of graphitic tubules based on realistic models.

Using the tight-binding molecular-dynamics (TBMD)[122-127] we have studied the structural and dynamical properties of the graphitic tubules.[128] The phonon dispersion and density of states of graphitic tubules with various helicities and diameters have been calculated. It has been found that, compared with graphite, phonon modes in tubules are softened by the curvature. While the overall spectra of graphitic tubule resemble that of graphite, especially when the tubule diameter is large, unique features are present in the low-frequency part of the spectra (< 50 meV). These low-frequency modes, which are the signature of the tubule structure, are sensitive to the diameter and the helical arrangement. The nature of the modes is analyzed and detailed comparison with unrolled graphite sheet is carried out.

In this paper, we will describe our recent work on parallel algorithms for the MD and QMD approaches and simulation results for structural properties of network glasses at high and low densities, C_{60} solids and graphite tubules. The outline of this paper is as follows. Parallel MD and QMD algorithms are discussed in Sec. 2. Simulation results for structural properties of highly densified and porous SiO_2 glasses are discussed in Secs. 3 and 4, respectively. Section 5 deals with the structural and vibrational properties of solid C_{60} and graphitic tubules. Section 6 describes the computing facilities in our laboratory. Concluding remarks are given in Sec. 7.

PARALLEL MOLECULAR DYNAMICS AND QUANTUM MOLECULAR DYNAMICS ALGORITHMS

Parallel Molecular Dynamics Algorithms

Molecular dynamics (MD) approach has played a key role in our understanding of classical and quantum microscopic processes in physical systems.[1,2] In the MD approach, one obtains the phase-space trajectories of particles from the numerical solution of Newton's equations. Physical properties of a system are calculated from phase-space trajectories of the constituent particles. The interparticle interaction energy is a vital input to MD simulations. This N-body term is commonly expressed as a combination of one-body, two-body, three-body potentials, etc.

For systems with a finite-range interparticle interaction, the total number of pairs contributing to the energy and forces is $N*N_b/2$, where N_b is the average number of particles within the range of the interaction, r_c. An efficient way to calculate interparticle interactions is to use the linked-list method.[1] In this approach, the simulation cell is divided into n^3 smaller cells, each with an edge L/n (L is the length of the MD cell) which is slightly larger than r_c, see Fig. 1. Using two integer arrays, a list of particles in each cell is constructed. The first array identifies the particle at the top of the list in each cell and the second array links particles belonging to the same cell. A major advantage of the linked-cell list technique is that the computation time is proportional to $14*N*(L/n)^3$, since each cell has 26 neighboring cells (Newton's third law is used to reduce the computation by a factor of 2). Furthermore, with the linked-list method the minimum-image convention can be implemented efficiently on distributed-memory MIMD machines.

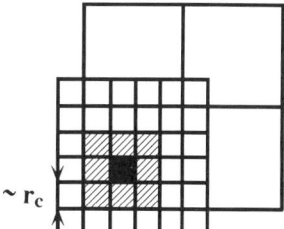

Figure 1. Linked-cell list method.

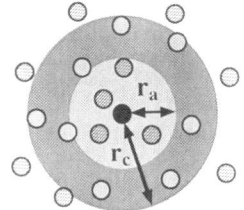

Figure 2. Multiple time-step method.

The computation of forces can be further reduced with the multiple time-step (MTS) approach.[3] The MTS approach exploits the fact that the force experienced by a particle can be separated into a rapidly varying primary component and a slowly varying secondary component. The primary interaction arises from nearest neighbors of a particle (within a range of r_a), whereas the secondary forces are due to other particles within its range of interaction, see Fig. 2. The primary component is calculated at every MD step. On the other hand, the secondary component is calculated at intervals of 5 and 15 steps. In between, the secondary component is extrapolated according to the Taylor series.

We have used the divide-and-conquer strategy based on domain decomposition[4] to implement the parallel MTS-MD algorithm.[5-7] The total volume of the system is divided into p subsystems of equal volume, and each subsystem is assigned to a node, see Fig. 3. The data associated with particles of a subsystem are assigned to the corresponding node.

The message-passing strategy we have used is shown schematically in Fig. 4.[5] First the data from node 0 are sent to node 1, data from node 1 to node 2, ..., and data from node p-1 to node 0 synchronously. Then using the linked-cell list method the contributions to secondary forces, potential energy, and their time derivatives are calculated. Node 0 sends back the calculated contributions to node p-1 while receiving the contributions calculated at node 1. Similar message-passing of calculated contributions takes place synchronously at other nodes as well. Next node 0 receives data from node p-2, node 1 from node p-1, ..., and node p-1 from node p-3. The contributions to forces, potential energy, and their time derivatives are calculated synchronously and the results are sent back to the nodes from which the data had been received. This procedure is continued until all the necessary interactions have been computed.

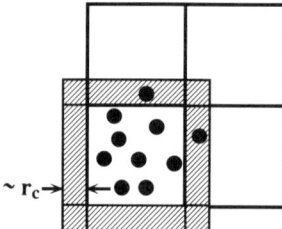

Figure 3. Domain decomposition.

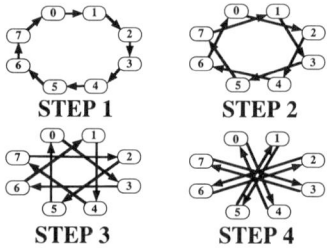

Figure 4. Internode communication strategy for secondary forces with 8 processors. Arrows indicate the direction of data motion.

The three-body force calculation is a time consuming part of the MTS-MD algorithm. Speed-up of the three-body force calculation is achieved by decomposing the three-body potential into a separable form.[6,7] Due to the decomposition, the evaluation of the potential energy and forces becomes similar to that of pair potentials. We accelerate our MTS-MD algorithm further using the separable three-body force calculation.

The most prohibitive computational problem is asociated with the Coulomb potential. Because of its long range, each atom interacts with all the other atoms in the system. Therefore the evaluation of the Coulomb potential for an N-particle system requires $O(N^2)$ operations, which makes large-scale MD simulations difficult. Recently, we have also implemented MD simulations involving the Ewald summation for Coulomb interaction on distributed-memory MIMD machines.[5] The Ewald summation[9] is the most widely used approach for the long-range Coulomb interaction in bulk systems. In this approach the interaction is written as a sum of a constant term, a sum in the Fourier space, and a sum of "short range" terms in real space. The parallel algorithm we have designed for the Ewald summation reduces the computational complexity from $O(N^{3/2})$ to $O(N)$. This is achieved by ensuring that both the real-space and Fourier-space contributions scale linearly with the size of the system. In real space, the potential energy and force calculations are truncated at $r_c = 5r_0$, where r_0 is the ion-sphere radius, $r_0 = (3/4\pi\rho)^{1/3}$, and ρ is the number density. This cutoff maintains the desired level of precision -- 0.01% for all system sizes. The real-space contributions are then calculated with the domain decomposition and the linked-list methods which scale as $O(N)$. The computation of Fourier-space contributions reveals that only a certain number of wave vectors need to be included. We performed simulations as a function of the number of Fourier components and system sizes. It is found that an increase in the number of wave vectors from 309 to 2,192 produces a change of 0.01% in the total potential energy. The computation time for the k-space calculation increases linearly as the number of wave vectors increases. Thus, the total execution time for the Ewald sum scales linearly with the number of particles (see Fig. 5) and is inversely proportional to the number of processors.

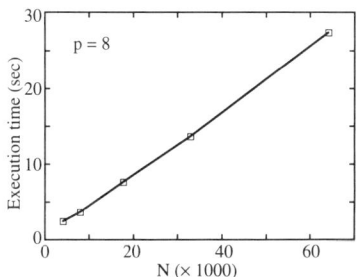

Figure 5. Execution time as a function of the number of particles in a bulk Coulombic system.

Recent hierarchical algorithms[10,11] have revolutionized the computation of the Coulomb potential. The fast multipole method (FMM) uses the truncated multipole expansion and local Taylor expansion for the Coulomb potential field.[11] By computing both expansions recursively on a hierarchy of cells, the Coulomb potential is computed with $O(N)$ operations. In many materials simulations, periodic boundary conditions are used to minimize surface effects. The summation over infinitely repeated image charges must be carried out to compute the Coulomb potential. Ding, Karasawa, and Goddard have developed the reduced cell multipole method (RCMM) which makes the computation of the Coulomb potential feasible for multimillion-particle systems with periodic boundary conditions.[12] In RCMM, distant images of multimillion particles are replaced by a small number of fictitious particles with the same leading multipoles as the original system. With little computational effort, the Ewald summation is applied to these reduced images. We have developed a highly efficient MD algorithm based on multiresolutions in both space and time.[7] The long-range Coulomb potentials in periodic systems are calculated with the cell-based RCMM and FMM, while the intermediate-range, non-Coulombic potentials are calculated by the list-based MTS method, see Fig. 6. The three-body potentials are calculated using the separable, tensor decomposition scheme.

Performance of the multiresolution algorithm is tested for SiO_2 systems on the 512-node Intel Touchstone Delta machine at Caltech and the 128-node IBM SP-1 system at

Argonne National Laboratory.[7] Figure 7 shows the execution time per MD step as a function of the number of processors, p. Number of particles is taken to be 8,232 p. For a 4.2 million particle system, the program requires only 4.84 second per MD step on the 512-node Delta. Communication accounts for only 8 % of the total elapsed time. On the IBM SP-1, the computation part runs 4.8 times faster than on the Delta, while the communication performs at about the same speed. As a result, the communication overhead is slightly larger on the SP-1.

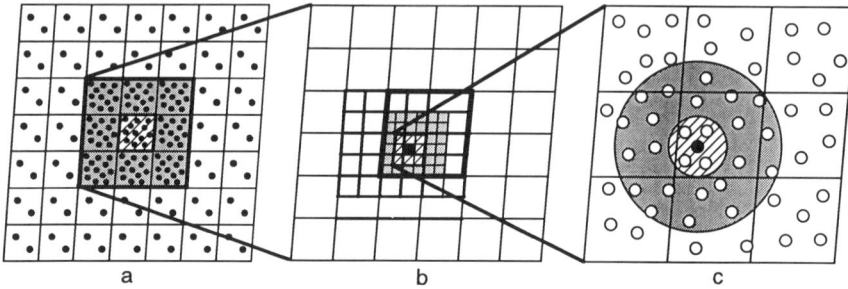

Figure 6. Multiresolution in space. (a) Periodically repeated images of the original MD box. Replacing far images by a small number of particles with the same multipole expansion up to a certain order reduces the computation enormously while maintaining the necessary accuracy. (b) A hierarchy of cells used in the fast multipole method. (c) The near-field force on a particle is due to primary, secondary, and tertiary neighbor particles.

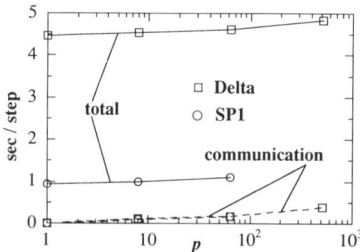

Figure 7. Execution time (solid curves) and communication time (dotted) per MD time step for SiO_2. Circles and squares represent the results on the IBM SP-1 and Intel Touchstone Delta, respectively. Here p is the number of nodes. The size of the system, N, increases as 8232 p.

Parallel Quantum Dynamics

In Quantum Dynamics (QD), we simulate the dynamics of coupled systems of electrons and phonons.[13-17] In this scheme, the electron-electron interaction is included in the framework of the time-dependent density functional theory.[18] The time evolution of electron orbitals, $\psi_i(\mathbf{r},t)$, is given by the time-dependent Kohn-Sham equations:[18,19]

$$[i\hbar\frac{\partial}{\partial t} - \frac{1}{2m}\left(\frac{\hbar\nabla}{i} + \frac{e}{c}[A(r,t) + A_{xc}(r,t)]\right)^2 - \frac{e^2}{2mc^2}\left(A^2(r,t) - [A(r,t) + A_{xc}(r,t)]^2\right) \quad (2)$$
$$- v_0(r,t) - v_{eI}(r,t) - \int dr' \frac{e^2 n(r',t)}{|r-r'|} - v_{xc}(r,t)]\psi_i(r,t) = 0,$$

where $v_0(r,t)$ is an external potential, $v_{xc}(r,t)$ is the exchange-correlation potential,[18] and $A_{xc}(r,t)$ is the exchange-correlation vector potential.[19] The electron density $n(r,t)$ is calculated from

$$n(r,t) = \sum_{i=1}^{N_e} |\psi_i(r,t)|^2, \quad (3)$$

where N_e is the number of electrons.

We study coupled systems of electrons and phonons in a mean-field approximation.[20] In this approximation, $v_{eI}(r,t)$ in Eq. (2) is expressed as,

$$v_{eI}(r,t) = \sum_k \left[Q_k e^{ik\cdot r} \sum_v \sqrt{v+1} \phi_k^{(v)*}(t) \phi_k^{(v+1)}(t) + c.c. \right], \quad (4)$$

where $\phi_k^{(v)}$ is the phonon wave function for the kth phonon mode in the number representation and Q_k is the corresponding electron-phonon interaction coefficient. Phonon wave functions evolve in time following the time-dependent Schrödinger equations,[15]

$$\left(i\hbar\frac{\partial}{\partial t} - v\hbar\omega_k\right)\phi_k^{(v)}(t) = Q_k n_k^*(t)\sqrt{v+1}\phi_k^{(v+1)}(t) + Q_k^* n_k(t)\sqrt{v}\phi_k^{(v-1)}(t), \quad (5)$$

where $n_k(t)$ is the Fourier transform of $n(r,t)$.

One of the consequences of the coupling between electrons and acoustic-phonon modes is to dampen quantum motion and drive the system toward equilibrium. Such dissipation effects are important to sustain a steady nonequilibrium state, and it is desirable to explicitly include dissipation in QD simulations. We have developed a scheme to incorporate dissipation by solving the Langevin equation for the center-of-mass motion of electrons.[16]

We have implemented the QD algorithms on a SIMD (single instruction multiple data) computer, an 8,192-node MasPar in the CCLMS.[17]

Conventional solutions of the time-dependent Schrödinger equation are based on the spectral method (SM).[21] The complexity of the SM is $O(M\log M)$, where M is the number of grid points used for discretization. On an array of M processors, the SM requires $O(\log M)$ parallel operations. Communication distance grows as $O(M^{1/D})$, where D is the spatial dimension. Recently, the space-splitting method (SSM) has been developed to solve the time-dependent Schrödinger equation.[22] The SSM is based on the decomposition of a tridiagonal matrix, related to the kinetic energy operator in the Schrödinger equation, into direct sums of 2×2 matrices. This decomposition provides an explicit scheme to propagate wave functions in time with $O(1)$ parallel operations. In addition, the operations in the SSM are local, and therefore only the nearest-neighbor communications are needed.

Figure 8 (a) shows the performance in million floating point operations per second (MFlops) of the SSM and SM for the solution of the time-dependent Schrödinger equation of an electron in two dimensions under a magnetic field on the MasPar. The number of grid points is taken to be 128 × 64. The SSM achieves 479 MFlops (64 % of the theoretical peak speed, 750 MFlops), while the SM achieves 22.3 % of the theoretical peak. Because the SSM involves only the nearest-neighbor communications, it executes 2.9 times faster than the SM.

In our example, the SSM requires 144M floating point operations. On the other hand, the SM requires $(16\log_2 M+18)M$ operations. For M larger than 230, the SSM involves less operations than the SM. Combined with the smaller communication overhead, it makes the SSM much faster than the SM. Figure 8 (b) shows the performance of the SSM and SM for the same problem in time steps executed per second. The SSM is 4.7 times faster than the SM and the performance of the SSM becomes even superior for larger problems.

On massively parallel processor arrays, conventional algorithms such as the fast Poisson solver (FPS)[23] and the multigrid method (MGM)[24,25] solve the Poisson equation in O(logM) parallel operations. In addition, communications in Poisson solvers vary with distance as $O(M^{1/D})$.

When the Poisson equation is coupled to dynamical equations, successive calls to a Poisson solver are correlated with each other. In such a case, it is possible to design a

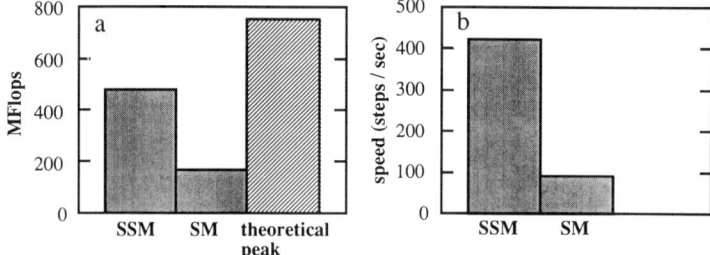

Figure 8. (a) Performance in MFlops of the SSM and the SM for the solution of the time-dependent Schrödinger equation for an electron in two dimensions under a magnetic field. The calculation employs 128 × 64 grid points on an 8,192-node MasPar MP-1. Measured performances (solid) in MFlops are compared with the theoretical peak speed (hatched) of the MasPar. (b) Performance of the SSM and SM for the same problem in time steps per second.

highly efficient algorithm by introducing fictitious dynamics for the electrostatic potential in the Poisson equation.[26] This dynamical-simulated-annealing (DSA) method is implemented with O(1) parallel time complexity. In contrast to conventional Poisson solvers, the DSA involves only the nearest-neighbor communications.

Figure 9 (a) compares the performance in MFlops of various Poisson solvers on the MP-1. We consider (1) dynamical-simulated-annealing (DSA) method, (2) fast Poisson solver (FPS), and (3) multigrid method (MGM). For the DSA and FPS, the results are obtained with 128 × 64 grid points. For the MGM, the result on 63 × 63 nodes are extrapolated to 8,192 nodes. The DSA runs at 460.0 MFlops (61.3 % of the theoretical peak, 750 MFlops), while the FPS and the MGM run at 160.3 MFlops (21.4 % of the theoretical peak) and 63.1 MFlops (8.4 %), respectively. The high speed of the DSA is due to the small communication overhead.

Figure 9 (b) shows the speed of various Poisson solvers in steps executed per second. Here 22.2 DSA steps are executed in the same time as one FPS step. Suppose that in a quantum dynamical simulation program with the DSA Poisson solver, we execute 22 time steps of the temporal evolution of the Hartree potential between two successive steps for the time-dependent Kohn-Sham equations. Then we get the same speed as that with the FPS Poisson solver. For larger systems, the DSA algorithm becomes much faster than the

FPS algorithm. For systems with nonuniform grid points, where the FPS is not applicable, the DSA is already faster than the MGM at M = 8,192. (With the ratio between time steps for electrons and the Hartree potential chosen as 22, the DSA is 8 times faster than the MGM.)

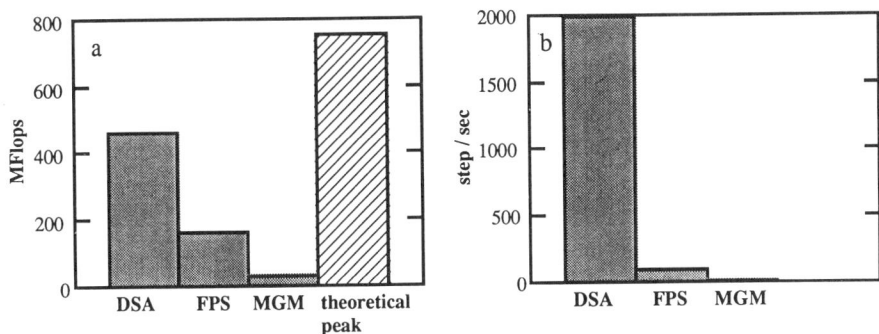

Figure 9. (a) Performances in MFlops of the DSA, FPS, and MGM on an 8,192-node MasPar MP-1 compared with the theoretical peak speed. (b) Speed of the (1) DSA, (2) FPS, and (3) MGM Poisson solvers for the same problem in steps executed per second on the MP-1.

Quantum Molecular Dynamics

Equations (2) and (5) are suitable for the electron transport at very low temperatures, where the quantum nature of lattice vibrations is important. On the other hand, in disordered materials at relatively high temperatures, it is more relevant to treat the motion of ions classically.[27-30]

For a coupled system of electrons and classical ions, $v_{eI}(r,t) = V_{eI}(r,\{R_J(t)\})$ in Eq. (2) is the interaction potential between an electron and ions, where $R_J(t)$ is the coordinate of the Jth ion. We solve Newton's equations for classical ions,

$$M_J \frac{d^2 R_J(t)}{dt^2} = - \frac{\partial V_I(\{R_J(t)\})}{\partial R_J(t)} - \int dr\, n(r,t) \frac{\partial V_{eI}(r, \{R_J(t)\})}{\partial R_J(t)}, \qquad (6)$$

concurrently with Eq. (2). In Eq. (6) M_J is the mass of the Jth particle and $V_I(\{R_J(t)\})$ is the interatomic potential for classical ions. The coupled equations, Eqs. (2) - (6), for quantum and classical particles are called the Quantum Molecular Dynamics (QMD).[27-30]

The parallel algorithms discussed above for MD and QD can be naturally combined to construct efficient parallel algorithms for QMD simulations on SIMD or MIMD machines.

SIO₂ GLASS AT HIGH PRESSURES

Recently Meade, Hemley, and Mao[46] have carried out in situ high-pressure (8 - 42 GPa) x-ray diffraction experiments on SiO_2 glass. These measurements reveal significant changes in the short-range and intermediate-range order (IRO). The position of the first sharp diffraction peak (FSDP) in x-ray structure factor, the fingerprint of IRO, changes from 1.55 Å^{-1} at 8 GPa to 2.37 Å^{-1} at 42 GPa. At the same time, there is a significant decrease in the height and an increase in the width of the FSDP. Furthermore, the pair correlation function shows that the nearest-neighbor (nn) tetrahedral coordination of Si-O changes to octahedral coordination as the pressure is increased from 8 to 42 GPa. Raman and infrared spectra at pressures greater than 28 GPa reveal the absence of tetrahedral vibrational modes.[43,44]

Molecular dynamics simulations were performed with interatomic potentials comprising two- and three-body terms.[54] The two-body potentials combine steric repulsion, long-range Coulomb interaction due to charge-transfer effects, and charge-dipole interaction due to large electronic polarizability of O^{2-} ions. The three-body covalent interactions include the effects of bond bending and stretching.[39] The MD simulations were carried out at normal mass density $\rho_0 = 2.20$ g/cm^3, and at pressures 0.9, 5.4, 22.7, and 42.3 GPa corresponding to densities ρ = 2.64, 2.94, 3.53, and 4.28 g/cm^3, respectively. Periodic boundary conditions were employed and the long-range nature of the Coulomb interaction was taken into account by Ewald's summation.

The SiO$_2$ glasses were generated by quenching well equilibrated liquids at high temperatures (~ 4,000 K).[39,54] High pressure SiO$_2$ glasses were also prepared in a similar fashion except that the length of the MD cell and atomic positions were first appropriately scaled in high-temperature liquid states. Low-temperature glasses were further relaxed with the aid of the steepest-descent quench.[39] At each temperature and density, structural and dynamical correlations were calculated with MD trajectories of at least 30 ps.

In Fig. 10, we present the MD results for the density dependence of the static structure factor, S(q). In the normal density glass, the FSDP is located at 1.6 Å^{-1}. With an increase in the density, the height of the FSDP decreases, its width increases, and its position shifts to higher values of q. Note that simple elastic compression [i.e., $(\rho/\rho_0)^{1/3}$]

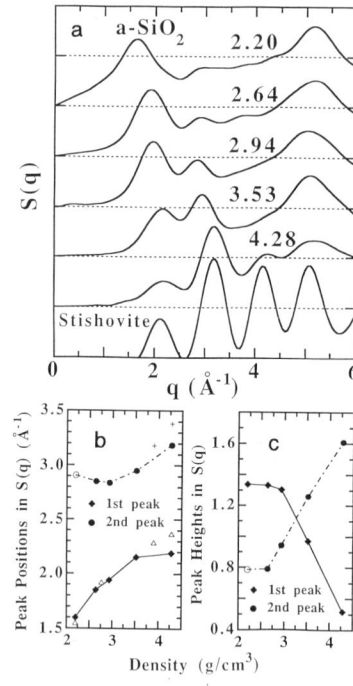

Figure 10. (a) Molecular dynamics results for the static structure factor, S(q), of normal and high density SiO$_2$ glasses at 300 K; (b) density dependencies of positions of the first two peaks in S(q); (c) density dependencies of heights of the first two peaks in S(q). In (b) and (c), open circle (O) at normal density are meant to indicate that the second peak is broad and has a small amplitude. Triangles (Δ) and crosses (+) in (b) represent the experimental data estimated from Ref. 46.

cannot account for the observed shift in the position of the FSDP. Elastic compression corresponding to density increases of 20%, 33%, 60%, and 95% would shift the FSDP to 1.71, 1.77, 1.88, and 2.01 Å^{-1} whereas the simulation results reveal higher values for the position of the FSDP: 1.85, 1.94, 2.15, and 2.19 Å^{-1}. The high-pressure x-ray measurements by Meade, Hemley, and Mao[46] reveal similar behavior for the FSDP.

Figure 10 also shows a new peak in the static structure factor. It appears when the density of the normal system is increased by 20%. Located at 2.85 Å^{-1}, the peak grows with further increase in the density. However, its position shifts only slightly at higher pressures [Fig. 10(b)]. These results are again well supported by the recent x-ray measurements at high pressures.[46]

Partial pair-distribution functions, $g_{\alpha\beta}(r)$, at the normal and the highest densities are shown in Fig. 11. The position of the first peak in $g_{Si-O}(r)$ and the corresponding Si-O coordination remain unchanged up to 3.53 g/cm^3. At a pressure of 42.3 GPa where the glass density (4.28 g/cm^3) reaches the stishovite density, the first peak in $g_{Si-O}(r)$ occurs at 1.67 Å instead of 1.61 Å and the Si-O coordination increases from 4 to 5.8. In stishovite, the highest-density crystalline phase of SiO$_2$, Si-O bond lengths are 1.76 and 1.81 Å and the Si-O coordination is 6. In the glass at 4.28 g/cm^3, the second peak in $g_{Si-O}(r)$ is at 3.15 Å, close to the next-nearest-neighbor (nnn) Si-O distance (~ 3.20 Å) in the stishovite.

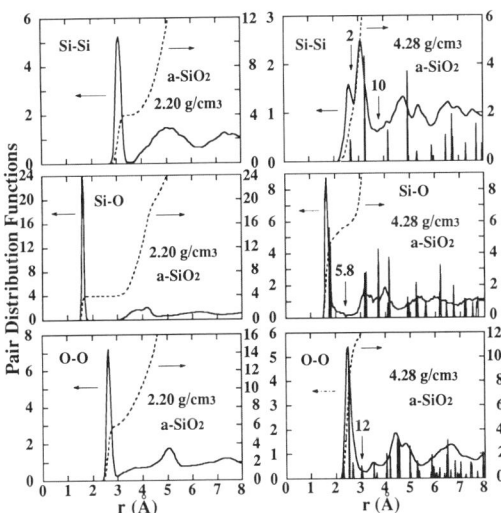

Figure 11. MD partial pair-distribution functions (solid lines) and coordination numbers (dotted lines) for SiO$_2$ glasses at normal and stishovite densities at 300 K. Sharp peaks in the figure at 4.28 g/cm^3 correspond to pair-distribution functions for crystalline stishovite.

Figure 11 also shows how the Si-Si pair-distribution function changes upon densification. The first peak splits into two peaks when the density increases to 4.28 g/cm^3. One of these peaks is located at 2.59 Å, close to the nn Si-Si distance (2.67 Å) in the stishovite. The second peak appears at 3.07 Å which is close to the nnn Si-Si distance (3.24 Å) in the stishovite. The area under the first peak gives a coordination of 2 while the area under the first two peaks is 10. At normal density, the nn O-O coordination is 6. It increases to 10 at 3.53 g/cm^3 and to 12 at 4.28 g/cm^3. In stishovite crystal, the O-O coordination is 12.

Figure 12 displays the MD results for O-Si-O and Si-O-Si bond-angle distributions in SiO$_2$ glasses at different densities. As the density increases, the peaks in these distributions broaden and also shift to lower angles because of increased distortions of Si(O$_{1/2}$)$_4$ tetrahedra. At normal density, the O-Si-O distribution has a peak at 109° with a full width at half maximum (FWHM) of 10°. With a 20% increase in the density, this peak moves to 107° and the FWHM increases to 12°. Further increase of 40% in the density shifts the peak to 104° and increases the FWHM to 17°. However, there is a dramatic change in the distribution when the glass density reaches the stishovite density: The O-Si-O distribution has broad peaks at 90° and 171°. In the crystalline stishovite on the other hand, the O-Si-O angles are 81.35°, 90°, 98.65°, and 180°.

In the normal density SiO$_2$ glass, the Si-O-Si bond-angle distribution has a peak at 142° and the FWHM of this peak is 26° (Fig. 12). Both of these results are in excellent agreement with NMR measurements.[37] With a density increase of 33%, this peak shifts gradually to 137°. With 60% densification, the peak moves to 127° and a broad shoulder appears between 135° and 150°. At the stishovite density, the Si-O-Si bond-angle in the glass has broad peaks around 95° and 128°. These values are close to the Si-O-Si angles, 98.65° and 130.67°, in the stishovite crystal. Thus, the results for pair-distribution functions and bond-angle distributions at 4.28 g/cm^3 contain strong evidence for distorted Si(O$_{1/3}$)$_6$ octahedra in the glass, joined at corners and sharing edges as well.

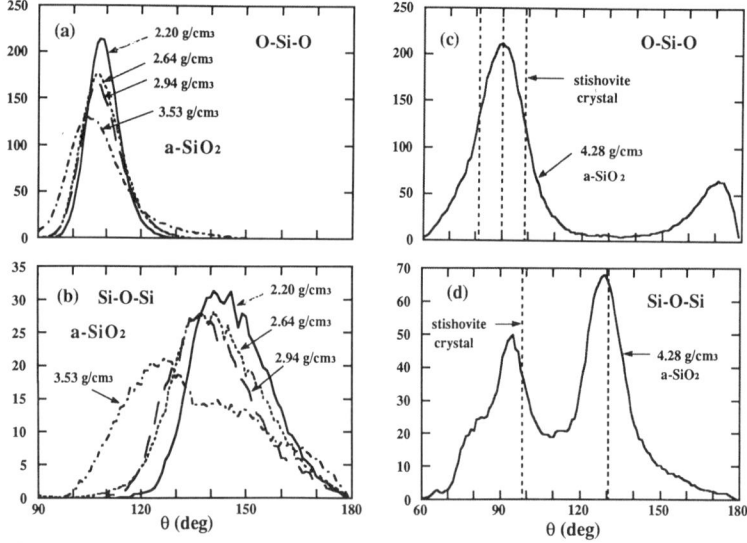

Figure 12. O-Si-O and Si-O-Si bond-angle distributions for normal and high density SiO$_2$ glasses.

The ring statistics in SiO$_2$ glass has been calculated to gain insight into network topology. For a given Si atom, the shortest closed path of Si-O bonds defines a ring. In the normal density SiO$_2$ glass, there are no two-fold rings (edge sharing tetrahedra) and the distribution is nearly symmetric with a peak at 6-fold rings. As the density increases, there is a decrease in the number of six-fold rings and an increase in the population of smaller rings. At the stishovite density, there are only two-, three-, four-, and five-fold rings in the glass, similar to the ring distribution in stishovite crystal which has only two-, three-, and four-fold rings.

MD SIMULATIONS OF POROUS SILICA

Structural correlations in porous silica span many hierarchical regimes. The short-range (< 4 Å) correlations are known to arise from the structure of the SiO_4 tetrahedral unit.[39] The intermediate-range (4 - 8 Å) correlations, manifested as the first sharp diffraction peak (FSDP) in neutron- and x-ray diffraction experiments, arise from the connectivity of the tetrahedral units.[39] Both these correlations exist at normal density as well as in low-density amorphous silica. Beyond the intermediate range, small-angle neutron scattering (SANS)[56,57] and small-angle x-ray scattering (SAXS)[58,59] experiments on porous silica reveal a fractal structure.

Molecular dynamics simulations of porous SiO_2 were pioneered by Kieffer and Angell.[60] We present the results of our MD simulations of porous SiO_2 at densities in the range of 2.2 - 0.1 g/cm^3.[61] These simulations have been performed on an 8-node Intel iPSC/860. The systems we have simulated consist of 41,472 Si and O atoms. Even at the lowest density, 0.1 g/cm^3, the length of the MD box (240 Å) covers all the hierarchical correlation regimes mentioned above. Simulations reported here took 1,200 hours on the 8-node iPSC/860 system. The simulations are based on an effective interatomic potential which combines two-body and three-body interactions.[39,61]

The porous SiO_2 systems are prepared as follows.[60,61] Starting with a well thermalized glass at the normal glass density 2.2 g/cm^3 and room temperature, porous glasses are obtained by successive expansions. At each expansion step, the coordinates of all the particles in the system are uniformly enlarged by a factor of 1.02 - 1.26 and subsequently the system is thermalized for 30,000 time steps. During thermalization, the temperature is kept at 300 K by removing heat from the system. Next, with a conjugate gradient scheme, the system is brought into a local minimum-energy configuration. In the relaxed configuration, particles are assigned random velocities according to the Maxwell distribution centered around 200 K. The equipartition of the energy is rapidly achieved (within 1,000 time steps) and the temperature drops to 100 K. At each density, statistical averages are calculated over 9,000 time steps.

Figure 13 displays snapshots of atomic positions of porous silica. At the condensed amorphous phase above 1.6 g/cm^3, the amorphous system possesses only short- and intermediate-range correlations, see Figs. 13 (a) and (b). However, as the density is lowered below 1.6 g/cm^3, density fluctuations that give rise to pores of various sizes set in.

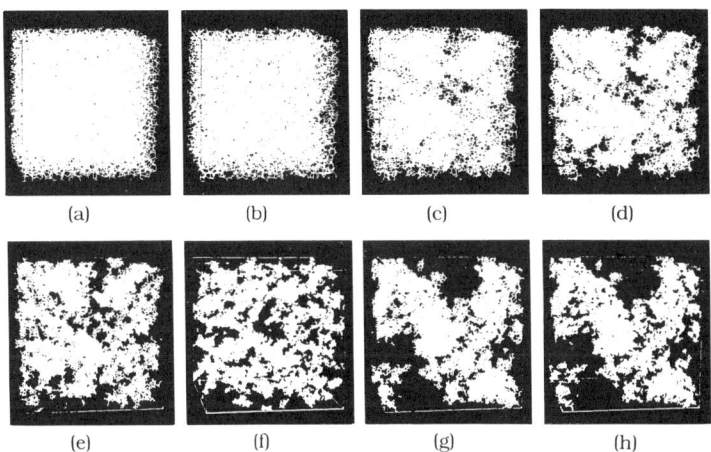

Figure 13. Snapshots of MD porous SiO_2 glasses at densities (a) 2.2, (b) 1.6, (c) 0.8, (d) 0.4, (e) 0.2, and (f) 0.1 g/cm^3 prepared at 300 K, and (g) 0.2 and (h) 0.1 g/cm^3 prepared at 1,000 K. Lines represent Si-O bonds.

A close examination of these snapshots (Figs. 13 (c) - (f)) reveals self-similarity at length scales between 5 - 25 Å.

In Fig. 14, we show a log-log plot of the pair distribution function g(r) at various densities. Short-range correlations manifest themselves as peaks at distances less than 5 Å. Some of the peaks split at lower densities, but the peak positions change very little over the entire range of density. At the normal density, the first peak is located at 1.62 Å. However, at lower densities, there is an additional peak at 1.58 Å due to the Si-O bond inside a triangle unit. The number of these triangular units grows as the density is lowered. In the range 5 - 25 Å, a power-law decay is superimposed on the peak structures. From the power-law, the fractal dimension, d_f, is calculated as $d_f = 3 + d \log[g(r)] / d \log(r)$.[61]

In real materials, the value of d_f depends on the aggregation process and sample preparation conditions such as pH value.[56-59] To investigate the effect of kinetic processes, we performed another set of MD simulations where the temperature was kept at 1,000 K instead of 300 K during the expansion process. Figures 13 (g) and (h) show the snapshots of the resulting glasses at densities 0.2 and 0.1 g/cm^3, respectively. Larger d_f is observed. Kinetic processes during the expansion determine the structure of the resulting glass. After an expansion, each particle or a cluster of particles searches an energy minimum through its

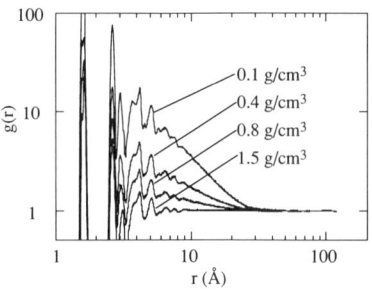

Figure 14. Log-log plot of pair distribution functions, g(r), of silica glasses at densities at 0.1, 0.4, 0.8, and 1.5 g/cm^3 at 300 K.

diffusive motion subject to the local interparticle correlation. Here, the balance between diffusion and local correlation is the most crucial factor which determines the structure. For higher temperatures, larger diffusion overcomes the correlation in immediate neighbors and more global configuration space is searched. As a result, energetically favored packed networks with larger d_f are formed. By controlling the balance between diffusion and correlation via temperature and expansion schedule, various dissimilar porous glasses with different d_f can be produced in MD simulations.

We have also performed large-scale MD calculations on 1.12-million particle amorphous silica systems, investigating the growth of pores with a decrease in the density of the system.[67] The low-density MD glasses were obtained by uniformly expanding the normal-density glass. The pores begin to form when the density of the system is reduced to 1.8 g/cm^3. Further decrease in the density of the system causes an increase in the number of pores and also the pores coalesce to form larger entities. There is a dramatic increase in the size of pores when the mass density is reduced to the critical value, $\rho_c = 1.4$ g/cm^3. At the critical density, some pores percolate through the entire system by catastrophic growth. In Fig. 15 we show one of the surfaces of the percolating pore.

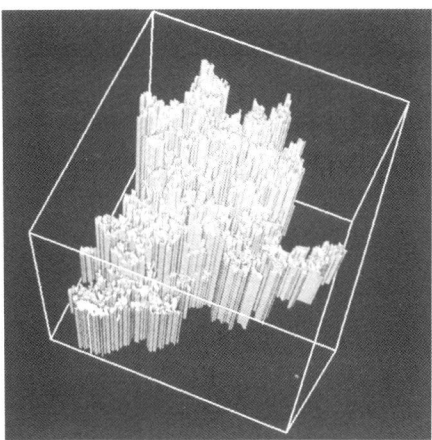

Figure 15. Snapshot of a fracture surface resulting from a percolating pore in silica glass at a mass density of 1.4 g/cm^3.

The roughness of this fracture surface is calculated from the height-height correlation function,[67] $G(\sigma) = \langle [h(y+y_0, z+z_0) - h(y_0, z_0)]^2 \rangle^{1/2}$, where $\sigma = (y^2 + z^2)^{1/2}$ and $h(y, z)$ is the highest vertical coordinate at the point (y, z). Figure 16 shows that the MD results for $G(\sigma)$ are well-described by the relation, $G(\sigma) \sim \sigma^\alpha$ with $\alpha = 0.87 \pm 0.02$ for $\sigma < 100$ Å. Experimental measurements on bakelite, concrete, steel, and aluminum alloys indicate that the roughness exponent α has a universal value 0.8.[65] The MD results for the roughness exponent agree with experimental measurements, thus lending further support to claims that the roughness exponent of fracture surfaces is a material-independent quantity. Furthermore, the MD results indicate that the universality of the roughness exponent may prevail even at length scales $\leq$ 10 nm.

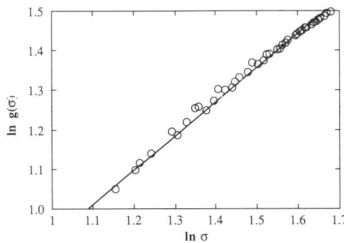

Figure 16. Height-height correlation function (open circles) versus the in-plane distance, σ, for the fracture surface shown in Fig. 15. The solid curve is the best fit, $G(\sigma) \sim \sigma^\alpha$, with $\alpha = 0.87 \pm 0.02$ for $\sigma < 100$ Å.

TBMD STUDY OF SOLID C$_{60}$ AND GRAPHITIC TUBULES

Model

Carbon is unique among all the elements in that it can form strong covalent bonds with various coordination numbers. Owing to the quantum nature of the directional chemical bonding, the description of carbon system by a classical potential becomes inadequate. Although first-principles methods (local-density-functional theory) are available, at present time their applications are still limited to small systems (~100 particles) because of enormous computing resource needed for such calculations. For systems in the range of 200 to 500 atoms (800 to 2000 electrons), an alternative approach is the tight-binding approximation.

Recently, by fitting to the results of first-principles calculations of electronic band structure and volume-dependent binding energies of various crystalline carbon phases, Xu, Wang, Chen, and Ho have proposed a transferable tight-binding potential for carbon.[99] In an empirical tight-binding approximation, the carbon-carbon interaction energy can be written as

$$E_{intra} = E_{bs} + E_{rep}, \qquad (7)$$

$$E_{bs} = \sum_{n}^{occ} \langle \psi_n | H_{tb} | \psi_n \rangle . \qquad (8)$$

The tight-binding Hamiltonian H_{tb} is constructed with a minimal basis of sp^3 orbitals:

$$H_{tb} = \sum_{i} \sum_{\alpha=s,p} \varepsilon_\alpha a^+_{i,\alpha} a_{i,\alpha} + \sum_{i,j} \sum_{\alpha,\beta=s,p} v_{\alpha,\beta}(r_{ij}) a^+_{i,\alpha} a_{j,\beta}, \qquad (9)$$

where the operator $a_{i,\alpha}$ annihilates an electron in orbital α at atom i. r_{ij} is the distance between atoms i and j. ε_s and ε_p are the on-site energies of s and p orbitals, respectively. $v_{ss\sigma}(r_{ij})$, $v_{sp\sigma}(r_{ij})$, $v_{pp\sigma}(r_{ij})$, and $v_{pp\pi}(r_{ij})$ are the hopping integrals. E_{bs} is the band-structure energy. E_{rep} is a short-ranged repulsive potential energy representing the ion-ion repulsion and the correction to the double counting of the electron-electron interaction in E_{bs},

$$E_{rep} = \sum_{i} f\left(\sum_{j} \varphi(r_{ij})\right) \qquad (10)$$

where $\phi(r_{ij})$ is a pair-wise potential between atoms i and j, and the function f describes the saturation of the repulsive energy for large values of $\Sigma_j \phi(r_{ij})$. Detailed description of the XWCH transferable tight-binding potential for carbon has been given by Xu et al. .[99]

Such a tight-binding potential model not only reproduces well the binding energies and bond lengths of the crystalline carbon with different coordination numbers, but also describes well the properties of several complex carbon systems far away from the ground state. These include micro-clusters, liquid, and amorphous carbon.[99] Therefore, this model should provide a good approximation for describing C_{60} and graphitic tubules.

Since the minimum separation (~3 Å) between any pair of atoms on different C_{60} clusters is large compared with the C-C covalent bond length (~1.4 Å), solid C_{60} can be regarded as a typical molecular crystal. This assumption is valid at least when the temperature and pressure are not too high. For a molecular crystal, the total energy is a sum of the intra- and intermolecular interaction energies.

In our unified approach, we use the XWCH model[99] to describe the intramolecular interaction and adapt a modified bond-bond charge model[97] for the intermolecular interaction in solid C_{60}. The interaction between molecules μ and ν is written as follows:

$$V(\mu,\nu) = \sum_{i,j=1}^{60} 4\varepsilon \left\{ \left[\frac{\sigma}{|r_{i\mu}-r_{j\nu}|}\right]^{12} - \left[\frac{\sigma}{|r_{i\mu}-r_{j\nu}|}\right]^{6} \right\} + \sum_{m,n=1}^{90} \frac{q_m q_n}{|b_{m\mu}-b_{n\nu}|} \operatorname{erfc}\left(\frac{|b_{m\mu}-b_{n\nu}|}{b_0}\right) \qquad (11)$$

where $r_{i\mu}$, $b_{m\mu}$ are the coordinates of the i-th C-atom and the m-th bond center of molecule μ, respectively. q_m is the effective charge of the m-th bond and a Coulomb interaction in conjunction with an error function is used to account for the interaction between bond charges. Parameters ε and q_L are determined by fitting to the lattice constant (a=14.05 Å) of the low temperature simple cubic structure and the bulk modulus (B=18 GPa) of solid C_{60}.[86] The parameters of the potential are: σ=3.407 Å, ε=2.8486 meV, b_0 =14.0 Å and q_L=0.30 |e|, where e is the electron charge. Because we use a different parametrization procedure, our parameters are slightly different from those used by Lu et al..[97] In Lu et al.'s bond-charge model, the bulk modulus is calculated to be about 9.6 GPa,[103] which is much lower than the experimental measurement.[86] T. Yildirim and A. B. Harris[103] have shown that the atom-bond model of Sprik et al.[98] results in an even lower bulk modulus of about 8.7 GPa.

In this unified model, all the degrees of freedom of C_{60} molecule in solid are treated explicitly. This approach not only permits the calculation of the intermolecular phonon dispersion and density of states, but also predicts the modifications of the intramolecular vibrational modes when the molecules form a solid and when the solid is under pressure.

Phonon Dispersion and Density of States of Solid C_{60}

Once the interaction in the unified model is defined, the ground-state properties can be completely determined. Using the tight-binding molecular dynamics (TBMD)[122-127] and the conjugate-gradient method, we have confirmed that the unified interaction model gives the correct $Pa\overline{3}$ symmetry of the ground-state structure of solid C_{60}. In the ground state, C_{60} molecules centered on the fcc Bravais lattice sites are oriented so that the molecule at (000) is rotated clockwise about the [111] axis by an angle Γ = 23.45° with respect to an initial standard orientation in which twofold axes are aligned along <100> directions, as are the other three molecules in the unit cell but about different <111> axes. This orientation angle agrees very well with experimental values (22° and 26°).[73,77]

Dispersion and Density of States of Orientationally Ordered Solid C_{60}. The phonon dispersion curves along some high symmetry directions in the first Brillouin zone are shown in Fig. 17. Since there are four C_{60} molecules per unit cell, there are 720 branches of dispersion curves. The lowest 24 branches correspond to the inter-molecular phonon modes and the remaining higher-energy branches correspond to the intra-molecular vibrational modes. Figure 17 shows that inter-molecular and intra-molecular phonon modes are well separated. This validates the rigid-molecule approximation. It is, however, surprising that not only the inter-molecular phonon modes but also the intra-molecular phonons show significant dispersions, especially those with energy below 70 meV. For comparison, we have plotted the vibrational spectrum of an isolated C_{60} molecule together with the symmetry notation and degeneracy calculated with the XWCH model.[99] Due to the I_h point-group symmetry of the C_{60} molecule, 174 vibrational modes of the C_{60} molecule span only 46 distinct vibrational frequencies. Among them, 4 modes (T_{1u}) are infrared-active and 10 modes (A_g and H_g) are Raman-active. Evidently in the solid most of the low-energy vibrational modes are split into bands except for a few modes. The intra-molecular vibrational modes are also shifted upward due to the inter-molecular interaction. Note that most of the Raman active modes (with H_g and A_g symmetry) show strong dispersions. For example, the lowest H_g Raman active mode splits into a broad band with a band-width of about 2.5 meV (20.1 cm^{-1}).

In Fig. 18, the calculated intermolecular phonon dispersion and density of states are compared with the neutron scattering experiments.[83] The calculated intermolecular phonon density of states shows strong features around 2.3 and 3.7 meV. The highest intermolecular phonon mode is around 7.6 meV. The calculated phonon spectrum agrees very well with the experimental measurements.

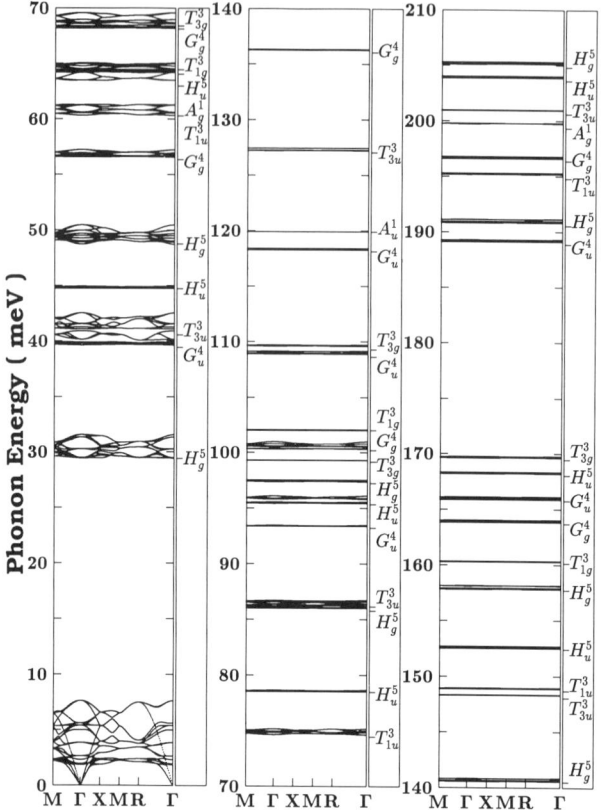

Figure 17. Phonon dispersion curves of solid C_{60} along some high symmetry directions in the first Brillouin zone at zero pressures and the vibrational frequencies of an isolated C_{60} molecule. The symmetry of the C_{60} molecule is indicated by the icosahedral I_h group label, the number on the label's shoulder indicates the degeneracy of the corresponding group representation.

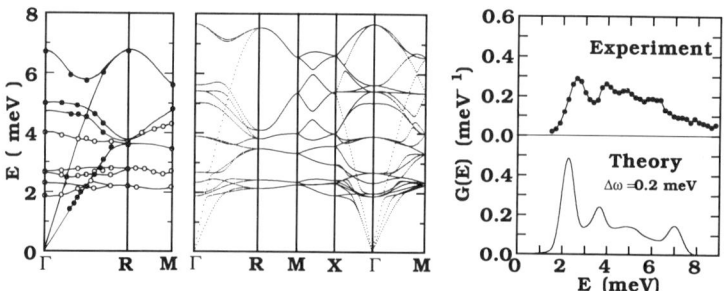

Figure 18. Comparison of the calculated intermolecular phonon dispersion and density of states of solid C_{60} with experiment measurements. (The experimental data are from Pintschovius et al. 1993)

Once the phonon density of states has been calculated, the specific heat can be evaluated directly. The temperature dependence of specific heat is shown in Fig. 19. The specific heat increases as T increases. Between 30 K to 70 K, C_v increases slowly with T because the contribution of the inter-molecular phonon modes has saturated. Above 100 K, C_v becomes a linear function of T. Also shown in Fig. 19 is the low-temperature specific heat measurements.[85] The agreement between theory and experiment is quite good. Notice that no fitting parameter has been used in the calculation of the specific heat.

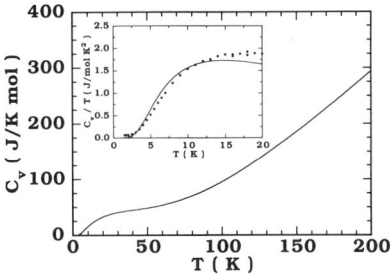

Figure 19. The specific heat C_V of solid C_{60} as a function of temperature T. In inset C_V/T vs. T (solid curve) is compared with the experimental measurements (solid circles) (Ref. 85).

Effect of Orientational Disorder. In a recent neutron diffraction experiment[82] on powder samples below T_c (250 K), there is evidence for two possible orientations of each C_{60} molecule. These orientations correspond to $\Gamma = \Gamma_1$ and $\Gamma = \Gamma_1 + 60° = \Gamma_2$ respectively, where $\Gamma_1 = 22°$. Here Γ is an orientation parameter. The probability that a molecule has an orientation Γ_1 is found to be 0.853, independent of the temperature as long as it is below 90 K.

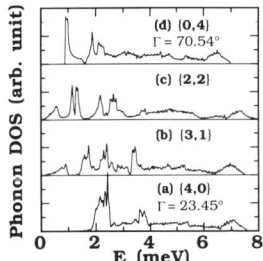

Figure 20. Intermolecular phonon density of states for four orientational configurations: (a) {4,0}; (b) {3,1}; (c) {2,2}; and (d) {0,4}.

To simulate the effect of the orientational disorder, we start with i molecules in the Γ_2 orientation and the remaining 4-i molecules in the unit cell in the Γ_1 orientation (the configuration is denoted by {4-i, i}, i = 0,1,2,3,4). This initial configuration is thermalized and then relaxed with the conjugate-gradient method. The configuration {4,0} is the ground state with $\Gamma = 23.45°$. The system with configuration {1,3} is found to be unstable and relaxes to the orientationally ordered ground state {4,0}. The configuration {0,4} initially with $\Gamma = 82°$ relaxes to a final configuration corresponding to $\Gamma = 70.54°$. In this configuration, hexagons and single (long) bonds of a C_{60} molecule face single bonds and hexagons of the nearest neighbor molecules, thereby maximizing bond-charge interactions.

We have calculated the phonon dispersion relation and density of states for each of the final configurations obtained by starting from the initial configuration {4,0}, {3,1}, {2,2} and {0,4}. In Fig. 20, we display the phonon density of states of the systems with different orientational orders: (a) {4,0}; (b) {3,1}; (c) {2,2}; and (d) {0,4}. Clearly, the intermolecular phonon spectrum is sensitive to the degree of orientational disordering which softens libron modes.

The Effects of Pressure. In experiments, pressure is the only variable which allows the delicate and continuous tuning of the strength of the inter-molecular potentials and rotational barriers which control the motions of the C_{60} molecules. Using the unified model, we have calculated the phonon dispersion and density of states of solid C_{60} at various pressures. For a given pressure or a given lattice constant, the system with four C_{60} molecules per unit cell is thermalized with TBMD and then the final equilibrium configuration is obtained by conjugate gradient method. The stability of the system is examined by checking the positivity of the eigenvalues of the dynamical matrix. In the pressure range (0 to 56 kbar) which we have studied, the simple cubic structure is found to be stable.

In Fig. 21, we show the phonon dispersion of solid C_{60} with a lattice constant of 13.50 Å, which correspond to a pressure of 56 kbar. Compared with Fig. 17 which corresponds to the case of zero pressure, it is evident that the phonon spectra are broadened and shifted upward in energy when the external pressure is applied. The pressure dependent phonon density of states of solid C_{60} is shown in Fig. 22. The libron modes shift to higher frequencies at a rate of about 0.05 meV/kbar (0.4 cm^{-1}/kbar). Intramolecular phonon modes

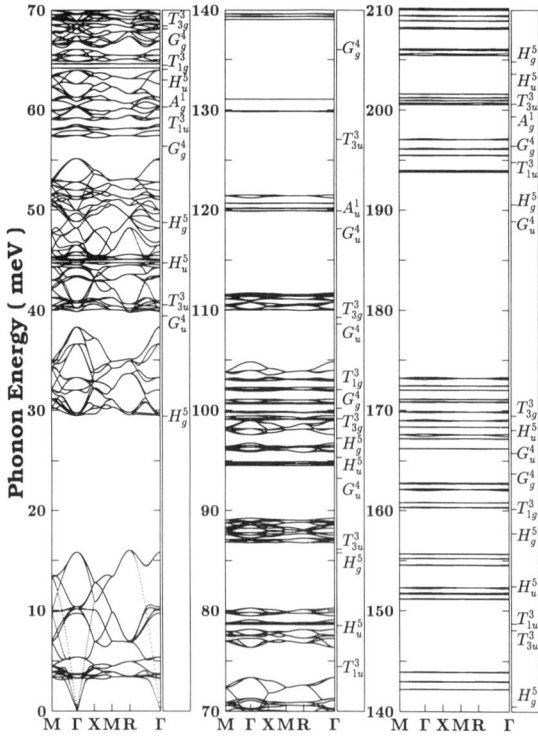

Figure 21. Phonon dispersion curves of solid C_{60} along some high symmetry directions in the first Brillouin zone when the system is under a pressure of 56 kbar (lattice constant a = 13.50 Å).

also show strong pressure dependence. When pressure is applied, the lower frequency modes are broadened into bands, and the higher frequency modes are split and shifted toward higher energy. Intramolecular phonon modes shift toward higher frequencies at a rate of up to 0.11 meV/kbar (0.88 cm^{-1}/kbar). Most Raman and IR active modes show strong pressure dependence. The high energy Raman (H_g) modes and the IR active (T_{1u}) modes are shifted upward and split into doublet or triplet. The splittings increase with increasing pressure.

Recently high-pressure IR spectra of solid C_{60} have been measured up to 32 kbar by Huang et al.[87] They have observed the splitting of the IR active T_{1u} modes when the pressure is above 14 kbar. Most of the IR modes show positive pressure dependence with slope ranging from 0.25 to 0.44 cm^{-1}/kbar. Our theoretical calculation agrees reasonably well with the experimental observation.

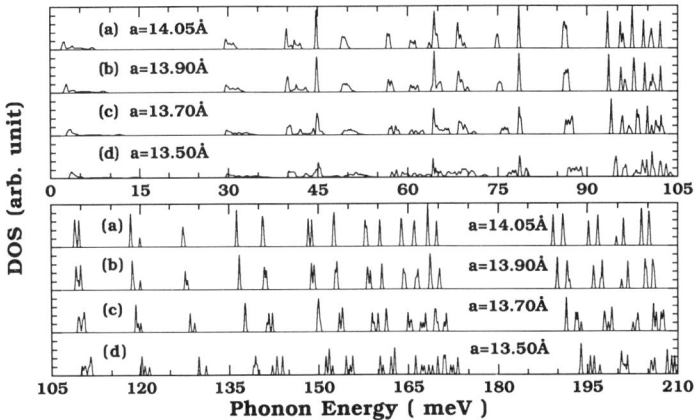

Figure 22. The phonon density of states of solid C_{60} at four different pressures: (a) p=0 kbar, a=14.05Å; (b) p=8 kbar, a=13.90Å; (c) p=26 kbar, a=13.70Å; and (d) p=56 kbar, a=13.50Å.

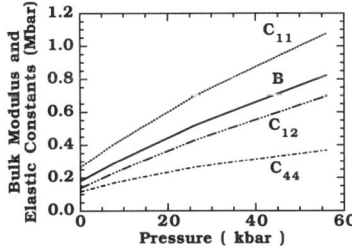

Figure 23. The elastic constants C_{11}, C_{12}, C_{44} and bulk modulus B vs lattice constant.

The pressure dependence of sound velocities and elastic constants and bulk modulus have also been calculated. In Fig. 23, the calculated elastic constants and bulk modulus are shown as functions of the lattice constant.

Our calculated phonon frequencies for solid C_{60} are within about 10% of the experimental values. For a better understanding of the effect of pressure, one should compare the value of shift per kbar or the percentage of shift in the frequency as a function of pressure rather than the absolute value. It should be emphasized that complete description of the inter- and intra-molecular vibrational modes of solid C_{60} is possible only in a unified model where C_{60} is not treated as a rigid ball.

Structural and Dynamical Correlations in Graphitic Tubules

The structure of a graphitic tubule can be described by choosing two lattice points on a graphite sheet. Denoting a fixed lattice point by $\mathbf{O}(0, 0)$, a specification of another lattice point, $\mathbf{R}_n(n_1, n_2)$, which will fold onto $\mathbf{O}$ uniquely defines the tubule structure. (See Fig.24). Alternatively, the lattice vector $\mathbf{R}_n$ or the index (n_1, n_2) can be used to specify the graphitic tubule. Here we denote it by $T(n_1, n_2)$ or $T(l_1, l_2)N$, where $n_1 = l_1 \cdot N$, $n_2 = l_2 \cdot N$, and N is the largest common divisor among n_1 and n_2. (l_1, l_2) determine the helicity of the tubule and N determines the rotational symmetry of the tubule (the tubule axis is a C_N axis). For a given vector $\mathbf{R}_n$ one can always find another vector $\mathbf{R}_m$ (m_1, m_2) such that $\mathbf{R}_n \cdot \mathbf{R}_m = 0$. $\mathbf{R}_n$ and $\mathbf{R}_m$ determine the tubule diameter and the unit cell length, respectively. For one-to-one correspondence, the parameters are confined by $n_1 \geq n_2 \geq 0$, and there is no common divisor among m_1 and m_2. The diameter (D), the unit cell length (L), and the number of carbon atoms (N_c) per unit cell are given by:

$$D = \frac{|\mathbf{R}_n|}{\pi} = \frac{a}{\pi}\sqrt{n_1^2 + n_1 n_2 + n_2^2}, \tag{12}$$

$$L = |\mathbf{R}_m| = a\sqrt{m_1^2 + m_1 m_2 + m_2^2}, \tag{13}$$

$$N_c = 2(n_1 m_2 - n_2 m_1) = 2N(l_1 m_2 - l_2 m_1), \tag{14}$$

where a is the lattice constant of the graphite, and (m_1, m_2) satisfy:

$$\frac{m_1}{m_2} = -\frac{n_1 + 2n_2}{2n_1 + n_2} = -\frac{l_1 + 2l_2}{2l_1 + l_2}. \tag{15}$$

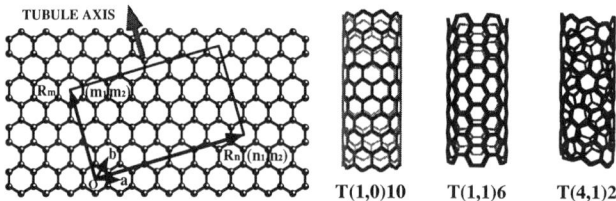

Figure 24. Graphite sheet and graphitic tubules. Vector $\mathbf{R}_n$ defines the tubule helicity and the diameter of the tubule. Vector $\mathbf{R}_m$ indicates the direction of the tubule, and $|\mathbf{R}_m|$ determines the length of a unit cell of the tubule. Also shown are some examples of the structure of graphitic tubules: T(1,0)10; T(1,1)6; and T(4,1)2.

Two special cases are the tubules: T(1,0)N and T(1,1)N. In the tubule T(1,0)N, there exist C-C bonds which are parallel to the tubule axis, while in tubule T(1,1)N there are bonds which are perpendicular to the tubule axis. Some examples of graphitic tubules with different helical structures are also shown in Fig. 24.

Although there are infinite atomic structures possible for graphitic tubules, the most interesting properties will emerge only in the smaller, nanometer-size diameters, where quantum effects become important. It has been demonstrated that the electronic property of graphitic tubule is sensitive to its helical structure. Theoretical studies predict that the tubule T(1,1)N are metallic while others are semiconducting.[108,109] Recently magnetoresistance and Hall coefficient measurements on a bundle of graphite nanotubules indicate that graphitic tubules behave as semi-metals.[120] We focus our attention on the structural and dynamical properties of graphitic tubules. Using the TBMD method, we have calculated the phonon spectra of graphitic tubes with various helical structures and diameters.[128] Helical and diameter dependence of Young's modulus of graphitic tube has been calculated. The systems under study include graphitic tubules with different chiral symmetry; coaxial microtubes with the same or different helical combinations; carbon cages; onion-like carbon clusters; doughnut-like graphitic tubule ring; and graphitic tubule arrays. Tight-binding molecular dynamics, dynamical stimulated annealing, and quantum molecular dynamics methods are used to investigate the structural stability, mechanical properties at finite temperature, structural transformations under external stresses -- compressing, stretching and twisting; the process of crack propagation and fragmentation.

In this section, we report the results of the phonon spectra and sound propagation in graphitic tubules. For a given graphitic tubule, first we thermalize it with TBMD and then quench it to the ground-state configuration with the conjugate-gradient method. Once the ground-state configuration is obtained, the force matrix is directly constructed by numerically differentiating the forces calculated by the Hellmann-Feynman theorem. The phonon spectra of graphitic tubules are calculated as functions of tubule diameter and helicity.

Figures 25 and 26 show the 1-D phonon dispersion and density-of-states of the graphitic tubules T(1, 1)6 and T(1, 0)10, respectively. For tubules with different helicities the phonon dispersion curves behave quite differently. This difference can be easily understood, because it mainly reflects the difference of dispersions of graphite along different directions in the 2-D graphite Brillouin zone. However, as a result of the real periodic boundary condition along the circumference of the tubule, a finite number of peaks emerge from the low-frequency part of the spectrum of graphitic tubules which depends on the size of the tubule diameter.

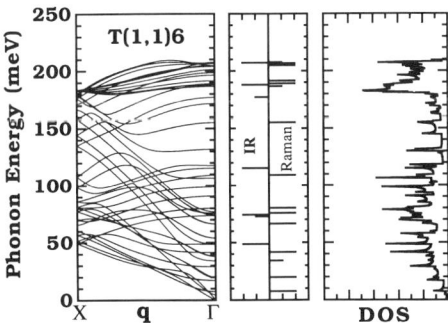

Figure 25. The phonon dispersion and density of states of the graphitic tubule T(1, 1)6. The IR-active and Raman active modes are explicitly marked with the lengths proportional to mode degeneracies.

To compare with graphite, we have also calculated the phonon spectra of the corresponding unrolled graphite sheet with a finite width that is equal to the circumference of the tubule. One example of the comparison is shown in Fig. 26. There is a one-to-one correspondence of the phonon modes between the graphitic tubule and the finite unrolled graphite sheet. Main features of the spectrum of the graphitic tubule resemble that of the corresponding finite graphite sheet. The difference in the spectra demonstrates the effect of the finite curvature of the tubule which softens the vibrational modes at Γ and hardens the lower-frequency modes at X. In a graphitic tubule, two acoustic modes with amplitudes perpendicular to the tubule axis become degenerate with a finite slope at Γ, while for the unrolled graphite, the acoustic modes are not degenerate and the lowest mode has a zero slope at Γ which corresponds to the translational motion perpendicular to the graphite plane. Compared with graphite, the finite curvature of graphitic tubule has two major effects: one is the hybridization of the π bonds with the σ bonds, and the other is the repulsion between the π orbitals inside the graphitic tubule. As in the case of a C_{60} molecule, the effect of the hybridization exceeds that of the repulsion between the π orbitals in graphitic tubules.

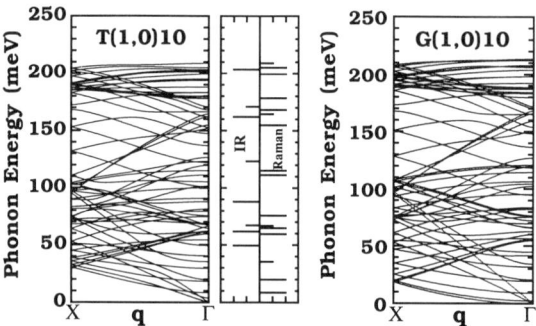

Figure 26. The phonon dispersion of the graphitic tubule T(1, 0)10 compared with that of an unrolled graphite G(1, 0)10 which has a finite width equal to the circumference of the tubule.

For a graphitic tubule with Nc particle per unit cell, there are 3Nc modes. Among them 4 modes have vanishing frequency at Γ, corresponding to 3 translational modes and one rotational mode (about the tubule axis). There are (3Nc+6)/2 distinct nonvanishing vibrational frequencies at Γ, among them 10 modes are nondegenerate and others are double degenerate.

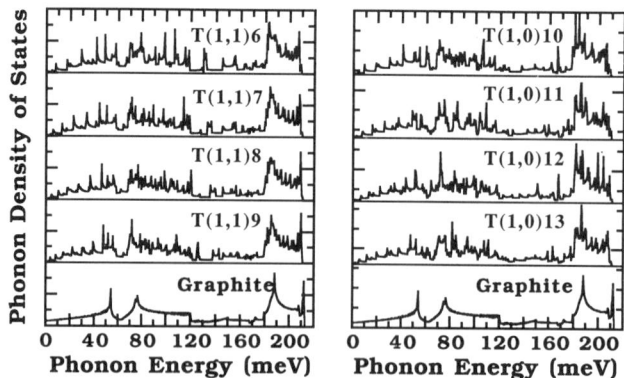

Figure 27. The phonon density of states of graphitic tubules T(1, 0)N and T(1, 1)N as a function of tubule diameters, and comparison with the density of states of an infinite graphite.

Figure 27 shows the phonon densities of states of graphitic tubules T(1, 0)N and T(1, 1)N as a function of tubule diameters and a comparison with the spectrum of infinite graphite. Evidently, the number of peaks in the low-frequency region of the spectra of graphitic tubules is proportional to the diameter of the tubules. As the diameter increases, the tendency of the spectra toward that of graphite is clearly evident. Again, the high-frequency modes are softened by the curvature.

Recently, Jishi, et al. have studied the symmetry properties of chiral graphitic tubules.[111] Here we analyze the phonon modes in the higher symmetry graphitic tubules, T(1, 1)N and T(1, 0)N, which represent two different limits of general helical graphitic tubules.

The factor group of the one-dimensional space group of the graphitic tubules T(1, 1)N and T(1, 0)N is isomorphic to the point group D_{Nd}. The character table for this group is given in Tables 1 and 2 for N = even and N = odd, respectively.

Table 1. The Character Table of $D_{(2n)d}$

$D_{(2n)d}$	E	$2(S_{2n})^{(2j-1)}$ (j=1,...,n)	$2(C_{2n})^{(2j)}$ (j=1,...,n-1)	C_2	$(2n)C_2'$	$(2n)\sigma_d$
A_1	1	1	1	1	1	1
A_2	1	1	1	1	-1	-1
B_1	1	1	1	1	1	1
B_2	1	1	1	1	-1	1
E_1	2	$2\cos((2j-1)\pi/n)$	$2\cos((2j)\pi/n)$	-2	0	0
E_2	2	$2\cos(2(2j-1)\pi/n)$	$2\cos(2(2j)\pi/n)$	2	0	0
...	...	...	...	...	...	...
E_k	2	$2\cos(k(2j-1)\pi/n)$	$2\cos(k(2j)\pi/n)$	$(-1)^k 2$	0	0
...	...	...	...	...	...	...
$E_{(2n-1)}$	2	$2\cos((2n-1)(2j-1)\pi/n)$	$2\cos((2n-1)(2j)\pi/n)$	-2	0	0

Table 2. The Character Table of $D_{(2n+1)d}$

$D_{(2n+1)d}$	E	$2(C_{2n+1})^{(j)}$ (j=1,...,n)	$(2n+1)C_2'$	i	$2(S_{2n+1})^{(j)}$ (j=1,...,n)	$(2n+1)\sigma_d$
A_{1g}	1	1	1	1	1	1
A_{2g}	1	1	-1	1	1	-1
E_{1g}	2	$2\cos(j2\pi/(2n+1))$	0	2	$2\cos(j2\pi/(2n+1))$	0
E_{2g}	2	$2\cos(2j2\pi/(2n+1))$	0	2	$2\cos(2j2\pi/(2n+1))$	0
...	...	...	...	...	...	...
E_{ng}	2	$2\cos(nj2\pi/(2n+1))$	0	2	$2\cos(nj2\pi/(2n+1))$	0
A_{1u}	1	1	1	-1	-1	-1
A_{2u}	1	1	-1	-1	-1	1
E_{1u}	2	$2\cos(j2\pi/(2n+1))$	0	-2	$2\cos(j2\pi/(2n+1))$	0
E_{2u}	2	$2\cos(2j2\pi/(2n+1))$	0	-2	$2\cos(2j2\pi/(2n+1))$	0
...	...	...	...	...	...	...
E_{nu}	2	$2\cos(nj2\pi/(2n+1))$	0	-2	$2\cos(nj2\pi/(2n+1))$	0

For tubule T(1, 1)N and T(1, 0)N, there are N_c = 4N particles per unit cell and there are $3N_c$ = 12N vibrational modes. They are decomposed according to the following irreducible representations:

(a) T(1, 1)N:

$$\Gamma^{vib} = 3A_1 + 3A_2 + 3B_1 + 3B_2 + 6E_1 + 6E_2 + 6E_3 + ... + 6E_{(N-1)}, \quad (N = \text{even}) \quad (16)$$

$$\Gamma^{vib} = 3A_{1g} + 3A_{2g} + 6E_{1g} + 6E_{2g} + ... + 6E_{(\frac{N-1}{2})g}$$
$$+ 3A_{1u} + 3A_{2u} + 6E_{1u} + 6E_{2u} + ... + 6E_{(\frac{N-1}{2})u} \quad (N = \text{odd}) \quad (17)$$

(b) T(1,0)N:

$$\Gamma^{vib} = 4A_1 + 2A_2 + 2B_1 + 4B_2 + 6E_1 + 6E_2 + 6E_3 + ... + 6E_{(N-1)}, \quad (N = \text{even}) \quad (18)$$

$$\Gamma^{vib} = 4A_{1g} + 2A_{2g} + 6E_{1g} + 6E_{2g} + ... + 6E_{(\frac{N-1}{2})g}$$
$$+ 2A_{1u} + 4A_{2u} + 6E_{1u} + 6E_{2u} + ... + 6E_{(\frac{N-1}{2})u} \quad (N = \text{odd}) \quad (19)$$

The translational modes have E_1 and B_2 (E_{1u} and A_{2u}) symmetries and the rotational mode has A_2 (A_{2g}) symmetry for even values of N (odd N). The modes that transform according to A or B irreducible representations are nondegenerate, and the modes that transform according to E irreducible representations are doubly degenerate. The zone-center infrared active modes and Raman active modes have the following symmetries:

(a) T(1,1)N:

$$\Gamma^{IR} = 2B_2 + 5E_1; \quad \Gamma^{Raman} = 3A_1 + 6E_2 + 6E_{(N-1)}, \quad (N = \text{even}) \quad (20)$$

$$\Gamma^{IR} = 2A_{2u} + 5E_{1u}; \quad \Gamma^{Raman} = 3A_{1g} + 6E_{1g} + 6E_{2g}, \quad (N = \text{odd}) \quad (21)$$

(b) T(1,0)N:

$$\Gamma^{IR} = 3B_2 + 5E_1; \quad \Gamma^{Raman} = 4A_1 + 6E_2 + 6E_{(N-1)}, \quad (N = \text{even}) \quad (22)$$

$$\Gamma^{IR} = 3A_{2u} + 5E_{1u}; \quad \Gamma^{Raman} = 4A_{1g} + 6E_{1g} + 6E_{2g}, \quad (N = \text{odd}) \quad (23)$$

respectively. In graphitic tubules T(1, 1)N, there are only 7 distinct infrared frequencies and 15 distinct Raman frequencies; in graphitic tubules T(1, 0)N, there are 8 distinct infrared frequencies and 16 distinct Raman frequencies; the symmetries of the infrared active modes

Table 3. The vibrational modes of the graphitic tubule T(1,1)6 at Γ (symmetry group: D_{6d})

i	mode	frequency (meV)	IR	Raman	i	mode	frequency (meV)	IR	Raman
0	B_2	0.00			19	E_3	106.67		
0	E_1	0.00			20	E_2	109.31		active
0	A_2	0.00			21	E_1	115.50	active	
1	E_2	7.61		active	22	A_2	117.76		
2	E_3	17.23			23	E_4	132.01		
3	E_5	19.77		active	24	E_5	155.25		active
4	E_4	29.23			25	B_1	171.81		
5	A_1	34.29		active	26	B_2	177.20	active	
6	E_4	38.23			27	A_1	186.76		active
7	E_5	41.61		active	28	E_1	188.06	active	
8	B_1	48.69			29	E_5	189.04		active
9	E_1	48.80	active		30	E_2	191.59		active
10	E_3	54.17			31	E_3	196.42		
11	E_2	66.51		active	32	E_4	198.61		
12	B_2	73.75	active		33	E_4	201.28		
13	E_1	74.37	active		34	E_3	204.05		
14	E_2	75.82		active	35	B_1	204.82		
15	A_2	77.07			36	E_5	204.87		active
16	E_5	80.23		active	37	E_2	206.62		active
17	E_4	90.95			38	E_1	207.10	active	
18	E_3	99.02			39	A_1	207.59		active

and Raman active modes are different. In a general chiral graphitic tubule there are 9 distinct infrared frequencies ($4A + 5E_1$) and 15 distinct Raman frequencies ($4A + 5E_1 + 6E_2$), and the infrared active modes are also Raman active modes.[111]

We have identified the symmetry of each vibrational mode in graphitic tubules. In Fig. 25 the infrared active modes and Raman active modes are explicitly marked by the left-going and right-going arrows for the graphitic tubule T(1, 1)6. The symmetries and frequencies of the vibrational modes of the graphitic tubules T(1,1)6 and T(1,0)10 at Γ are listed in Tables 3 and 4, respectively.

Table 4. The vibrational modes of the graphitic tubule T(1,0)10 at Γ (symmetry group: D_{10d})

i	mode	frequency (meV)	IR	Raman	i	mode	frequency (meV)	IR	Raman
0	B_2	0.00			31	E_9	115.25		active
0	E_1	0.00			32	E_4	119.74		
0	A_2	0.00			33	E_4	122.26		
1	E_2	8.57		active	34	B_2	123.41	active	
2	E_3	18.41			35	E_3	126.73		
3	E_9	19.88		active	36	E_5	137.12		
4	E_4	29.67			37	E_6	147.61		
5	A_1	35.49		active	38	E_3	152.39		
6	E_5	40.49			39	E_7	152.61		
7	E_8	40.64			40	E_2	154.89		active
8	E_1	49.53	active		41	E_1	162.08	active	
9	E_6	54.50			42	E_8	162.21		
10	E_2	59.53		active	43	A_1	164.54		active
11	E_7	61.44			44	E_4	166.56		
12	E_1	61.73	active		45	E_9	168.12		active
13	E_3	62.08			46	B_2	170.90	active	
14	E_9	64.64		active	47	E_7	177.96		
15	A_1	66.73		active	48	B_1	177.97		
16	B_2	66.97	active		49	E_9	178.41		active
17	E_7	67.90			50	E_8	179.81		
18	E_8	68.15			51	E_6	180.21		
19	E_4	68.93			52	E_5	183.13		
20	A_2	74.72			53	E_5	183.38		
21	E_2	75.59		active	54	E_4	188.04		
22	E_5	79.66			55	E_3	193.85		
23	E_6	81.68			56	E_6	195.05		
24	E_1	88.08	active		57	E_2	199.48		active
25	E_6	91.07			58	E_7	201.62		
26	E_7	100.37			59	B_1	202.47		
27	E_3	101.14			60	E_8	202.83		
28	E_5	101.72			61	E_1	203.58	active	
29	E_8	105.45			62	E_9	205.18		active
30	E_2	111.40		active	63	A_1	209.02		active

According to the vibrational eigenvectors, the normal modes of a graphitic tubule can be approximately classified into radial modes and tangential modes. Correspondingly, the phonon spectrum of a graphitic tubule can be divided into three regions:

(a) In the low-frequency region (< 50 meV), the phonon modes are dominated by the radial modes whose number depends on the tubule diameter.

(b) In the medium-frequency region (50 -- 120 meV), both radial modes and tangential modes are present, and the detailed structure of the spectrum depends on tubule diameter and helicity.

(c) In the high-frequency region (> 120 meV), the phonon modes belong to tangential modes which are sensitive to the helicity of the graphitic tubule. The spectra of graphitic tubules deviate strongly from that of graphite and carry unique information of the tubule helical structures in the region of 120 -- 180 meV.

The breathing mode is within the low-frequency region. It has A_1 (A_{1g}) symmetry, and it is Raman active. Figure 28 shows that the frequency of the breathing mode decreases as the tubule diameter increases.

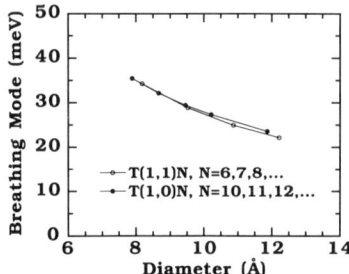

Figure 28. The frequency of the breathing mode as a function of tubule diameters.

Figure 29 shows the low-frequency (< 50 meV) normal modes in the graphitic tubules with same helicity but different diameters: T(1,1)6 and T(1,1)7. It is evident that the phonon modes in the low-frequency region are dominated by the radial modes. The number of vibrational modes is determined by the tubule diameter. Figure 30 is the pictorial view of some of the infrared and Raman modes in the graphitic tubule T(1,1)6.

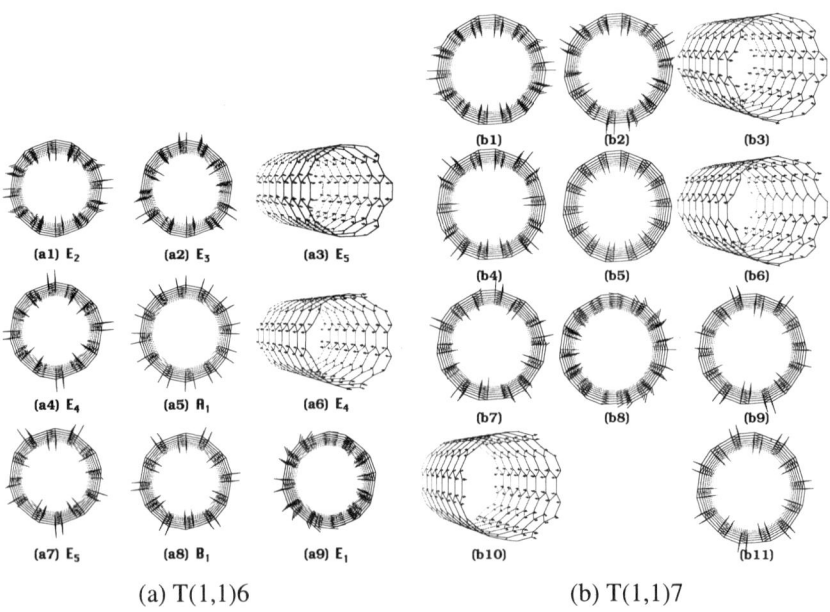

Figure 29. Eigenvector representation of the low-frequency (< 50 meV) modes of tubules: (a1-a9): T(1,1)6 and (b1-b11): T(1,1)7.

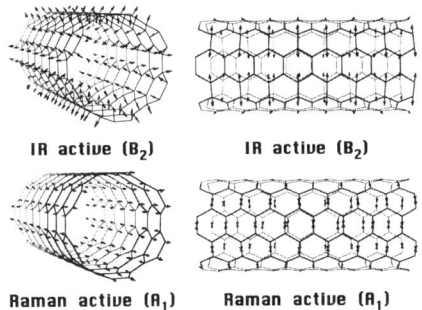

Figure 30. Eigenvector representation of some infrared and Raman active modes of the graphitic tubule T(1,1)6.

We have also calculated the phonon spectra for graphitic tubules with general chiral symmetries. One example is shown in Fig. 31 for graphitic tubule T'(6, 5). Again the lower-frequency part of the spectrum corresponds to the radial modes with the number of peaks depending on the diameter of the tubule. Unlike the higher symmetry cases, the higher-frequency tangential modes in the graphitic tubules with a general helicity cannot be classified into pure z-modes and θ-modes according to their vibrational eigenvectors. The higher-frequency part of the spectra carries unique information about the tubule helical structures.

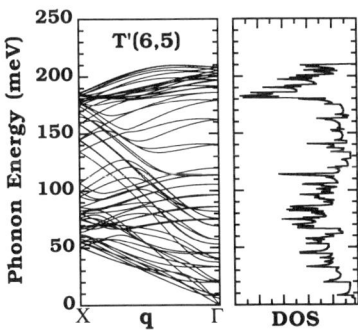

Figure 31. The phonon dispersion and density of states of the graphitic tubule T'(6, 5).

From the slopes of the acoustic-phonon branches in the long-wavelength limit, one can calculate the velocity of sound. Figure 32(a) shows the speed of sound propagating along the tubule axis as a function of the tubule diameter for two different helicities T(1, 1)N and T(1, 0)N. The two transverse sound velocities are degenerate. The longitudinal velocity of sound is lower than that in graphite. Both the transverse and longitudinal sound velocities are helicity dependent. However, for a given helicity, the sound velocities depend only weakly on the tubule diameter. Fig. 32(b) shows the effect of an external stress on sound velocities (as a function of tubule unit cell length). The strong helicity dependence of sound velocities on pressure represents, in response to external stresses, the different characteristic of the C-C bond compressing or stretching and bond angle bending. Compressing or stretching changes the bond length in T(1, 0)N and bond angles in tubule T(1, 1)N. Tubules with helicity T(1, 0)N that have bonds parallel to the tubule axis are easy to compress but hard to stretch, while tubules with helicity T(1, 1)N are hard to compress but easy to stretch. T(1, 0)N and T(1, 1)N are the two limiting cases: the behavior of tubules with other helicities lies in between that of tubules T(1, 0)N and T(1, 1)N. As the diameter increases, the difference between different helicities diminishes. Detailed results for the effects of external stress on the phonon spectra, and structural and dynamical properties of graphitic tubules will be reported elsewhere.

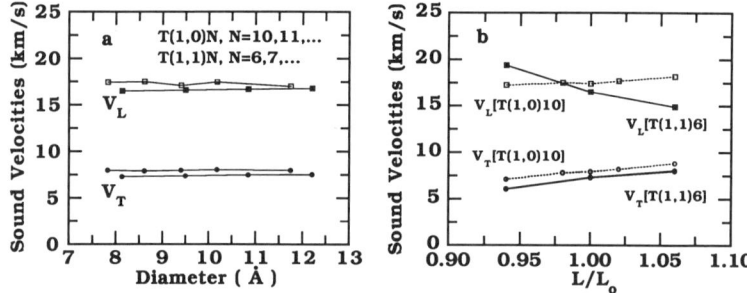

Figure 32 (a) Sound velocities as functions of tubule diameters and helicities at zero pressure; (b) The effect of compressing and stretching on the sound velocities in graphitic tubules. (Solid circles and squares for tubules T(1, 0)N and open circles and squares for tubules T(1, 1)N. Solid lines are drawn to guide the eye)

Recently, Iijima *et al.* have demonstrated that single graphitic tubule about 700Å long can be obtained experimentally.[121] Our model calculations will be useful in guiding infrared, Raman, and sound measurements in characterizing fullerene tubules.

CONCURRENT COMPUTING LABORATORY FOR MATERIALS SIMULATIONS

In the past two and a half years, we have received from the Louisiana Board of Regents three equipment enhancement grants totaling more than $2.5M. With the first grant we have purchased a MasPar MP-1, an 8-node Intel iPSC/860 system, a SPARCserver, and SPARC workstations. These facilities are housed in a laboratory in Nicholson Hall. The second grant has been used to purchase a 64-cell iWarp machine, a Silicon Graphics Power Center 4D/380VGX and associated visualization equipment, multiprocessor Sun S670, workstations, and X-terminals. These computing facilities are housed in a laboratory in an adjacent building (Coates Hall). The laboratory has a dedicated systems manager. From the third grant we plan to purchase a 32-node IBM SP-1 system. A brief description of each machine in the *Concurrent Computing Laboratory for Materials Simulations* (CCLMS) is given below and the overall setup is shown in Fig. 33.

Plans are in place to connect these machines by FDDI (Fiber Distributed Data Interface) to form a distributed multiparallel processing network so that different parts of a large-scale simulation can run concurrently on different parallel architectures along with real-time visualization.

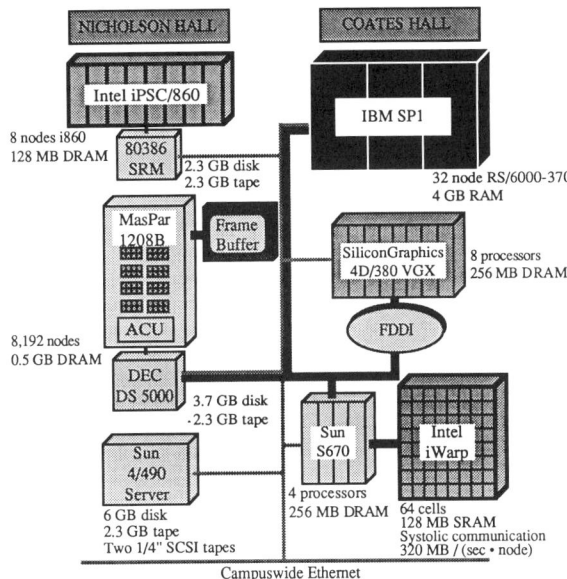

Figure 33. CCLMS computing facilities at Louisiana State University.

[1] **Intel iWarp**: iWarp is a distributed-memory MIMD (multiple instruction multiple data) machine which supports fine-grain systolic communication. In a systolic system, data flows among processors rhythmically without returning to memory. Each iWarp cell contains both a 20 MHz computation processor (20 MFlops for 32-bit precision) and a high throughput (320 MB/sec), low latency (100-150 µs) communication engine. iWarp cells are connected in a two-dimensional mesh topology. CCLMS's system has 64 cells with the theoretical peak performance of 1.28 GFlops for 32-bit precision.

[2] **MasPar**: MasPar MP-1 is a SIMD (single instruction multiple data) machine. Each processor element has a 12.5 MHz, 4-bit control processor (1.8 MIPS for 32-bit integer instruction), forty 32-bit registers, and 64KBytes of RAM. The peak speed of the 8,192-node system in the CCLMS is 750 MFlops and the communication bandwidth is 12 GB/sec. The system can be extended up to 16,384 nodes. Processor elements are connected in a two-dimensional mesh topology.

[3] **Silicon Graphics Power Center**: Silicon Graphics 4D/380VGX is a shared-memory MIMD machine with 8 processors. It has VGX graphics and a video creator. Each processor is a 33 MHz MIPS R3000 RISC processor which operates at 13 MFlops for 64-bit precision.

[4] **Multiprocessor SUN**: CCLMS's Sun S670MP is a shared-memory MIMD machine with four SPARC RISC processors. It is connected to the iWarp via a VME bus. The SUN acts as front end for the iWarp as well as a general purpose server for X-terminals.

[5] **Intel iPSC/860**: iPSC/860 is a distributed-memory MIMD machine. Each node is a 40 MHz i860 processor with the theoretical peak of 80 MFlops for 32-bit precision. Processors are connected in a hypercube topology. CCLMS's iPSC/860 has 8 nodes with the theoretical peak of 640 MFlops (32-bit).

[6] **IBM SP-1**: Each node of SP-1 is an IBM RISC System/6000 model 370 (theoretical peak is 125 MFlops for 64-bit precision). With a high-performance switch, communication bandwidth is 8.5 MB/sec with 50 µsec latency.

CONCLUSION

Algorithms have been designed to implement molecular dynamics (MD) and quantum molecular dynamics (QMD) simulations on emerging concurrent architectures. Simulations have been performed for densified and porous SiO_2 glasses, solid C_{60} and graphite tubes, and electron transport in nanodevices and disordered materials. We believe that for the next ten years, there is an enormous opportunity to develop new and efficient algorithms for parallel computers to solve grand challenge problems in materials science.

ACKNOWLEDGMENTS

This work was supported by the U.S. Department of Energy, Grant No. DE-FG05-92ER45477 and National Science Foundation Grants No. ASC-9109906 and ASC-9310314. The computations were performed using the facilities in the Concurrent Computing Laboratory for Materials Simulations (CCLMS) at Louisiana State University. The facilities in the CCLMS were acquired with Equipment Enhancement Grants awarded by the Louisiana Board of Regents through Louisiana Education Quality Support Fund (LEQSF). Computations were also performed using the Touchstone Delta and iPSC/860 machines operated by Caltech on behalf of the Concurrent Supercomputing Consortium.

REFERENCES

1. M. P. Allen and D. J. Tildesley, "Computer Simulation of Liquids" Oxford University Press, Oxford (1990).
2. R. Car and M. Parrinello, *Phys. Rev. Lett.* 55:2471 (1985).
3. W. B. Streett, D. J. Tildesley, and G. Saville, *Mol. Phys.* 35:639 (1978).
4. D. C. Rapaport, *Comput. Phys. Commun.* 62:217 (1991).
5. R. K. Kalia, S. W. de Leeuw, A. Nakano, P. Vashishta, *Comp. Phys. Comm.* 74:316 (1993).
6. R. K. Kalia, S. W. de Leeuw, A. Nakano, D. L. Greenwell, and P. Vashishta, *Supercomputer* 54(X-2):11 (1993); A. Nakano, P. Vashishta, and R. K. Kalia, *Comput. Phys. Commun.* 77:303 (1993).
7. A. Nakano, P. Vashishta, and R. K. Kalia, *Comput. Phys. Commun.*, submitted; W. Li, R. K. Kalia, S. W. de Leeuw, A. Nakano, D. L. Greenwell, and P. Vashishta, *Mat. Res. Soc. Symp. Proc.* 291:267 (1993).
8. D. Frenkel, *in*: "Simple Molecular Systems at Very High Density", A. Polian and P. Loubeyre, eds., Plenum, New York (1989).
9. S. W. de Leeuw, J. W. Perram, and E. R. Smith, *Proc. R. Soc. London A* 373:27 (1980).
10. J. Barnes and P. Hut, *Nature* 324:446 (1986).
11. L. Greengard and V. Rokhlin, *J. Comput. Phys.* 73: 325 (1987).
12. H.-Q. Ding, N. Karasawa, and W. A. Goddard, *Chem. Phys. Lett.* 196:6 (1992); *J. Chem. Phys.* 97:4309 (1992).
13. A. Nakano, P. Vashishta, and R. K. Kalia, *Phys. Rev. B* 43:9066 (1991).
14. A. Nakano, R. K. Kalia, and P. Vashishta, *Phys. Rev. B* 44:8121 (1991).
15. A. Nakano, R. K. Kalia, and P. Vashishta, *Appl. Phys. Lett.* 62:3470 (1993).

16. A. Nakano, R. K. Kalia, and P. Vashishta, *Appl. Phys. Lett.* 64:2569 (1994).
17. A. Nakano, P. Vashishta, and R. K. Kalia, *Comput. Phys. Commun.*, submitted.
18. E. K. U. Gross and W. Kohn, *in*: "Advances in Quantum Chemistry", Vol. 21, S. B. Trickey, ed., Academic Press, San Diego (1990) p. 255.
19. G. Vignale and M. Rasolt, *Phys. Rev. Lett.* 59:2360 (1987).
20. B. Jackson, *Comput. Phys. Commun.* 63:154 (1991).
21. M. D. Feit, J. A. Fleck, and A. Steiger, *J. Comput. Phys.* 47:412 (1982).
22. J. L. Richardson, *Comput. Phys. Commun.* 63:84 (1991); H. de Raedt, *Comput. Phys. Rep.* 7:1 (1987).
23. R. W. Hockney and J. W. Eastwood, "Computer Simulation Using Particles", Adam Hilger, New York (1988).
24. A. Brandt, *Math. Comput.* 31:333 (1977).
25. P. O. Frederickson and O. A. McBryan, *in*: "Multigrid Methods", S. F. McCormick, ed., Marcel Dekker, New York (1988) p. 195.
26. R. Car and M. Parrinello, *Solid State Commun.* 62:403 (1987).
27. A. Selloni, P. Carnevali, R. Car, and M. Parrinello, *Phys. Rev. Lett.* 59:823 (1987).
28. R. N. Barnett, U. Landman, and A. Nitzan, *Phys. Rev. Lett.* 62:106 (1989).
29. R. K. Kalia, P. Vashishta, L. H. Yang, F. Dech, and J. Rowlan, *Int. J. Supercomputer Applications* 4:22 (1990).
30. A. Nakano, P. Vashishta, R. K. Kalia, and L. H. Yang, *Phys. Rev. B* 45:8363 (1992).
31. "The Physics and Technology of Amorphous SiO_2", R. A. B. Devine, ed., Plenum, New York (1988).
32. S. C. Moss and D. L. Price, *in*: "Physics of Disordered Materials", D. Adler, H. Fritzsche, and S. R. Ovshinsky, eds., Plenum, New York (1985) p. 77.
33. R. J. Bell and D. C. Hibbins-Butler, *J. Phys. C* 8:787 (1975).
34. P. A. V. Johnson, A. C. Wright, and R. N. Sinclair, *J. Non-Cryst. Solids* 58:109 (1983).
35. P. N. Sen and M. F. Thorpe, *Phys. Rev. B* 15:4030 (1977).
36. J. M. Carpenter and D. L. Price, *Phys. Rev. Lett.* 54:441 (1985).
37. R. Dupree and R. F. Pettifer, *Nature* 308:523 (1984).
38. B. Feuston and S. H. Garofalini, *J. Chem. Phys.* 89:5818 (1988).
39. P. Vashishta, R. K. Kalia, J. P. Rino, and I. Ebbsjö, *Phys. Rev. B* 41:12197 (1990).
40. S. R. Elliot, *Nature* 354:445 (1991); *Phys. Rev. Lett.* 67:711 (1991).
41. M. Grimsditch, *Phys. Rev. Lett.* 52:2379 (1984); *Phys. Rev. B* 34:4372 (1986); A. Polian and M. Grimsditch, *Phys. Rev. B* 41:6086 (1990).
42. P. McMillan, B. Piriou, and R. Couty, *J. Chem. Phys.* 81:4234 (1984).
43. R. J. Hemley, H. K. Mao, P. M. Bell, and B.O. Mysen, *Phys. Rev. Lett.* 57:747 (1986).
44. Q. Williams and R. Jeanloz, *Science* 239:902 (1988).
45. S. Susman, et al., *Phys. Rev. B* 43:1194 (1991).
46. C. Meade, R. J. Hemley, and H. K. Mao, *Phys. Rev. Lett.* 69:1387 (1992).
47. L. V. Woodcock, C. A. Angell, and P. Cheeseman, *J. Chem. Phys.* 65:1565 (1976).
48. R. A. Murray and W. Y. Ching, *Phys. Rev. B* 39:1320 (1989).
49. L. Stixrude and M. S. T. Bukowinski, *Phys. Rev. B* 44:2523 (1991).
50. S. Tsuneyuki, Y. Matsui, H. Aoki, and M. Tsukada, *Nature* 339:209 (1989).
51. Y. Tsuchida and T. Yagi, *Nature* 340:217 (1989).
52. J. S. Tse and D. D. Klug, *Phys. Rev. Lett.* 67:3559 (1991); J. S. Tse, D. D. Klug, and Y. Le Page, *Phys. Rev. Lett.* 69:3647 (1992); N. Binggeli and J. R. Chelikowsky, *Nature* 353:344 (1991); *Phys. Rev. Lett.* 69:2220 (1992).
53. L. E. McNeil and M. Grimsditch, *Phys. Rev. Lett.* 68:83 (1992).
54. W. Jin, R. K. Kalia, P. Vashishta, and J. P. Rino, *Phys. Rev. Lett.* 71:3146 (1993); *Phys. Rev. B*, in press; W. Jin, R. K. Kalia, and P. Vashishta, *Mat. Res. Soc. Symp. Proc.* 293:225 (1993).
55. J. Fricke, *J. Non-Cryst. Solids* 121:188 (1990); ibid. 147&148:356 (1992).
56. T. Freltoft et al., *Phys. Rev. B* 33:269 (1986).
57. R. Vacher et al., *Phys. Rev. B* 37:6500 (1988); *J. Non-Cryst. Solids* 106:161 (1988); J. Pelous et al., ibid. 145:63 (1992).
58. D. W. Shaefer and K. D. Keefer, *Phys. Rev. Lett.* 56:2199 (1986).
59. T. Lours et al., *J. Non-Cryst. Solids*, 121:216 (1990).
60. J. Kieffer and C. A. Angell, *J. Non-Cryst. Solids* 106:336 (1988).
61. A. Nakano, L. Bi, R. K. Kalia, and P. Vashishta, *Phys. Rev. Lett.* 71:85 (1993); *Phys. Rev. B* 49:9441 (1994).
62. F. Family and T. Vicsek, *J. Phys. A* 18:L75 (1985); M. Kardar, G. Parisi, and Y. C. Zhang, *Phys. Rev. Lett.* 56:889 (1986); J. Villain, *J. Phys. I (France)* 1:19 (1991); J. M. Kim and S. Das Sarma, *Phys. Rev. Lett.* 72:2903 (1994).

63. "Dynamics of Fractal Surfaces", F. Family and T. Vicsek, eds., World Scientific, Singapore (1991); F. Family, *Physica A* 168:561 (1990); F. Family and R. Pandey, *J. Phys. A* 25:L745 (1992); J.-F. Gouyet, M. Rosso, and B. Sapoval, *in:* "Fractals and Disordered Systems", A. Bunde and S. Havlin, eds., Springer Verlag, Berlin (1991) p. 229.
64. E. Bouchaud, G. Lapasset, and J. Planès, *Europhys. Lett.* 13:73 (1990).
65. K. J. Måløy, A. Hansen, E. L. Hinrichsen, and S. Roux, *Phys. Rev. Lett.* 68:213 (1992).
66. V. Y. Milman, R. Blumenfeld, N. A. Stelmashenko, and R. C. Ball, *Phys. Rev. Lett.* 71:204 (1993).
67. A. Nakano, R. K. Kalia, and P. Vashishta, *Phys. Rev. Lett.*, submitted.
68. H. W. Kroto, J. R. Heath, S. C. O'Brien, R. F. Curl, and R. E. Smalley, *Nature* 318: 162 (1985).
69. W. Kratschmer, L. D. Lamb, K. Fostiropoulos, and D. R. Huffman, *Nature* 347: 354 (1990).
70. H. Hebard et al., *Nature* 350: 600 (1991); M. Rosseinsky et al., *Phys. Rev. Lett.* 66: 2830 (1991); K. Holczer et al., *Science* 252: 1154 (1991).
71. C. S. Yannoni, et al., *J. Phys. Chem.* 95: 9 (1991); R. D. Johnson, et al., *Science* 255: 1235 (1992).
72. R. Tycko, et al., *J. Phys. Chem.* 95: 518 (1991); R. Tycko, et al., *Phys. Rev. Lett.* 67: 1886 (1991).
73. P. A. Heiney, et al., *Phys. Rev. Lett.* 66: 2911 (1991).
74. J. E. Fischer, et al., *Science* 252: 1288 (1991).
75. R. Sachidanandam and A. B. Harris, *Phys. Rev. Lett.* 67: 1467 (1991); P. A. Heiney, et al., *Phys. Rev. Lett.* 67: 1468 (1991).
76. R. M. Fleming, et al., *Mater. Res. Soc. Symp. Proc.* 206: 691 (1991).
77. W. I. F. David, et al., *Nature* 353: 147 (1991).
78. R. L. Cappelletti, et al., *Phys. Rev. Lett.* 66: 3261 (1991).
79. D. A. Neumann, et al., *Phys. Rev. Lett.* 67: 3808 (1991).
80. K. Prassides, et al., *Chem. Phys.* 187: 455 (1991); K. Prassides, et al., *Nature* 354: 462 (1991).
81. J. R. D. Copley, D. A. Neumann, R. L. Cappelletti, and W. A. Kamitakahara, *in:* "The Fullerenes", H. W. Kroto, J. E. Fischer, and D. E. Cox, ed., Pergamon Press, Oxford, (1993).
82. W. I. F. David, et al., *Europhys. Lett.* 18: 219 (1992).
83. L. Pintschovius, B. Renker, F. Gompf, R. Heid, S. L. Chaplot, M. Haluska, and H. Kuzmany. *Phys. Rev. Lett.* 69: 2662 (1992); L. Pintschovius, et al., *in* "Proc. Int. Winterschool on Electronic Properties of Novel Materials", H. Kuzmany, ed., Springer Series on Solid State Science, (1993).
84. C. Coulombeau, et al., *J. Phys. Chem.* 96: 22 (1992).
85. W. P. Beyermann, et al., *Phys. Rev. Lett.* 68: 2046 (1992).
86. S. J. Duclos, et al., *Nature* 351:380 (1991).
87. Y. Huang, D. F. R. Gilson, and I. S. Butler, *J. Phys. Chem.* 95: 5723 (1991).
88. Fred Moshary, Nancy H. Chen, Isaac F. Silvera, *Phys. Rev. Lett.* 69: 466 (1992).
89. D. W. Snoke, et al., *Phys. Rev. B* 47: 4146 (1993).
90. G. A. Samara, et al., *Phys. Rev. B* 47: 4756 (1993).
91. Q. Zhang, Jae-Yel Yi, and J. Bernholc, *Phys. Rev. Lett.* 66: 2633 (1991).
92. B. P. Feuston, W. Andreoni, M. Parrinello, and E. Clementi, *Phys. Rev. B* 44: 4056 (1991).
93. J. L. Feldman, et al., *Phys. Rev. B* 46: 12731 (1992).
94. Y. Guo, N. Karasawa, and W. A. Goddard III, *Nature* 351: 464 (1991).
95. W. Y. Ching, M-Z. Huang, Y-N. Xu, W. G. Harter, and F. T. Chan, *Phys. Rev. Lett.* 67: 2045 (1991); Y.-N. Xu, M.-Z. Huang, and W. Y. Ching, *Phys. Rev. B* 46: 4241 (1992).
96. A. Cheng and M. L. Klein, *J. Phys. Chem.* 95: 6750 (1991); A. Cheng and M. L. Klein, *Phys. Rev. B* 45: 1889 (1992).
97. J. Lu, X. Li, and R. M. Martin, *Phys. Rev. Lett.* 68: 1551 (1992); X. Li, J. Lu, and R. M. Martin, *Phys. Rev. B* 46: 4301 (1992).
98. M. Sprik, A. Cheng, M. L. Klein, *J. Phys. Chem.* 96: 2027 (1992).
99. C. H. Xu, C. Z. Wang, C. T. Chan, and K. M. Ho, *J. Phys: Condens. Matter* 4: 6047 (1992); C. Z. Wang, C. H. Xu, C. T. Chan, and K. M. Ho, *J. Phys. Chem.* 96: 3562 (1992); C. Z. Wang, K. M. Ho, and C. T. Chan, *Phys. Rev. Lett.* 70: 611 (1993).
100. M. B. Walker, *Phys. Rev. B* 45: 13849 (1992).
101. Z. Gamba, *J. Chem. Phys.* 97: 553 (1992).
102. E. Burgos, E. Halac, and H. Bonadeo, *Phys. Rev. B* 47: 7542 (1993); *ibid.* 47: 13903 (1993).
103. T. Yildirim and A. B. Harris, *Phys. Rev. B* 46: 7878 (1992).
104. J. Yu, R. K. Kalia, and P. Vashishta, *Appl. Phys. Lett.* 63: 3152 (1993).
105. J. Yu, R. K. Kalia, and P. Vashishta, *Phys. Rev. B* 49: 5008 (1994).
106. S. Iijima, *Nature* 354: 56 (1991).
107. T. W. Ebbesen and P. M. Ajayan, *Nature* 358: 220 (1992).
108. J. W. Mintmire, B. I. Dunlap, and C. T. White, *Phys. Rev. Lett.* 68: 631 (1992).
109. N. Hamada, S. Sawada, and A. Oshiyama, *Phys. Rev. Lett.* 68: 1579 (1992).

110. R. Saito, M. Fujita, G. Dresselhaus, and M. S. Dresselhaus, *Appl. Phys. Lett.* 60: 2204 (1992); M. S. Dresselhaus, G. Dresselhaus, and R. Saito, *Phys. Rev. B* 45: 6234 (1992).
111. R. A. Jishi and M. S. Dresselhaus, *Phys. Rev. B* 45: 11305 (1992); R. A. Jishi, M. S. Dresselhaus, and G. Dresselhaus, *Phys. Rev. B* 47: 16671 (1993).
112. G. B. Adams, O. F. Sankey, J. B. Page, M. O'Keeffe, and D. A. Drabold, *Science* 256: 1792 (1992).
113. J.-C. Charlier and J.-P. Michenaud, *Phys. Rev. Lett.* 70: 1858 (1993).
114. M. R. Pederson and J. Q. Broughton, *Phys. Rev. Lett.* 69: 2689 (1992).
115. D. H. Robertson, D. W. Brenner, and J. W. Mintmire, *Phys. Rev. B* 45: 12592 (1992).
116. C. T. White, D. H. Robertson, and J. W. Mintmire, *Phys. Rev. B* 47: 5485 (1993).
117. Jae-Yel Yi and J. Bernholc, *Phys. Rev. B* 47: 1708 (1993).
118. X. Blase, L. X. Benedict, E. L. Shirley, and S. G. Louie, *Phys. Rev. Lett.* 72: 1878 (1994).
119. D. Ugarte, *Nature* 359: 707 (1992).
120. S. N. Song, X. K. Wang, R. P. H. Chang, and J. B. Ketterson, *Phys. Rev. Lett.* 72: 697 (1994).
121. S. Iijima and T. Ichihashi, *Nature* 363: 603 (1993).
122. C. Z. Wang, C. T. Chan, and K. M. Ho, *Phys. Rev. B* 39: 8586 (1989); C. Z. Wang, C. T. Chan, and K. M. Ho, *ibid*. 40: 3390 (1990); C. Z. Wang, C. T. Chan, and K. M. Ho, *ibid*. 42: 11276 (1990); C. Z. Wang, C. T. Chan, and K. M. Ho, *Phys. Rev. Lett.* 66: 189 (1991).
123. F. S. Khan and J. Q. Broughton, *Phys. Rev. B* 39: 3688 (1989); J. Q. Broughton and F. S. Khan, *ibid*. 40: 12098 (1989); F. S. Khan and J. Q. Broughton, *ibid*. 43: 11754 (1991).
124. D. Tomanek and M. A. Schluter, *Phys. Rev. Lett.* 56: 1055 (1986); D. Tomanek and M. A. Schluter, *Phys. Rev. B* 36: 1208 (1987).
125. O. F. Sankey and D. J. Niklewski, *Phys. Rev. B* 40: 3979 (1989); O. F. Sankey, et al., *ibid*. 41: 12750 (1990); D. A. Drabold, et al., *ibid*. 42: 5135 (1990); D. A. Drabold, et al., *ibid*. 43: 5132 (1991); G. B. Adams, et al., *ibid*. 44: 4052 (1991).
126. K. Lassonen and R. M. Nieminen, *J. Phys. C* 2: 1509 (1990).
127. R. E. Allen and M. Menon, *Phys. Rev. B* 33: 5611 (1986); M. Menon and R. E. Allen, *ibid*. 33: 7099 (1986) ; M. Menon and R. E. Allen, *ibid*. 38: 6196 (1988); M. Menon and K. R. Subbaswamy, *Phys. Rev. Lett.* 67: 3487 (1991).
128. J. Yu, R. K. Kalia, and P. Vashishta, *Phys. Rev. Lett.* , submitted.

MOLECULAR DYNAMICS ON A MASSIVELY PARALLEL COMPUTER FOR APPLICATION TO SURFACE SYSTEMS

S. Pickering and I. Snook

Department Of Applied Physics
Royal Melbourne University Of Technology
P.O. BOX 2476V, Melbourne 3001
Victoria, Australia

INTRODUCTION

The surface of materials and the interface between phases is interesting to both basic science and to a wide range of applied science and engineering discipline areas such a electronics, crystal growth, catalysis, corrosion, friction, lubrication and wear, electrode reactions and biological cell function. One facet of this field involves the understanding of the properties of surfaces and interfaces at the atomic level. Thus, the theoretical study of these systems at this level is a very active area of research. One aspect of this theoretical work involves the use of the computational techniques of numerical statistical mechanics[1]. These methods have been applied to the study of the surfaces, for example to investigate the surface structure of crystals and liquids[2-4], to treat surface diffusion[5] and to investigate how gases and liquids are adsorbed on surfaces[3,6,7].

However, in most cases relatively small systems have been simulated, typically containing a few hundred to a few thousand particles ie. with surfaces of only a few tens of nm squared[1-7]. This is adequate to calculate many interesting properties eg the surface tension of simple liquids and the adsorption of gases and liquids on single, low index crystal planes. However, most surfaces are very irregular, containing atomic steps, ledges and terraces and many of the most interesting and useful surface phenomena (eg. catalysis) only occur on such heterogeneous surfaces. In order to simulate these systems much larger surface areas are needed, of the order of hundreds to thousands of nm^2 which involves the use of simulations treating tens of thousands to millions of atoms.

One of the most powerful atomic simulation methods is the Molecular Dynamics (MD) technique[1] and in this chapter we outline an approach to using the MD method for simulating systems of millions of atoms. This makes use of a Massively Parallel Computer and as this is a very new area of research and since these machines are extremely difficult to program efficiently we concentrate on describing how this is done in some detail instead of merely discussing applications. The ideas described here will then, hopefully, be seen to have a much wider application in Computational Condensed Matter Physics.

MOLECULAR DYNAMICS

As outlined in other articles in this proceedings, the basis of Molecular Dynamics, is solving Newton's equations of motion for a collection of N particles, with masses m_i and coordinates r_i by a finite difference scheme. Where F_i is the total force on particle i.

$$m_i \frac{dr_i^2}{dt^2} = F_i$$

Molecular Dynamics is an important tool for molecular analysis, as it allows the modelling and analysis of many properties on an atomic level. This detailed molecular information, can be used to provide more accurate models of the behaviour of not just the bulk of a sample, but the many smaller atomic features such as positioning of implanted species, nucleation and ion bombardment effects.

Solving the equations for r_i and v_i for many sequential time steps, produces information about the time evolution of the particle system. The potential model used here is pairwise additive, and the forces F_i are calculated from the potential function.

Many finite difference schemes exist for solving Newton's equations and thus to calculate the motion of the particles over small time intervals. The method used in this simulation is that of Verlet[7], with constant temperature scaling added to the algorithm. The algorithm is a two-steps method, requiring the previous two sets of positions, to calculate the next one.

Program Outline

The program is the basic MD algorithm, with many extra modelling features added. Periodic boundary conditions have been implemented to allow the modelling of samples which are usually solids, within infinite crystals or with a free surface. Various velocity scaling algorithms are available to produce such effects as constant temperature systems, constant energy systems, temperature gradients or thermal impulses. Heating and cooling stages have been implemented to allow the observation of phase changes and annealing effects.

The program consists of an initialisation section, that sets up all of the working variables, and then a series of loops that solve the equations of motion for the atoms in the simulation. As the simulation evolves, time-averaged thermodynamic data is accumulated. This data can be used to calculate many macroscopic properties of the sample. The positions and velocities are also recorded as a function of time, for subsequent analysis of the time-correlation functions.

NEIGHBOUR LISTS

"Neighbour Lists" are an optimisation technique that can be used in conjunction with Molecular Dynamics, to reduce the number of interaction-pairs that need to be calculated. The number of interacting particles is limited by the cut-off radius of the potential, as used by Verlet[8]. This is achieved by each atom maintaining a list of only those atoms that it interacts with, due to the short range nature of the potential being used. The neighbour lists are periodically updated as the atoms diffuse within the sample.

Neighbour List Algorithms

Two of the major methods for devising neighbour lists will be given as examples, showing how with a small modification to the selection of interacting pairs, a third method can easily be developed.

Without Newton's Third Law

The first method is the brute force method of calculation. The main time consuming step is in the calculation of the forces, which is an N^2 process. Each atom is tested against every other atom in the system. If the distance between the atoms is within the interaction range, the indices of the atoms are recorded in an array. This method actually records each interaction twice, as it doesn't utilise Newton's Third Law[9] between the interacting pairs. Once the array is completed, only those specified interacting pairs contribute to the calculations. Approximately every 20 time steps, the neighbour lists are recalculated.

With Newton's Law (I < J Indexed Pairs)

By using Newton's Third Law, the total number of interacting pairs that need to be stored, is halved as we only need to calculate those pairs where the numerical index of atom I is less than that of atom J (I < J). The advantage of this method is that the same data is stored, but in half the time and therefore, the execution time of the force calculations is also halved. The disadvantage of this method though is that the array that stores the data is now filled like an upper-half triangular matrix. The utilisation of the array is poor, as atom 1 has the most neighbours, whereas atom N has none. The array must be declared at twice the size of what is actually used.

Height Indexed Pairs ($Z_I < Z_J$)

The third method alleviates this problem by redistributing the pairs more evenly throughout the array. Searching for the neighbours is similar to the first method, with an extra condition added. The Z coordinate of atom J (or its periodic image) must be greater than that of atom I. Assuming an isotropic system, the number of neighbours for each atom is determined by the size of the interaction sphere that surrounds the atom. As this is the same for all atoms, the number of neighbours should be the same. The only differences would be caused by local density variations. eg. by the presence of a surface. The array is now evenly filled for each atom, and contains the same pairs that would have been found in the first two methods.

Neighbour List Performance

The increase in performance of a neighbour list algorithm over the serial implementation depends on how often the neighbour list data is updated. By assuming that the neighbour list data is accumulated as the forces are calculated and that the lists are updated every 20 time steps, the total number of interactions can be calculated. The number of force calculations for creating the neighbour lists is N(N-1)/2, whereas the number of calculations when using the neighbour lists is approximately 30*N, for each time step (assuming that the average number of interacting neighbouring atoms per atom is approximately 30).

Serial Version $\quad 20N(N-1)/2$

Neighbour List Version $\quad (19*30*N)+(N(N-1)/2)$

Analysis of these results shows that the serial version is mainly proportional to N^2, whereas the neighbour list version is mainly linear with N with a much smaller N^2 term. For systems with more than 64 atoms, the neighbour list method would be faster.

MASSIVELY PARALLEL COMPUTER (MASPAR)

The massively parallel computer used for these simulations was a DECmpp 12000, with 16,384 processors[10]. It is a SIMD (Single Instruction, Multiple Data) local memory computer. The same instruction is executed on all active processors concurrently, where each processor potentially stores different data.

Hardware Layout

The MasPar consists of two sections, the Array Control Unit (ACU) and the Data Processing Unit (DPU). The ACU generates all of the timing and control signals for the processors. It contains 128kb of data memory and 4Gb of instruction memory. The DPU contains the actual processors. The processors can be arranged in either a square (1024, 4096, 16384) or rectangular layout (2048, 8192). Each processor contains 64kb of local memory. Toroidal Wrap Topology[10] is used to connect neighbouring processors for faster communication paths. Each processor can be accessed individually or as a member of the active set of processors.

Inter-Processor Communication

The MasPar supports two major methods of inter-processor communication, "XNET" and "Global Router". Each method has its own uses, advantages and disadvantages. Both methods perform similar tasks but they are not always interchangeable. Which communication method to choose is highly dependent on the application, how the data is to be shared and moved amongst the processors and more importantly the programming language which the application is written in. Choosing the wrong method can result in the application running many times slower than optimal, or that which can be achieved.

XNET Communication Paths

The XNET commands are used to access processors (PEs) that are a set distance and direction from the active PEs. The directions of communication are the eight major compass directions N, NE, E, SE, S, SW, W and NW [11].

An example of an XNET command would be as follows,

i = xnetN[1].j

All active processors store the value of their first North neighbour processor's variable j in their own variable i.

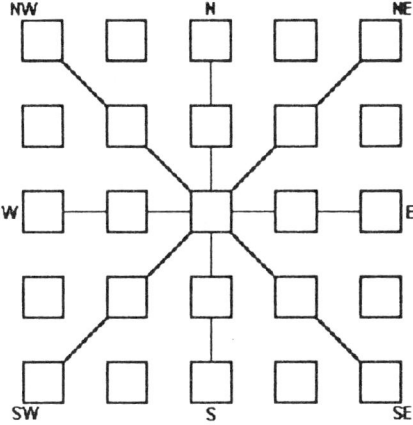

Figure 1. Possible XNET directions for communicating processors.

Toroidal wrapping is used to wrap the processors on the boundaries of the processor grid, around to the opposite edge of the grid. XNET is usually the fastest communication method, as it is independent of the number of processors participating. Communication time is proportional to the distance between the communicating processors.

Global Router

The Global Router is used when the relationship between the sending and receiving processors is random. Each 16 PEs are assigned to a PE cluster. Only one PE per cluster can send or receive at a time, but one can send while another is receiving. If more than one PE wishes to send or receive, a data collision occurs, and the transfers become serialised. The performance of the Router is very data dependent, and is usually much slower than using XNET. It is limited by the number of collisions that occur, but is independent of the distance between the communicating processors.

Compiler Languages

The two programming languages available on the MasPar are Fortran 90[15] and MPL[11] (MasPar Application Language). Both have distinct advantages and disadvantages, depending on the application that they are used for.

Fortran 90 (F90)

Fortran 90 is an extension of Fortran 77[15,16], with parallel loop constructs and data mapping commands. Each iteration of a loop can be assigned to an individual processor, so that many iterations of the same loop can be executed at once. If insufficient processors exist, virtual processors are created, with each real processor doing the work of many processors. Array data can be spread across the processors using many different mapping techniques, depending on how the data is to be accessed. Access to the processors, or control over the communication methods between the processors is unavailable, and can only be altered by adjusting the data mapping directives to achieve the desired result.

Fortran 90 is probably the most commonly used language on the MasPar, as it is identical to F77, converting the code is straightforward, and it is also available on many other types of machines, eg. the Cray computers. The performance increases obtainable for small systems, can be proportional to the number of processors, but problems arise when virtual processors[12,13] are created. The difference between a 16,384 and a 16,385 system on a 16,384 processor machine could be a factor of many times slower, depending on the types of operations being performed. This processor boundary effect is predominant for many vector and matrix operations on the MasPar as is shown in Figure 2.

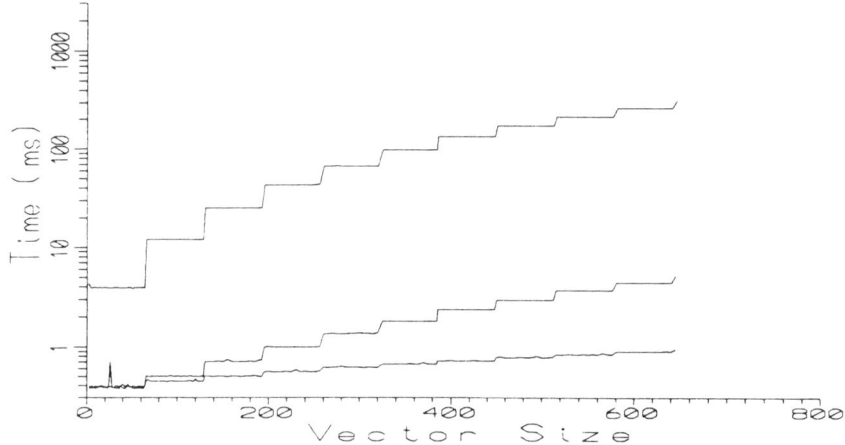

Figure 2. Graph of a range of Vector/Matrix operations showing processor boundary effects on a 64x64 processor grid [12]. Note the effect at each multiple of 64 along the major axis.

MPL (MasPar Application Language)

The MasPar Application Language[11] is a super-set of "C", where a new data type modifier "plural" has been added which allows the declaration of a variable that can contain a different value on every processor. Low level processor commands allow direct access to the mechanisms that control which processors participate in any of the instructions, and how the processors communicate and exchange data.

It is more difficult to use Fortran 90, as the spreading of data amongst the processors must be implemented by the programmer, but this allows better allocation of the data structures, depending on their intended use. The design of the processor grid (square or rectangular) is two dimensional, whereas most simulations are inherently three dimensional. This allows easy mapping of any the multi-dimensional data structures onto the processor grid, with the extra dimensions being accommodated in the processor's memory.

MASSIVELY PARALLEL IMPLEMENTATION

We chose MPL to write the simulation code in, since knowledge of the algorithm showed that by using MPL, many features of the architecture of the MasPar could be exploited whereas Fortran 90 only had the advantage of having many processors to help solve the problem.

The samples of interest were usually rectangular or square along two of the axes, allowing easy mapping of the cartesian coordinates onto the processor grid. Periodic boundaries in the X and Y directions would be facilitated by using the Toroidal Wrap Topology of the processors. As MD is mainly concerned with local area interactions, neighbouring processors could contain the data for the neighbouring atoms, reducing the communication costs to transfer data, as it would always be close by. By using the XNET commands, all processors could shift most of their data simultaneously.

The solution to the MD calculations could be reduced from solving many equations once each, to solving one equation many times in parallel, if the data for each processor could be isolated. This would produce a performance increase proportional to the number of processors in the system. By careful implementation of the algorithms, the problems associated with the model size crossing processor boundaries would be removed. This resulted in linearly increasing and not stepped execution times, resulting in a system that is independent of the number of processors in the computer.

Data Structures

Each processor contains information about a local area of the sample. The sample is sub-divided into columns of data similar to the used by Lin[14], with each processors controlling one column of information each. The physical length, width and height of a processor is determined from the structure. For those atoms assigned to a particular processor, their positions, velocities and accelerations are all stored on that processor. The data is stored in arrays, which can all be accessed concurrently.

Periodic boundaries in the Z direction were easily implemented by accessing each array as a circular queue. If the surface of the sample is uneven, eg. a stepped crystal, the processors could contain differing numbers of atoms, with the execution time being proportional to the length of the longest array. Lists of the neighbouring atoms would not need to be stored, as they can be found easily by looking at the neighbouring processors' array of positions. This is one of the major differences between this implementation and other neighbour list implementations. ie. This method doesn't require any extra memory storage to maintain the lists of interacting neighbours, as it follows directly from the machines architecture.

Neighbour Searching Algorithms

Finding the neighbouring atoms is now extremely easy, as they must reside on the neighbouring processors. Every processor can be identified by three unique numbers, so finding the neighbour processors means searching around in the XNET directions from the central processor. The number of processors to search in any direction can be determined from the physical dimensions of the processors and the cut-off radius of the potential. The number of processors to search in the X,Y directions can be calculated from the width and length of the processors. The number of layers# to search vertically can be calculated from the distance between the crystal planes.

\# The length, width and height of a processor is determined by the density of the system, and the number of processors. eg. For a cubic crystal of reduced density 1.0, the dimensions of the processors would be (1.0, 1.0, 1.0).

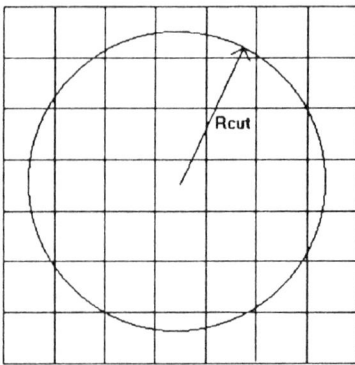

Figure 3. A schematic diagram showing which processors are included within the cut-off radius of the interactions.

Each square represents a single processor. The circle specifies the range of interactions from the central processor in the X,Y plane. Those processors partially or fully contained within the circle, would contain some atoms that are within range of the atoms in the central processor. Processors outside the circle are still included in the search, but may or may not contain valid interactions. The same is also true for moving vertically through the layers[Ψ] in the sample.

Four slightly different searching algorithms were developed. Each a significant improvement over the previous version, by reducing the interaction volume (number of interactions that need to be calculated) for each atom. The number of processors and layers to search is determined by a set of indices that determine the extent of the search. An example of these searching dimensions (N_X, N_Y, N_Z) would be (3, 3, 5). Three processors from the central processor in the X direction, three processors in the Y direction, and five layers up through the sample.

Every Layer Search

Every processor simultaneously searches one of the XNET[Δ] directions and retrieves the atom from the current layer contained on that processor. All atoms on the originating processor interact with that fetched atom, and the resulting accelerations are stored on the original processor, and then sent back to the processor where the atom was fetched from. This is repeated for all of the XNET directions, and all of the layers in the sample.

The total number of interactions is given by.

Total Interactions $\quad N_L^2((2N_X+1)(2N_Y+1)) - N_L$

Interaction Volume $\quad N_L(2N_X+1)(2N_Y+1)$

[Ψ] The definition of the term "layers" has been altered. It does not refer to the number of physically distinct layers in the sample, but the number of array locations occupied by the data.

[Δ] XNET refers only to the eight major compass directions. The other minor compass directions (eg. NNW, SSE) can be achieved by using two XNET instructions, one after the other.

The interaction volume is a tall rectangle extending from the bottom to the top of the sample, with width and length ($2N_X + 1$) and ($2N_Y + 1$) respectively. The first method is still an N^2 process, as is serial molecular dynamics without using a neighbour list algorithm.

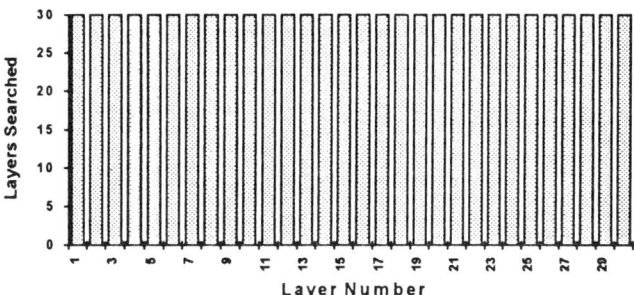

Figure 4. Layers searched for the "Every Layer Search" algorithm.

Layers Above Search

This second method is analogous to the second neighbour list method. The number of interactions can be practically halved by not repeating the same interactions twice, and in this case, only layers on the same level or above the current search layer are tested for interactions.

Total Interactions $\quad \dfrac{N_L^2}{2}\left((2N_X+1)(2N_Y+1)\right) - \dfrac{N_L}{2}$

Average Interaction Volume $\quad \dfrac{N_L+1}{2}\left((2N_X+1)(2N_Y+1)\right)$

The second method is still N^2, but is nearly twice as fast, which was expected by utilising Newton's Third Law.

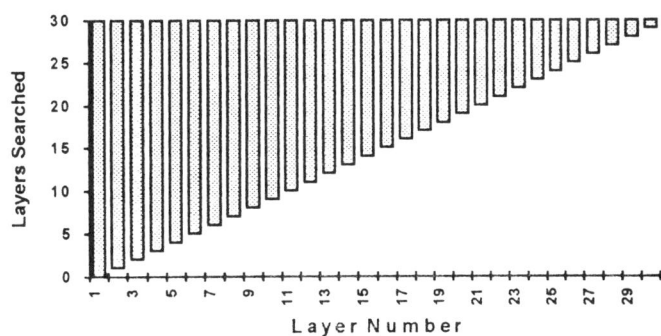

Figure 5. Layers searched for the "Above Layers Search" algorithm.

Limited Layers Above Search

Now we also know that the potential has a finite interaction range for typical short range forces, thus for thick samples, many layers are being searched above the current layer, even though the distance between the atoms may be outside the cut-off radius of the potential. The number of layers above to be searched can be limited by this interaction range. Instead of searching to the top of the sample, only a finite number of layers N_Z are searched above each layer.

Total Interactions $\quad N_L\left(N_Y + N_X(2N_Y + 1) + (N_Z - 1)(2N_X + 1)(2N_Y + 1)\right)$

Average Interaction Volume $\quad N_Z(2N_X + 1)(2N_Y + 1)$

This process is now linearly proportional to N, and is consistent with other neighbour lists algorithms.

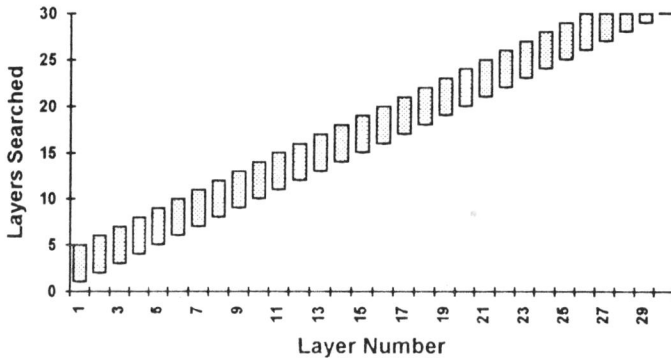

Figure 6. Layers searched for the "Fixed Layers Above Search" algorithm.

Hemispherical Search

The fact that the potential is spherically symmetric means that the interaction volume is really a sphere (ellipsoid). Since we are only concerned with layers above the current layer, the volume can be reduced to a hemisphere. The hemispherical search envelope varies with the local density fluctuations in the sample. The hemisphere can be represented by a collection of histograms with height proportional to their location in the X,Y plane. The histograms that have no height, represent those processors outside the circle of interaction which would contain no valid interactions.

Each of the directions enclosed in the square of interaction is tested in order. Each processor searches up through the layers in the specified XNET directions until the interactions are out of range, terminating the search as early as possible. The number of layers to search upwards is determined by the mean distance between the physical layers in the sample and how the layers are stored in memory.

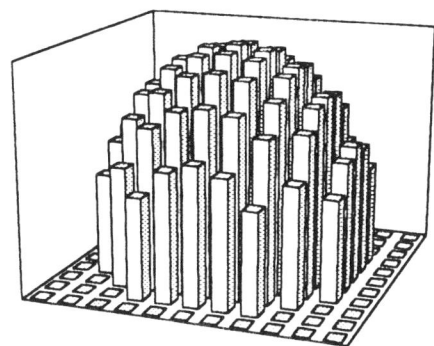

Figure 7. Layers searched for the "Hemispherical Search" algorithm.

Total Interactions   Dependent On Crystal Structure (≈ 40)

Interaction Volume $\dfrac{\pi}{6}(2N_X + 1)(2N_Y + 1)N_Z$

Analysis Of Algorithms

Each of the neighbour search algorithms is described by a set of four indices N_X, N_Y, N_Z and N_L. These represent the number of processors to search in the X and Y directions, the number of layers upwards to search and the number of layers in the sample, respectively. As result of the method that was used to construct the crystal structure, these indices can be calculated easily. For any known crystal structure, the dimensions of a processor can be determined from the density of the sample. Following is an example of the indices for the three major faces of a Face Centred Cubic crystal.

Table 1. Table describing search indices for various crystal structures.

Dominant Crystal Face	N_X	N_Y	N_Z
FCC (100)	2	2	8
FCC (110)	3	2	7
FCC (111)	3	2	7

Table 2. Table showing number of interactions for each of the search algorithms.

All Layers Search	$25N_L^2 - N_L$
Layers Above Search	$(25N_L^2 - N_L)/2$
Limited Layers Above Search	$187 N_L$
Hemispherical Search	$44 N_L$

As a result, the total number of interactions for any search method can be calculated.

The first two methods are N^2 processes, as expected from typical MD algorithms. The last two methods are proportional to N, as other Neighbour List methods are. For the Hemispherical Search, the number 44 is the average number of interactions that any atom found within the hemisphere of interaction.

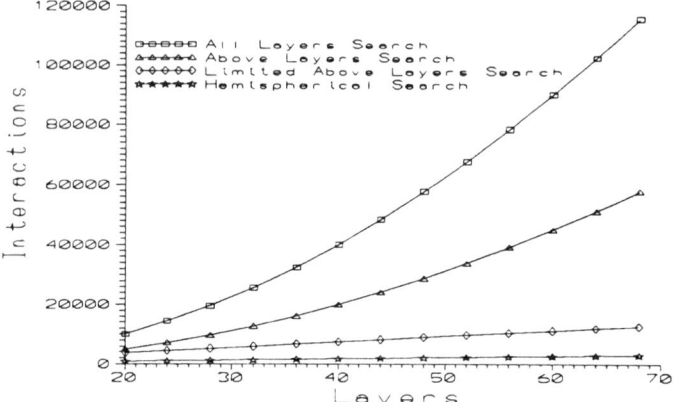

Figure 8. Number of interactions for each of the search algorithms.

As diffusion in the sample causes the originally stated number of layers to alter, there may be an increase of up to 50% in the number of layers in the sample. As the sample equilibrates, the maximum number of layers decreases back towards the original value. The original value is never reached, but the maximum values decreases to approximately 10% above the original value. There is usually an increase of between 2 and 4 in the total number of layers. The distribution of the number of layers is gaussian around the original number of layers. As all of these results are dependent on the maximum number of layers in the sample, as stated before these numbers of interactions are the minimum values that could be calculated. The time taken is therefore proportional to the maximum number of layers in the structure.

For an MD system, the total number of interactions found is N(N-1)/2, but for a neighbour list algorithm, this is reduced to approximately 30N, where 30 is the average number of neighbour per atom. So for a sample with 327680 atoms, the total number of interactions that needs to be calculated for a typical MD algorithm would be 5.3687E+10 and a neighbour list algorithm would be 9.83E+6. Using the MasPar Hemispherical Search algorithm (20 layers), a minimum total number of interactions of 880 is achieved. This may increase to a maximum number of interactions calculated at 1,056.

The number of interactions calculated on the MasPar (minimum or maximum) are still much less than for the original MD algorithms. Assuming that these algorithms were executed on similar type processors, and that the execution time is proportional to the number of interactions, this represents a speed increase of nearly 10,000 times faster. This number is important, as it shows that most of the processors are being utilised most of the time. If full utilisation had occurred and the processor types were identical, this speed increase would have been 16,384, the number of processors in the MasPar.

PERFORMANCE AND TIMINGS

As the times are proportional to the number of interactions, the graph of the timings for the four search methods will be practically identical to the graph displaying the number of interactions. The variation in the number of layers during the simulation is average out during the total execution time of the simulation. Timings for a neighbour list algorithms were formed on a Cray Y-MP EL for comparison. The results are as follows.

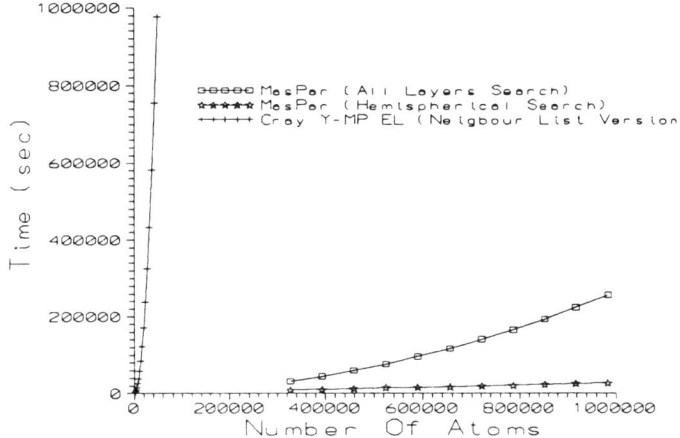

Figure 9. Comparison of execution time for different implementations of the MD algorithms.

SUMMARY OF RESULTS

It is demonstrated the Molecular Dynamics on a massively parallel computer is vastly superior to a similar implementation on a serial or vector computer. The large number of processors available has greatly increased the sizes of the systems that can be modelled. By efficient programming, a speed up proportional to the number of processors in the computer has been obtained. This method can be applied to many other types of systems, that require large sample sizes. eg. Crack Propagation, Stepped Surfaces and Surface Melting.

ACKNOWLEDGMENTS

We would like to thank Mr. Sonny Trinder of the RMIT Computer Centre for his many helpful discussions on the idiosyncrasies of the MasPar. We also wish to acknowledge the financial assistance of the Australian Research Council in a variety of areas and in particular for funding the Ormond Supercomputer Centre.

REFERENCES

[1] M.P. Allen and D.J. Tildesley, *Computer Simulation Of Liquids*, (Clarendon, Oxford) 1987.
[2] D. Henderson, *Fundamentals of Inhomogeneous Fluids*, (Marcel Dekker, New York) 1992.
[3] D. Nicholson and N.G. Parsonage, *Computer Simulation and the Statistical Mechanics of Adsorption*, (Academic Press, London) 1982.
[4] J.K. Lee, J.A. Barker and G.M. Pound, Surface Structure and Surface Tension: Perturbation Theory and Monte Carlo Calculation, J. Chem. Phys. 60:1976 (1974).

[5] John C. Tully, George H. Gilmer, and Mary Shugard, Molecular dynamics of surface diffusion. I. The motion of adatoms and clusters, J. Chem. Phys, 71:1630 (1979).
[6] I.K. Snook and W. van Megen, Solvation Forces in Simple Dense Fluids I, J. Chem. Phys. 72:2907 (1980).
[7] W. van Megen and I.K. Snook, Physical Adsorption of Gases at High Pressure I The Critical Region, Mol. Phys. 45:629 (1982); II Effect of Temperature, ibid. 47:1417 (1982): Adsorption in Slit-Like Pores, ibid. 54:741 (1985).
[8] L. Verlet, Computer "Experiments" on Classical Fluids. I. Thermodynamical Properties of Lennard-Jones Molecules, Phys. Rev. 159:98 (1967).
[9] J.B. Marion and W.F. Hornyak, *Physics For Science and Engineering*, (Holt Saunders, Japan) 1982.
[10] MasPar System Overview and MPPE Manuals, (MasPar Computer Corporation) 1991.
[11] MasPar MPL Programming Manuals , (MasPar Computer Corporation) 1991.
[12] B. Trippit, *The Performance of MasPar Array Intrinsics*, Proceedings of Fifth Australian Super-Computing Conference, 1992.
[13] L.L. Boyer and P.J. Edwardson, *Application of Massively Parallel Machines to Molecular Dynamics Simulation of Free Clusters*, (IEEE, New York) 1988.
[14] C.S. Lin, A.L. Thring and J. Koga, A Parallel Particle-in-Cell Model for the Massively Parallel Processor, (IEEE, New York) 1988.
[15] MasPar Fortran Programming Manuals, (MasPar Computer Corporation) 1991.
[16] M. Metcalf and J. Reid, *Fortran 90 Explained*, (Oxford Science Publications, Oxford) 1990

FRIEDEL OSCILLATIONS IN CONDENSED MATTER CALCULATIONS

John F. Dobson

Faculty of Science and Technology
Griffith University
Nathan, Queensland, 4111, Australia

I. INTRODUCTION

Friedel oscillations[1] are spatial modulations of the electron density n(r) which occur in microscopic zero-temperature calculations of metallic properties, whenever the metal is subject to a local disturbance. Static Friedel oscillations take the form of a standing-wavelike perturbation in the electron density, falling off with distance from the disturbance. These oscillations are most clearly derived in the jellium model in which the positive metal ions are represented by a uniform positive background. This suppresses periodic crystalline density oscillations and leaves the Friedel oscillations plainly visible. One example is provided by the electron density profile n(z) in calculations[2] of the jellium metal surface (Fig. 1): here the surface itself constitutes the "disturbance", and the wavenumber of the oscillations is twice the Fermi momentum in the bulk metal, $2k_F$. The oscillations die as z^{-2} where -z is the depth into the metal.

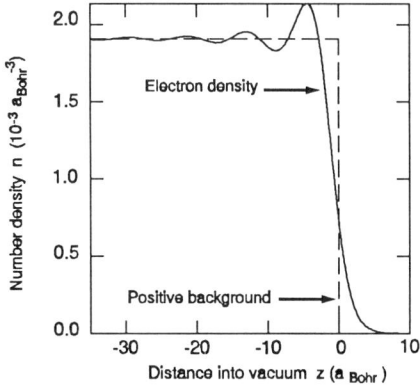

Figure 1. Groundstate density profile n(z) at a jellium surface (r_s=5, T=0K).

Computational Approaches to Novel Condensed Matter Systems
Edited by D. Neilson and M.P. Das, Plenum Press, New York, 1995

Similar effects occur near a point impurity[3] in a metal. Once more we see weakly decaying oscillations at wavenumber $2k_F$, this time falling off as $\sin(2k_F r - \phi)/r^3$.

The basic cause of static Friedel oscillations is the oscillatory nature of the electron wavefunctions at the Fermi surface of a metal: the effect is therefore not present in hydrodynamic or classical calculations which do not employ wavefunctions. In the summation over squared wavefunctions to obtain the density in the presence of a disturbance, oscillations due to lower wavefunctions largely cancel one another (phase averaging), so that the highest occupied states dominate at large spatial separations. This is tied to a strong decrease in density of occupied states at the Fermi surface. Since this effect depends on the sharpness of the Fermi surface, the Friedel oscillations become exponentially damped as the temperature is raised from zero.

There is a second, perturbative, way to understand the static oscillations, and to motivate the existence of generalized Friedel oscillations for time-periodic disturbances at frequency ω. One Fourier-decomposes the localized perturbation potential into plane-waves with wavenumber q: the electron gas can make transitions from occupied states with momentum $\mathbf{k}$ and energy $E(\mathbf{k})$ to unoccupied ones at $\mathbf{k}+\mathbf{q}$ and $E(\mathbf{k})+\hbar\omega$. The unperturbed and perturbed wavefunctions beat to form density waves at $\mathbf{k}-\mathbf{k'}= \mathbf{q}$, and phase averaging as described above means that the oscillations are dominated by extremal allowed values of $\mathbf{k}$ and $\mathbf{k}+\mathbf{q}$. For example, at $\omega=0$, extremal values of $\mathbf{k}$ and $\mathbf{k'}$ lie on opposite sides of the unperturbed Fermi surface. In the case of a spherical Fermi surface this leads to a perturbation density dominated by $q=2k_F$. In the case of real metals, Friedel oscillations are strongest for those with strongly "nested" Fermi surfaces, i.e. ones with large nearly flat parallel extremal portions[4].

The Kohn anomaly in phonon spectra of metals[5] can be regarded as a Friedel oscillation phenomenon, in this case expressed as a singularity of the electronic response at a Fermi-surface-spanning wavevector q. This strongly influences phonons of the same wavenumber: alternatively, we can think of the relatively slowly time-varying phonon distortion as a "static" external perturbation that causes a strong electronic response when it meshes with the spatial Friedel oscillations.

Where a spin-polarizing impurity occurs, the spin polarization density of the metallic electron gas also exhibits Friedel oscillations, leading to the oscillatory electron-mediated RKKY interaction[6] between magnetic impurities: this interaction is believed to be responsible in some cases for spin-glass behaviour. Perhaps the best direct experimental evidence for static Friedel oscillations comes from the case of sandwiches consisting of a spacer layer of nonmagnetic metal between two layers of magnetic metal[6]. The spin-polarised Friedel oscillations in the middle layer cause macroscopic magnetic properties of the sandwich to be oscillatory functions of the thickness of the nonmagnetic layer. Related phenomena are currently receiving considerable attention[7-10]. A Friedel-like explanation for an attractive pairing interaction in the cuprate high-Tc superconductors has also been suggested[11], based on spin models of doped CuO planes.

Static Friedel oscillations are also closely related to quantum size effects in thin metal films[12] or wide parabolic quantum wells in semiconductors[13]. For thin metal films Schulte[12] has shown that these Friedel or quantum size effects lead to a striking cusped behaviour of the work function as the thickness is varied.

For cases where the density would be spatially fairly slowly varying but not necessarily constant in a classical approximation, Kohn and Sham[14] have shown that Friedel oscillations still occur. They even interpret shell structure of atomic densities in this light.

In this paper we will discuss the linear response of an inhomogeneous metallic system to a small potential perturbation $\delta V(\underline{r})\exp(-i\omega t)$. This problem is relevant to the study, for example, of plasmons on the inhomogeneous electron gas at metal surfaces and slabs. At finite frequency it is known that, even in the uniform electron gas, the static Friedel oscillation at $2k_F$ is replaced by spatial oscillations with several different wavenumbers. Further complications occur because of new physics introduced by an inhomogeneity (e.g. a surface). We will show that these phenomena cause difficulties in calculations of plasmon properties on metal surfaces and wide quantum wells. A discussion will be given of various attempts to avoid these difficulties. New results are given for the case of non-zero surface-parallel wavevector $q_\parallel$.

II. LINEAR RESPONSE OF AN INHOMOGENEOUS ELECTRON GAS

Consider the effect of a space- and time-dependent electric field on a metallic sample. Specific examples include Electron Energy Loss Spectroscopy (EELS) experiments which have recently obtained the dispersion of surface plasmons on metals[15,16]. The same ideas apply to epitaxially grown GaAlAs well structures[13,17] in which case the external field is infrared radiation incident via a grating coupler.

We suppose that the unperturbed equilibrium state of a many-electron system has been solved, within the Kohn-Sham Local Density Functional theory[18] for example, to give a set of unperturbed effective wavefunctions $\Psi_I(r)\exp(-i\omega_I t)$ with eigenvalues $E_I = \hbar\omega_I$, a total effective single-particle potential $V^0_{eff}(r)$ and a total electron density

$$n_0(r) = \sum_I f_I |\Psi_I(r)|^2 \qquad (1)$$

where f_I is the Fermi distribution. We imagine a perturbation $\delta V_{eff}(r)\exp(st)$ occurring in the total self-consistent potential, where an imaginary frequency $\omega = is$ has been chosen for now, to ensure a real perturbation. The eigenfunctions acquire a linear perturbation $\delta\Psi_I(r)\exp[(-i\omega_I+s)]t$, where

$$[-\frac{\hbar^2}{2m}\nabla^2 + V^{eff}_0(r) - \hbar\omega_I - i\hbar s]\delta\Psi_I(r) = -\delta V^{eff}(r)\Psi_I(r) \qquad (2)$$

with suitable boundary conditions. Equation (2) can also be written

$$\delta\Psi_I(r) = -\int G(r,r',E_I+i\hbar s)\delta V_{eff}(r')\Psi_I(r')\,dr' \qquad (3)$$

where

$$G(r,r',E) = \sum_J \frac{\Psi^*_J(r)\Psi_J(r')}{E_J-E} = [G(r',r,E^*)]^* \qquad (4)$$

is the Green function for the one-body Schrodinger equation: it is the solution for a point source, satisfying

$$[\frac{-\hbar^2}{2m}\nabla^2 + V^0_{eff}(r) - E]G(r,r',E) = \delta(r-r') \quad , \qquad (5)$$

and satisfying spatial boundary conditions appropriate to the problem at hand. Note that the energy E cannot be exactly real, for then the denominator in (4) would make G ambiguous.

The resulting density perturbation is, to first order,

$$\delta n(\mathbf{r}, t) = 2\sum_{I} f_{I} [\Psi_{I}^{*}(\mathbf{r}) \delta\Psi_{I}(\mathbf{r}) + \Psi_{I}(\mathbf{r}) \delta\Psi_{I}^{*}(\mathbf{r})] \exp(st) \qquad (6)$$

where $f_I = f(E_I)$ is the Fermi distribution and we have assumed a spin-unpolarised state, accounting for spins by the initial factor 2. Using (3) we can write the density change in the form

$$\delta n(\mathbf{r}) = \int \chi_0(\mathbf{r}, \mathbf{r}', \omega) \delta V_{eff}(\mathbf{r}') d\mathbf{r}' \qquad (7)$$

where the BARE SUSCEPTIBILITY is

$$\chi_0(\mathbf{r}, \mathbf{r}', \omega = is) = -2\sum_{I} f_{I} [\Psi_{I}^{*}(\mathbf{r}) G(\mathbf{r}, \mathbf{r}', \hbar\omega_I + \hbar\omega) \Psi_{I}(\mathbf{r}') \\ + \Psi_{I}(\mathbf{r}) G(\mathbf{r}', \mathbf{r}, \hbar\omega_I - \hbar\omega) \Psi_{I}^{*}(\mathbf{r}')] \qquad (8)$$

The second part of (4) was used in the second term. Using (4) and interchanging dummy indices in the second term we can also obtain the possibly more familiar form

$$\chi_0(\mathbf{r}, \mathbf{r}', \omega) = 2\sum_{I,J} [f_I - f_J] \frac{\Psi_{I}^{*}(\mathbf{r}) \Psi_{J}^{*}(\mathbf{r}') \Psi_{I}(\mathbf{r}') \Psi_{J}(\mathbf{r})}{E_I - E_J + \hbar\omega} \qquad (9)$$

Similar expressions for the time-ordered version of our retarded function χ_0 can also be derived by evaluating the open-bubble Feynman diagram[19] in which each particle propagator includes the interaction to all orders with the effective external potential V^0_{eff}.

The quantity χ_0 represents the linear density perturbation at $\mathbf{r}$ due to a temporally varying point-localized (delta-function) disturbance to the effective potential at $\mathbf{r}'$. We shall see that χ_0 contains all the necessary information about Friedel oscillations. Although we used a pure imaginary frequency in deriving (8) and (9), (in order to have a real potential and to be able to write the density perturbation from each orbital as $\delta(\Psi^*\Psi)$), the structure of (8), (4) and (9) shows that χ_0 is in fact analytic for ω lying in the upper half plane: thus χ_0 is given by (8) or (9) for any complex frequency ω such that Im(ω) > 0. For experimental situations in which external excitation is applied, this means ω will lie just above the real axis. The small positive imaginary part of ω corresponds to an adiabatic switching-on process.

While (7) is formally the density perturbation of independent electrons, the many-body nature of the problem is acknowledged via the requirement that the effective potential perturbation must be consistent with the density perturbation, thus necessitating a self-consistency loop in general. This self-consistency condition, when linearized, can be written in the form of a screening equation:

$$\delta V_{eff}(\mathbf{r}) = \delta V_{ext}(\mathbf{r}) + \int [\frac{e^2}{|\mathbf{r}-\mathbf{r}'|} + f_{xc}(n_0(\mathbf{r}))] \delta(\mathbf{r}-\mathbf{r}')] \delta n(\mathbf{r}') d\mathbf{r}' \qquad (10)$$

where δV_{ext} is a given externally-imposed potential, and all perturbed quantities have been assumed to vary as exp(st), or more generally as exp(-iωt) where Im(ω) > 0. Equation (10) can be written in convolution notation as $\delta V^{eff} = \delta V^{ext} + (V_c + f_{xc}) * \delta n$. In similar notation we obtain the selfconsistent integral equation for the density perturbation;

$$\varepsilon * \delta n = \chi_0 * \delta V_{ext},$$
$$\varepsilon = 1 - \chi_0 * V_c - \chi_0 * F_{xc} \qquad (11)$$

Without the f_{xc} term, equations (7), (8) and (10) define the time-dependent Hartree (uncorrelated, product-wavefunction) approximation. The term involving f_{xc} is an exchange-correlation term accounting approximately for the fact that the many body wavefunctions must be antisymmetric under exchange of particle labels, plus the fact that electrons avoid one another via their mutual repulsion (Coulomb correlation). Here we take $f_{xc}(n) = d\mu_{xc}/dn$ where $\mu_{xc}(n)$ is the exchange-correlation contribution to the chemical potential of the homogeneous electron gas: this amounts to the "Time-dependent local density approximation" (TDLDA)[20,21]. More sophisticated approximations can be made[22,23] but their inclusion does not affect the conclusions we will draw here concerning Friedel oscillations.

When χ_0 has been found, solution of the spatial integral equation (11) yields the interacting density $\delta n(r)\exp(-i\omega t)$, leading to an understanding of the response of the system to an externally imposed perturbation potential $\delta V^{ext}(r)\exp(-i\omega t)$. Alternatively, solution of (11) with the external potential set to zero will yield the eigenfrequencies (bulk and surface plasmon frequencies) of the inhomogeneous metal: these lie on or below the real frequency axis, so analytic continuation is required in general[24]. In the course of such plasmon calculations, the Friedel oscillations present in χ_0 cause a serious computational difficulties. These are significantly worse than the problems inherent in a static calculation, because several different Friedel wavenumbers are involved in the dynamic case. The easiest way to motivate the existence of these multiple wavenumbers is to look first at the linear response of a strictly uniform electron gas: this follows in the next section.

III. FRIEDEL EFFECTS IN THE LINEAR RESPONSE OF UNIFORM JELLIUM

In the case of the homogeneous electron gas (uniform jellium), the space Fourier transform of the bare susceptibility (9) can be evaluated in closed form. In this case the unperturbed eigenfunctions $\Psi_I(r)$ are free-electron plane waves, labelled by momentum $\mathbf{k}$, and occupied up to the Fermi momentum k_F. χ_0 then takes the form

$$\chi_0(\mathbf{r},\mathbf{r}',\omega) = (2\pi)^{-3}\int \chi_0(q,\omega) e^{i\mathbf{q}\cdot(\mathbf{r}-\mathbf{r}')} d\mathbf{q}. \qquad (12)$$

where, after some algebra, one finds[19] at T = 0K for a frequency just above the real axis

$$\mathrm{Re}\chi_0(q,\omega+i0) = -\frac{mk_F}{2\pi^2\hbar^2}[1-\{\frac{1-(F-Q)^2}{4Q}\ln|\frac{1+F-Q}{1-F+Q}|\}-\{F\rightarrow F\}], \qquad (13)$$

$$Q = q/(2k_F), \qquad F = \hbar\omega/(4E_F Q).$$

When this Lindhard expression is transformed back into three-dimensional $\mathbf{r}$ space, it represents the density response at distance r from a time-varying point potential disturbance at the origin, in a zero-temperature Fermi gas of independent electrons of density $n = k_F^3/(3\pi_2)$:

$$\chi_0(r,\omega+i0) = \frac{1}{(2\pi)^3}\int \chi_0(q,\omega+i0) e^{i\mathbf{q}\cdot\mathbf{r}} d^3r$$
$$= \frac{1}{2\pi^2}\int q\sin(qr)\chi_0(q,\omega+i0) dq \qquad (14)$$

In performing the inverse transformation (14) of equation (13), at large values of r the integration is dominated by singularities (branch points) of the logarithms. The four logarithmic terms cause spatial oscillations with up to four periods, as discussed by Persson[25]. The corresponding wavenumbers k make one of 1+F-Q, 1-F+Q, 1-F-Q or

1+F+Q vanish. Using (13) we find that these generalized Friedel wavenumbers are those from the following set which are real:

$$k(\alpha, \beta) = \alpha k_F + \beta \sqrt{k_F^2 + \beta 2m\omega/\hbar} \qquad (15)$$

where α and β independently take the values ± 1. For $\omega = 0$, these roots are degenerate, two being zero and two taking the values $\pm 2k_F$, yielding the well-known static Friedel oscillations. For $\hbar\omega > E_F$ only two of these generalised Friedel wavenumbers are real. In the case $0 < \hbar\omega < E_F$ there are four real Friedel wavenumbers. For later purposes it is also convenient to label these wavenumbers via an alternative scheme:

$$\begin{aligned}
k_1 &= k(1,1) &&= (k_F^2 + 2m\omega/\hbar)^{1/2} + k_F \\
k_2 &= k(-1,1) &&= (k_F^2 + 2m\omega/\hbar)^{1/2} - k_F \\
k_3 &= k(1,-1) &&= k_F - (k_F^2 - 2m\omega/\hbar)^{1/2} \\
k_4 &= k(-1,-1) &&= -k_F - (k_F^2 - 2m\omega/\hbar)^{1/2}
\end{aligned} \qquad (16)$$

The above argument applied to the bare response, but in the present homogeneous gas case we can easily show that the TDLDA interacting response contains similar terms. Fourier-transformation of (11) turns the convolutions into simple products and we can solve for the density perturbation $\delta n(q,\omega)$ in terms of the external potential:

$$\delta n(q,\omega) = \frac{\chi_0(q,\omega)}{1 - (V_q + f_{xc})\chi_0(q,\omega)} \delta V_{ext}(q,\omega) \qquad (17)$$

and this has branch points and cuts at the same values of q as does χ_0, leading to Friedel oscillations in $\delta n(r,t)$ similar to those of $\chi_0(r,t)$, though with modified amplitude. As explained in ref. 25, the poles of $\delta n(q,\omega)$ due to the vanishing of the denominator in (17) lead to outgoing plasma waves in $\delta n(r,t)$. These fall off only as r^{-1} and are not Friedel oscillations: they do not appear in the noninteracting response χ_0. The resulting response found in ref. 25 is

$$\delta n(r,t) \underset{r\to\infty}{\sim} e^{-i\omega t} \sum_\mu A_\mu \frac{\exp(ik_\mu r)}{r^3} + (\text{plasmon term}) \qquad (18)$$

where the $\{A_\mu\}$ are constants. The same form (18) (with different coefficients A) was obtained whether the external source is a point oscillating potential, a more physical point oscillating charge, or a more general localised source.

It is well known[19] that the logarithmic singularities in $\chi_0(q,\omega)$ are related to the possiblity of exciting single electrons from momentum **k** to **k+q** with an energy change $\hbar\omega$: this is closely related to Landau damping of collective plasmons. Thus from Persson's approach it is clear that **dynamic Friedel oscillations are related to the onset of real (as opposed to virtual) single-particle excitations.**

We now discuss an alternative way of deriving the above results for χ_0. It is more closely related to the method we will shortly use for metal surface applications. The Green function for a free electron (the solution of the Schrodinger equation (5) with zero potential and a point source at **r'**) is an outgoing spherical wave:

$$G(\mathbf{r},\mathbf{r}',E) = \frac{2m}{\hbar^2} \frac{\exp(i(2mE/\hbar^2)^{1/2}|\mathbf{r}-\mathbf{r}'|)}{4\pi|\mathbf{r}-\mathbf{r}'|} \quad (19)$$

where E must contain a nonzero imaginary part and the sign of the complex square root is chosen to make the Green function remain bounded as the separation between **r** and **r'** goes to infinity. Putting (19) into (8) we find the bare susceptibility, representing the perturbed density at **r** due to an oscillating point potential disturbance at **r'**=0, is

$$\chi_0(\mathbf{r},0,\omega+i0) = -\frac{4m}{(2\pi)^3\hbar^2}\int_0^{k_F} 4\pi k^2 \frac{\sin(kr)}{kr}\left[\frac{e^{ir\sqrt{k^2+2m\bar{\omega}/\hbar}}}{4\pi r} + \frac{e^{-iK_2 r}}{4\pi r}\right]dk \quad (20)$$

Here K_2 takes the value $(k^2 - 2m\omega/\hbar)^{1/2}$ when $\hbar^2 k^2/(2m) > \hbar\omega$ and the value $-i(2m\omega/\hbar - k^2)^{1/2}$ when $\hbar^2 k^2/(2m) < \hbar\omega$. The latter condition always holds if $\hbar\omega > E_F$. The integral in (20) can be separated into four phase integrals of form

$$I(r) = \int_{-\infty}^{\infty} w(k) e^{iq(k)r} dk \quad (21)$$

where the phase function is $q(k) = \alpha k + \beta(k^2 + \beta 2m\omega/\hbar)^{1/2}$, α and β independently taking the values ±1. Such phase integrals will be pivotal throughout this review. As $r \to \infty$ *and when q is real* the rapid oscillation of the imaginary exponential as a function of k tends to make the integral vanish. At large r values the integral is thus dominated by those values of k where either

(i) the weight function w(k) has a singularity (in the present case w(k) has to represent the finite integration cutoffs in (20) and so has singularities at k=0 and k=k_F.)

or (ii) the phase function is stationary, dq/dk = 0. (This case can be converted to the first case by changing integration variable from k to q so that dk becomes dq/(dq/dk), where the last term is singular at the points of stationary phase.) Note that points where dq/dk → ∞ do **not** contribute as r → ∞.

The integrals of form (21) arising from (20) have no points of stationary phase so their values for large r are dominated by the endpoint singularities as in (i) above. Their effect is easily calculated by changing variable to q and using integration by parts:

$$\int_0^{k_F} e^{iq(k)r} k\, dk = \int_{q(0)}^{q(k_F)} e^{iqr} k(q) \frac{dk}{dq} dq$$

$$= \left[\frac{e^{iqr}}{ir} k(q) \frac{dk}{dq}\right]_{q(0)}^{q(k_F)} + O(r^{-2}) \quad (22)$$

Substitution of (22) into (20) then gives χ_0 for large r as a sum of up to four terms of form $A_\mu r^{-3}\exp(ik_\mu r)$ where the $\{A_\mu\}$ are constants and the $\{k_\mu\}$ are the wavenumbers given in (16). Thus we have rederived the results of Persson[25] for the bare response. The argument following equation (16) shows that the interacting response falls off in the same Friedel fashion as the bare response just obtained.

IV. LINEAR RESPONSE OF THE JELLIUM SURFACE

The calculations described here are motivated partly by EELS experiments to detect surface plasmons on metals[15,16], and by grating-coupled IR absorption experiments on artificial metal layers in GaAlAs[13,17]. In these cases one has to consider the effect, on a bounded metal, of temporally oscillating near-field components satisfying $\nabla \cdot \mathbf{E} \approx 0$, and describable by a scalar potential

$$\delta V_{ext}(\mathbf{r}, t) = V_1 \exp(i\mathbf{q}_\| \cdot \mathbf{r}_\| + q_\| z - i\omega t) \ . \tag{23}$$

Here the surface-parallel wavenumber $q_\|$ is determined by the grating coupler in the IR experiment, or by the observed electron momentum loss parallel to the surface in the EELS experiment. The aim is to calculate, within linear response theory, the induced oscillating electron density $\delta n(z)\exp(i\mathbf{q}_\| \cdot \mathbf{r} - i\omega t)$. Then using the continuity equation one can deduce the absorbed power Re $\int \mathbf{J} \cdot \mathbf{E} d^3 r \propto -\text{Im} \int \exp(q_\| z) \delta n(z) dz$, and hence predict peaks in the spectra at specific frequencies, permitting a study of the plasmon dispersion $\omega(q_\|)$.

The discussion here will be limited to the jellium model of a metal surface[2] in which the discrete positive ions are replaced by a uniform positive background occupying the half-space $z < 0$, leaving Friedel effects clearly discernible. Because of the many-body nature of the electronic system, calculations even in this jellium model are nontrivial. The groundstate properties of the jellium metal surface were calculated within the Local Density Functional theory by Lang and Kohn[2]. The electrons are free to move with momentum $\hbar \mathbf{k}_\|$ in the xy plane parallel to the surface so that the box-normalised effective single-particle wavefunctions for the groundstate are of form

$$\Psi_I(\mathbf{r}) \equiv \Psi_{\mathbf{k}_\|, k}(\mathbf{r}) = A^{-1/2} e^{i\mathbf{k}_\| \cdot \mathbf{r}_\|} \psi_k(z) \tag{24}$$

where $k = k_z$ and A is the surface area. The self-consistent effective potential $V^0_{eff}(z)$ is shown in Fig 2: the groundstate density $n_0(z)$ was given in Figure 1.

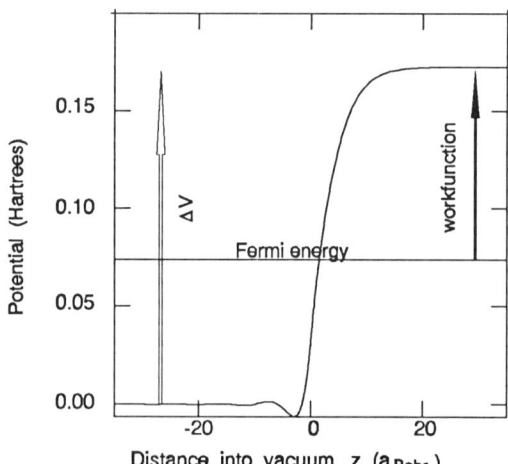

Figure 2. Kohn-Sham potential V_0^{eff} for jellium surface ($r_s=5$)

Deep in the metal V^0_{eff} becomes constant and the one-dimensional wavefunction becomes sinusoidal, with a phase-shift $\gamma_k = kZ_k$ determined by the detailed shape of the surface potential:

$$\psi_k(z) \sim (2/L)^{1/2} \sin(kz - \gamma(k)) \quad \text{as} \quad z \to \infty \quad . \tag{25}$$

Here the factor $(2/L)^{-1/2}$ ensures box normalization in a length L. It is shown in Ref. 2 that the phase shift $\gamma(k)$ of the kth wavefunction approaches zero linearly as $k \to 0$.

As discussed in the Introduction, Friedel oscillations are apparent in the groundstate density of Fig 1 and to a lesser degree[2] in the potential of Fig 2, but we shall see that far larger and more intractable spatial oscillations occur in the dynamic response, and these make predictions of surface plasmon dispersion more difficult. To analyze the dynamic case we first calculate the bare susceptibility χ_0 (equ. (8)) and then solve the selfconsistent screening problem (equs (11)).

For the present geometry we can use the unperturbed wavefunctions (24) in equation (4) to put the 3D Green function in (8) into the form

$$G(\mathbf{r}, \mathbf{r}', E) = \sum_{p_\parallel} e^{i\mathbf{p}_\parallel \cdot (\mathbf{r} - \mathbf{r}')} \sum_k \frac{\psi_k(z) \psi_k(z')}{\hbar^2 (2m)^{-1} (p_\parallel^2 + k^2) - E} \quad . \tag{26}$$

Now we can use equation (4) once more to recognise the sum over k in (26) as a *one-dimensional* Green function $g(z,z',E-\hbar^2 p_\parallel^2/2m)$ where g satisfies the 1D Schrodinger equation

$$[\frac{-\hbar^2}{2m} \frac{d^2}{dz^2} + V_{eff}^0(z) - \varepsilon] g(z, z', \varepsilon) = \delta(z - z') \quad . \tag{27}$$

It is well-known[26,27] that such a 1D Green function can be written as a product of two solutions of the homogeneous Schrodinger equation, thus avoiding an infinite summation over intermediate states:

$$g(z, z', \varepsilon) = \frac{2m}{\hbar^2} \frac{\psi^R(z_>, \varepsilon) \psi^L(z_<, \varepsilon)}{W(\psi^R, \psi^L)} \quad . \tag{28}$$

Here the "right" solution $\psi^R(z)$ satisfies the homogeneous Schrodinger equation (equ (27) with zero right-hand side) and vanishes as $z \to +\infty$. Similarly the "left" solution $\psi^L(z)$ satisfies the homogeneous equation and vanishes as $z \to -\infty$. $z_>$ and $z_<$ are the greater and lesser of z and z', while $W(\psi^R, \psi^L) = \psi^R d\psi^L/dz - \psi^L d\psi^R/dz$ is the Wronskian of the two solutions, which can be proved to be independent of z. The derivative discontinuity of (28) at $z = z'$, when acted on by d^2/dz^2, produces the delta function on the right side of equation (27). The 3D Green function (26) thus becomes

$$G(\mathbf{r}, \mathbf{r}', E) = \frac{2m}{\hbar^2} A^{-1} \sum_{p_\parallel} e^{i\mathbf{p}_\parallel \cdot (\mathbf{r} - \mathbf{r}')} \frac{\psi^R(z_>, \varepsilon) \psi^L(z_<, \varepsilon)}{W(\psi^R, \psi^L)} \tag{29}$$

where $\varepsilon = E - \hbar^2 p_\parallel^2/(2m)$. Putting (29) into (8), writing $\mathbf{k}_\parallel - \mathbf{p}_\parallel = \mathbf{q}_\parallel$ and using $\sum_{k_\parallel} = A(2\pi)^{-2} \int d^2 k_\parallel$, we find[28]

$$\chi_0(\mathbf{r}, \mathbf{r}', \omega) = (2\pi)^{-2} \int d^2 q_\parallel \, \chi_0(q_\parallel, z, z', \omega) \exp(i\mathbf{q}_\parallel \cdot (\mathbf{r} - \mathbf{r}')) \tag{30}$$

where the bare susceptibility, in coordinates appropriate to the jellium surface, is

$$\chi_0(q_\parallel, z, z', \omega) = -2m\pi^{-3}\hbar^{-2} \int dk \int dk_x \sqrt{k_F^2 - k^2 - k_x^2} \; \psi_k(z) \psi_k(z')$$
$$\times \left[\left\{ \frac{\psi^R(z_>, \varepsilon_+) \psi^L(z_<, \varepsilon_+)}{W(\psi_R, \psi_L, \varepsilon_+)} \right\} + \{\varepsilon_+ \to \varepsilon_-\} \right] \tag{31}$$

with $\varepsilon_\pm = \pm\omega + \hbar^2(2m)^{-1}(2k_x q_\parallel - q_\parallel^2 + k^2)$. Taking the limit of a frequency just above the real axis, we then find

$$\chi_0(q_\parallel, z, z', \omega + i0)$$
$$= 2m\pi^{-3}\hbar^{-2} \int dk \int dk_x \sqrt{k_F^2 - k^2 - k_x^2} \; \psi^+(k, z) \psi^+(k, z') \tag{32}$$
$$\times \left[\{|p_+|^{-1} \psi^+(p_+, z_>) \psi^L(p_+, z_<)\} + \{p_+ \to p_-\}^* \right], \text{ where}$$

$$p_\pm^2 = \pm 2m\frac{\omega}{\hbar} + 2k_x q_\parallel - q_\parallel^2 + k^2 \tag{33}$$

while ψ^+ is the right eigenfunction satisfying

$$\left[\frac{d^2}{dz^2} + p^2 - \frac{2m}{\hbar^2} (V_0^{eff}(z) - V_0^{eff}(-\infty)) \right] \psi^+(z) = 0, \tag{34}$$

and either vanishing as $z \to \infty$ if $p^2 - V_{eff}^0(+\infty) + V_{eff}^0(-\infty) \equiv \xi^2 < 0$, or behaving as const $\exp(i|\xi||z - \gamma(|\xi|)|)$ if $\xi^2 > 0$. The right eigenfunctions are normalised by prescribing their deep-metal behaviour, as follows:

$$\psi^+(p, z) \underset{z \to -\infty}{\sim} \begin{cases} \sin(|p|z - \gamma(|p|)), & p^2 > 0 \\ \sinh(|p|z - \bar\gamma(|p|)), & p^2 < 0. \end{cases} \tag{35}$$

The other solution also satisfies (34) but has the following behaviour deep in the metal:

$$\psi^L(p, z) \underset{z \to -\infty}{\sim} \begin{cases} \exp(-i|p|z + i\gamma(|p|)), & p^2 > 0 \\ \exp(+|p|z - \bar\gamma(|p|)), & p^2 < 0. \end{cases} \tag{36}$$

Note that the phase shift $\gamma(p)$ becomes complex for $p^2 > V_{eff}^0(+\infty) - V_{eff}^0(-\infty)$. This corresponds to unequal reflected and transmitted amplitudes when the electron can propagate freely into the vacuum. The Wronskian W of the left and right eigenfunctions, from (35) and (36) is $-|p|$, accounting for the prefactor in the braces in (32).

Equation (32) is the analogue, for a jellium metal surface at finite real frequency and finite surface-parallel wavevector $q_\parallel$, of the formula (20) derived above for the bare susceptibility of the uniform electron gas. These equations, or equivalent, have been evaluated numerically by a number of authors[24,26-32] as a first step towards predicting plasmon dispersion or absorption data on simple-metal surfaces. An example of such a χ_0 calculation is given in Fig 3. The perturbing plane of potential is at the jellium edge ($z' = 0$), and the frequency is close to the surface plasmon frequency. Also shown is the interacting response χ^{TDLDA}. We note that the slowest Friedel oscillation is somewhat reduced by the interactions, but the most rapid oscillation is hardly affected.

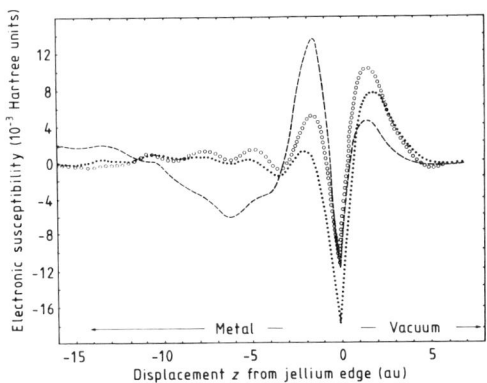

Figure 3. Bare susceptibility Re $\chi_0(q_\|,z,z'=0,\omega)$ of jellium surface for r_s = 2.07, $q_\|$=0.3 au, ω=0.5 a.u. (dashed line). Also shown are interacting susceptibilities $\chi=\varepsilon^{-1}\chi_0$ (dotted line RPA, circles TDLDA).

Typical features include a peak with derivative discontinuity at $z = z'$, and a combination of weakly decaying sinusoidal (generalized Friedel) oscillations as z moves into the metal, for fixed z'. From the graph, it will be seen that these oscillations in the quantities χ_0 can be a much larger fraction of the maximum value than are the static density oscillations occurring in the groundstate calculation (see Fig 1).

In the next two Sections we will show analytically that (32) contains a complicated array of Friedel phenomena when one or both of z,z' lie deep in the metal.

V. FRIEDEL ASYMPTOTICS OF THE BARE SURFACE RESPONSE WHEN ONLY ONE OF Z,Z' LIES DEEP IN THE METAL

We will show that a disturbance of planar geometry located at z' near the surface of a metal can lead to a bare response $\delta n(z) = \chi_0(q_\|,z,z',\omega)$ containing Friedel oscillations when z lies deep in the metal, with up to six distinct wavenumbers to leading order. Four of these wavenumbers are the same as for the oscillations we have already discussed for the uniform gas ("bulklike" Friedel oscillations, see equ.(16)), and two are "photoemission" Friedel oscillations occurring because of the presence of the surface. The amplitudes fall off slowly (z^{-2} for these planar disturbances compared with r^{-3} for a point source) and depend on the phase-shifts $\gamma(k) = kZ_k$ of the eigenfunctions of the surface problem.

Equation (32) gives χ_0 as an integral over k_x and k ($k \equiv k_z$) inside the Fermi circle $k_x^2 + k^2 = k_F^2$. We consider the case $z << -\lambda_F$, $z - z' << -\lambda_F$. Thus z lies deep in the metal, and we allow z' to lie anywhere provided that it is substantially less deep than z, so that we are in a Friedel asymptotic region. We will obtain the first term in an asymptotic expansion in the large variable z. For this case $z_< = z$ and the wavefunctions $\psi^+(k,z)$ and $\psi^+(p_\pm,z_<)$ in (32) can be replaced by their asymptotic forms (35) and (36).

Then, expressing $\psi^+(z)$ from (35) as a sum of two complex exponentials, we find that (32) consists of a sum of four 2D phase integrals (c.f. (21)), all of which have the form

$$\chi_0^{\alpha,\beta}(q_\parallel, z, z', \omega+i0)$$
$$= \int_{-k_r}^{k_r} dk \int_{-k_o}^{k_o} dk_x \, S_{\alpha,\beta}(k_x, k, q_\parallel, \omega, z') \sqrt{k_F^2 - k_x^2 - k^2} \, \exp[iq_{\alpha,\beta}(k_x,k)z], \tag{37}$$

$$S_{\alpha,\beta}(k_x, k, q_\parallel, \omega, z') = \frac{2m}{\pi^2 \hbar^3} \psi^+(k, z') \frac{\psi^+(p_\beta, z')}{p_\beta} \exp[-i(\alpha\gamma(k) + \beta\gamma(p_\beta))],$$

$$p_\beta = \sqrt{2m\beta\omega\hbar^{-1} + 2k_x q_\parallel - q_\parallel^2 + k^2} \tag{38}$$

$$q_{\alpha\beta}(k_x, k) = \alpha k + \beta p_\beta = \alpha k + \beta \sqrt{2m\beta\omega/\hbar + 2k_x q_\parallel - q_\parallel^2 + k^2}, \quad (\alpha = \pm 1, \beta = \pm 1) \tag{39}$$

and $k^0 = (k_F^2 - k^2)^{1/2}$. In (38) and (39), α and β independently take the values ± 1 corresponding to four contributions to the integral (32). In the parameter ranges leading to Friedel oscillations, $S_{\alpha\beta}(k_x,k,q_\parallel,\omega,z')$ is a smooth function of k and k_x except for a derivative discontinuity at the photoemission threshold to be discussed below. S vanishes at $k = 0$ and is continuous at the points where $p_\beta = 0$ (note that $\psi^+(p,z)$ vanishes as $p \to 0$).

Friedel behaviour of χ_0 for $q_\parallel = 0$, $z \to -\infty$, z' finite

We consider first the special case $q_\parallel = 0$. Then both p_α and $q_{\alpha,\beta}$ are independent of k_x, so that the k_x integration is trivial and (37) becomes

$$\chi_0^{\alpha,\beta}(q_\parallel=0, z, z', \omega+i0)$$
$$= \int_{-k_r}^{k_r} dk \, S_{\alpha,\beta}(k_x=0, k, q_\parallel=0, \omega, z') \pi(k_F^2 - k^2) \exp[iq_{\alpha,\beta}(k_x=0,k)z] \tag{40}$$

The phase function in this case is $q_{\alpha\beta} = \alpha k + \beta(2\beta m\omega/\hbar + k^2)^{1/2}$ with $-k_F < k < k_F$. Where this is imaginary there is no oscillatory integrand and no Friedel contribution (e.g. there is none when $\hbar\omega > E_F$). Where $q_{\alpha\beta}$ is real, $dq_{\alpha\beta}/dk = \alpha + \beta k p_\beta^{-1}$ never vanishes so there is no stationary phase point which could dominate. Thus the integral is dominated by any singularities of the weight function S and by the endpoints at $k = \pm k_F$. (See the discussion of phase integrals following equation (21)). To obtain the endpoint contributions we change variable from k to q ($\equiv q_{\alpha\beta}$) and perform integration by parts twice, thus introducing a factor z^{-2} from integrating $\exp(iqz)$. (Two integrations are required to obtain a nonzero asymptotic result because the factor $(k_F^2 - k^2)$ vanishes at the endpoints, amounting to a derivative discontinuity at $\pm k_F$.) We then obtain from (40)

$$\chi_0^{\alpha,\beta}(z, z', q_\parallel=0, \omega) \sim a_{\alpha,\beta} z^{-2} e^{iq_{\alpha,\beta}(k_x=0, k=k_F)z} + b_{\alpha,\beta} z^{-2} e^{iq_{\alpha\beta}(k_x=0, k=-k_F)z} \tag{41}$$

Note that $q_{\alpha,\beta}(k_x=0,-k_F) = q_{-\alpha,\beta}(k_x=0,+k_F)$ so that the $b_{\alpha,\beta}$ terms in (41) merely duplicate the $a_{\alpha,\beta}$ terms in the summation over α and β. Thus the bare susceptibility, which is a sum of the contributions (41) for all four possible values of the pair α,β, is given asymptotically as $z \to -\infty$ by

$$\chi_0^{Bulklike}(q_\parallel=0,z,z',\omega) \underset{z\to\infty}{\sim} \frac{1}{z^2}\sum_{\alpha,\beta} C_{\alpha,\beta} \exp[i(\alpha k_F + \beta\sqrt{k_F^2 + 2\beta m\omega\hbar^{-1}})z]$$

$$= \frac{1}{z^2}\sum_\mu C_\mu \exp(ik_\mu z) \qquad (42)$$

where α and β independently take the values ± 1. The coefficients $C_{\alpha\beta} \equiv C_\mu$ may depend on ω and z' but are independent of z. The Friedel wavenumbers k_μ appearing in (47) are just the wavenumbers $k_1..k_4$ which arose in our previous analysis of a localised finite-frequency disturbance in a uniform electron gas (see equation (16)).

We now turn to the possibility that the weight function S in (38) has a singularity at some point k in the integration range $[-k_F, k_F]$. This does occur for the case $V(\infty)-V(-\infty) > \hbar\omega > V(\infty)-V(-\infty) - E_F$, corresponding to the possibility of photoemission at driving frequency ω. Specifically[31-33] the right eigenfunction ψ^+ has a sharply varying derivative at a p value such that

$$\frac{\hbar^2 p^2}{2m} = \Delta V \equiv V_{eff}^0(z\to\infty) - V_{eff}^0(z\to-\infty) \qquad (43)$$

so that the particle is barely excited over the barrier into the vacuum, a process that has no analogue in the uniform electron gas (see vertical arrows in Fig. 2). This amounts to a derivative discontinuity in the prefactor S, similar to the behaviour at the endpoints discussed in the previous paragraph. Thus a power z^{-2} is expected from integrating by parts. From (43) and (38) with $q_\parallel=0$, this occurs at $k = \pm K_{photo} = \pm\hbar^{-1}[2m(\Delta V - \hbar\omega)]^{1/2}$. Thus the onset phenomenon associated with photoemission can give rise to two further Friedel terms in χ_0 as $z\to-\infty$:

$$\chi_0^{photo}(z,z') \sim \frac{1}{z^2}\sum_{\mu=5}^{6} C_\mu(z') e^{ik_\mu z} \qquad (44)$$

$$k_{5(6)} = -\sqrt{2m\Delta V/\hbar^2} \pm \sqrt{2m(\Delta V - \hbar\omega)/\hbar^2}$$

Note that the $\beta=-1$ term in the general contribution (37) to χ_0 does not lead to a photoemission process. The asymptotic Friedel behaviour of the bare susceptibility deep in the metal can now be written

$$\chi_0(q_\parallel=0,z,z',\omega) \underset{z\to\infty}{\sim} \frac{1}{z^2}\sum_{\mu=1}^{6} C_\mu(z',\omega) e^{ik_\mu z} \qquad (45)$$

where $k_1..k_4$ are the bulk dynamic Friedel oscillation wavenumbers from (16) and k_5, k_6 are the surface photoemission Friedel wavenumbers from (44). Note that C_3 and C_4 are zero if $\hbar\omega > E_F$, and C_5 and C_6 are zero unless $\Delta V - E_F < \hbar\omega < \Delta V$.

Friedel Behaviour of χ_0 for $q_\parallel \neq 0$, $z \to -\infty$, z' finite

The case of a jellium halfspace under a perturbation with a nonzero surface-parallel wavevector, $q_\parallel \ne 0$, is more involved. We first consider the inner integral over k_x in (37). For nonzero $q_\parallel$ at the large negative z values we are considering, the rapidly varying exp(iqz) factor makes this a phase integral of the type already discussed following equation (21). At large negative z values, the dominant contributions come from the endpoints $k_x = \pm k_0$, and in some cases from a derivative discontinuity in S due to the onset of photoemission. (There is no stationary phase point for finite $q_\parallel$ because, from (39), $\partial q/\partial k_x = \beta q_\parallel/p \ne 0$ where $\beta = \pm 1$.) The endpoint region $k_x \approx k_0$ of the integral can be evaluated by replacing all slowly varying factors by their value at $k_x = k_0$. This leaves an integral, corresponding to the $+k_F$ endpoint of (48), of form

$$S(k_0, k, q_\parallel, \omega, z') (2k_0)^{1/2} \int_{...}^{k_0} e^{iz[q(k_0, k) + \frac{\partial q}{\partial k_x}|_{k_0}(k_x - k_0) \cdots]} (k_0 - k_x)^{1/2} dk_x \quad (46)$$

By changing variable to $u = z[\partial q/\partial k_x|_{k_0}(k_0 - k_x)]^{1/2}$ and then bending the contour we reduce this integral to a gaussian integral of the form $\exp(-u^2) u^2 du$. Together with dimensional factors from the change of variable this reduces (46) to

$$S(k_0, k, q_\parallel, \omega, z') (2k_0)^{1/2} (\beta q_\parallel |z|/p_0)^{-3/2} e^{iq(k_0, k)z} \quad (47)$$

Still considering the inner k_x integration in (37), we find once more a derivative discontinuity in ψ^+ (and hence in S) due to "surface photoemission", similar to that discussed above for $q_\parallel = 0$. This discontinuity can occur where $p_\beta^2 = 2m\Delta V/\hbar^2$ i.e. for $k_x = (2q_\parallel)^{-1}[q_\parallel^2 - k^2 + 2m(\Delta V + \beta\hbar\omega)/\hbar^2]$. A little algebra shows that this k_x value lies within the integration range $-(k_F^2 - k^2)^{1/2} < k_x < (k_F^2 - k^2)^{1/2}$ provided

$$|k^2 - k_{phot}^2| < 2q_\parallel \sqrt{k_F^2 - 2m\hbar^{-2}(\Delta V - \hbar\omega)} \quad ,$$
$$k_{phot}^2 = K_{phot}^2 - q_\parallel^2 = 2m\hbar^{-2}(\Delta V - \hbar\omega) - q_\parallel^2 \quad (48)$$

For k in this range the k_x integral has a surface photoemission contribution

$$A_{photo}(z') z^{-2} \exp[-i(P_{photo} \pm k) z]$$
$$P_{photo} = \sqrt{2m\hbar^{-2}\Delta V} \quad , \quad (49)$$

in addition to the "endpoint" term (47).

It remains to integrate (49) and (47) with respect to k between k_F and $-k_F$, to give the contribution $\psi^{0\alpha,\beta}$ to the bare susceptibility as per equation (37). This is again a phase integral. Then because of the $k_0^{1/2}$ factor, the endpoint contributions are similar to those of the inner k_x integral, and so would yield an oscillatory term with a further $z^{-3/2}$ prefactor, making an overall power of z^{-3} for the bulklike terms. These are not the dominant dependences, however, as we will see in the following two paragraphs.

In the bulklike (endpoint) contribution (47), for some ranges of $q_\parallel$ and ω, the phase factor $q(k_0,k) \equiv q([k_F^2 - k^2]^{1/2}, k)$ has a stationary point which dominates the integral. After considerable algebra based on equation (39) plus the definition $k_0^2 = k_F^2 - k^2$, one finds that this stationary point (where $dq/dk = 0$) occurs at a k value, k_{00}, such that

$$Q(\alpha, \beta) = \sqrt{k(\alpha, \beta)^2 - q_\parallel^2} \equiv Q_\mu \quad (50)$$

where $k_{(\alpha,\beta)} = k_\mu$ ($\mu = 1,...,4$) is the same as for the case $q_\parallel = 0$ (see equation (16)). The result of the k integration in the limit $z \to -\infty$ can be found by Taylor-expanding the phase factor q to second order around its stationary point at $k = k_{00}$ and changing to the dimensionless variable u where $u^2 = |z|(d^2q/dk^2)(k-k_{00})^2$. Then one bends the contour so that it passes through the stationary point on a path of steepest descent, leaving a Gaussian integral of form $\exp(-u^2)$ du. The change of variable yields a dimensional prefactor $(d^2q/dk^2 |z|)^{-1/2}$, which multiplies the power $|z|^{-3/2}$ already found from the inner (k_x) integration (equ (47)). Thus the overall decay of the bulklike Friedel oscillations occurs with a power z^{-2}, just as for the case $q_\parallel = 0$:

$$\chi_0(q_\parallel, z, z', \omega) \underset{z \to \infty}{\sim} \frac{1}{z^2} \sum_\mu C_\mu e^{iQ_\mu z} \qquad (51)$$

where Q_μ is the wavenumber from equ (50).

For the surface photoemission term (49), the situation is not entirely clearcut when one performs the outer k integration. The phase factor (49) has no point of stationary phase in the limited range of k integration (48) applicable to this term. Thus, for substantial finite $q_\parallel$, the integral is reduced by phase averaging at large z, and so is dominated by the endpoints implied by (48). Integration by parts will then yield an negative power of z, multiplying the power z^{-2} already mandated by phase averaging in the k_x integration as in (49). Thus the photoemission Friedel term has a faster decay into the bulk than does the bulklike Friedel term (51). For very small $q_\parallel$ one will have to go into the bulk a distance on the order of $q_\parallel^{-1}$ in order for this to occur, however. Thus for small $q_\parallel$ it may be necessary to allow additional z^{-2} photoemission terms in (50), with wavenumbers $-p_{photo} \pm k_{photo}$ as in (48) and (49).

Summary of Friedel asymptotics of χ_0 for z deep in metal and z' finite

We can now summarize the asymptotics of χ_0 when z, but not z', lies deep in the metal. For general $q_\parallel$ the asymptotics are given by an equation of form (51). Each Friedel oscillation term is absent for $q_\parallel > k_\mu$ (see equs (50) and (16)) and two of the four are absent for $\hbar\omega > E_F$. In the special case $\omega \to 0$, $q_\parallel = 0$, two of the Friedel wavenumbers go to zero (with cancelling amplitudes) and two go to $\pm 2k_F$, corresponding to the well-known static Friedel period. For $\omega = 0$ and $0 < q_\parallel < 2k_F$, the Friedel wavenumber is $\pm(4k_F^2 - q_\parallel^2)^{1/2}$. For $q_\parallel$ near to zero and ω in the photoemission range $\Delta V > \hbar\omega > \Delta V - E_F$ one needs to add two further z^{-2} Friedel terms of the "photoemission" type as discussed in the paragraphs immediately above: thus there are 6 Friedel terms in this case, with the photoemission wavenumbers Q_5 and Q_6 given by (44) and (50).

Note also that the wavefunctions $\psi(p,z)$ do have $O(z^{-2})$ corrections deep in the metal, due to the $2k_F$ oscillations in the static potential. In the present case where z lies much deeper than z', these contribute terms in χ_0 of higher order than z^{-2}.

VI. FRIEDEL ASYMPTOTICS OF χ_0 WHEN BOTH z AND z' ARE DEEP IN THE METAL

When both z and z' are large and negative the asymptotic forms (35) and (36) can be used for all the wavefunctions ψ in (32), to lowest order. We will write the expression χ_0 using only the oscillatory forms of wavefunction (the upper cases in equs. (35), (36)), as these are what lead to Friedel oscillations:

$$\chi_0(q_\parallel, z, z', \omega+i0)$$
$$\approx 2m\pi^{-3}\hbar^{-2}\int dk \int dk_x \sqrt{k_F^2-k^2-k_x^2} \sin(kz-\gamma(k))\sin(kz'-\gamma(k)) \qquad (52)$$
$$\times [\{|p_+|^{-1}\sin(p_+z_>-\gamma(p_+))e^{-i(p_+z_<-\gamma(p_+))}\} + \{p_+\to p_-\}^*]\ .$$

Using product-to-sum trigonometric identities in (52) we can split the susceptibility into three terms:

(a) a "bulk" contribution $\chi_0^{bulk}(q_\parallel, |z-z'|, \omega)$ which depends only on separation and is the bare Lindhard response of the uniform electron gas, Fourier transformed to real space in the z direction;

(b) a term corresponding to "reflection" off the surface barrier, which is asymptotically of form

$$\chi_0^{refl} \sim \sum_\mu F_\mu \frac{\exp(i[Q_\mu(z+z')-\phi])}{(z+z')^2} \qquad (53)$$

where the $\{Q_\mu : \mu=1,...,6\}$ are the Friedel wavenumbers from (50), (44) and (16), and the phase shift Φ is independent of z and z'; and

(c) an "interference" term χ_0^{int} depending both on $|z-z'|$ and $(z+z')$ and falling off as $(z+z')^{-2}$. It has a derivative discontinuity at $z=z'$, which adds to that of χ_0^{bulk}. In addition to the Friedel wavenumbers $\{Q_\mu\}$ this term has a $2k_F$ oscillation which causes values of χ_0 near to the peak at $z=z'$ to have oscillations mimicking those of the groundstate density $n_0(z)$, and therefore falling off as z^{-2}. Similar effects also come from the first-order Friedel corrections to the wavefunctions.

These effects can be summarised for present purposes by

$$\chi_0 \underset{z,z'\to\infty}{\sim} \chi_0^{bulk}(q_\parallel, |z-z'|, \omega) + O[(z^{-2}, (z')^{-2})]\ . \qquad (54)$$

VII. FRIEDEL TERMS ARISING IN THE SCREENING OF THE BARE SURFACE RESPONSE

When an external field of the form (23) is applied to a metal surface, from experience with the uniform electron gas (see the discussion following equ. (17)), we expect that the Friedel oscillations in χ_0 will not be fully screened and will appear also in the selfconsistent density perturbation δn. This is a delicate matter and several papers have considered the subject[27,29-34], resulting in incomplete agreement. Here we attempt to analyze this as far as possible in a simple fashion. We denote $V_c * \chi_0$ and $F_{xc} * \chi_0$ by P_c and P_{xc} respectively, and write out the screening equation (10), (11) in more detail, assuming a perturbed density solution of form $\delta n(z)\exp(iq_\parallel \cdot r - \omega t)$:

$$\delta n(z) = \delta n_0(z) + \int P(z,z')\delta n(z')dz' \qquad (55)$$

where $P = P_c + P_{xc}$, and the Coulomb (Hartree) and exchange-correlation kernels are

$$P_c(z,z') = \frac{2\pi e^2}{q_\parallel}\int \chi_0(z,z'')e^{-q_\parallel |z''-z'|}dz'' \qquad (56)$$

$$P_{xc}(z,z') = \chi_0(z,z')\mu_{xc}(n_0(z'))\ .$$

The source term in (55) is

$$\delta n_0(z) = \int \chi_0(z,z') \exp(q_\| z') \, dz' \tag{57}$$

$$\underset{z\to-\infty}{\sim} \exp(q_\| z)[A+Bz^{-2}\cos(2k_F z-\phi_0)] + z^{-2}\sum_\mu D_\mu \exp(iQ_\mu z)$$

where the $\exp(q_\| z)$ term comes from the peak (derivative discontinuity) in χ_0 and includes a $2k_F$ Friedel ripple as discussed under (c) above: this term is only important for very small $q_\|$. The remaining terms in (57) come, via (51), from values of z' near the surface and are important for all $q_\|$ and ω such that the respective Friedel terms involving Q_μ appear in χ_0.

The deep-metal behaviour of $P_{xc}(z,z')$ is essentially the same as that of χ_0: there is a peak with derivative discontinuity at $z = z$,' at which the amplitude is almost constant but has an $\exp(i2k_F z)/z^2$ ripple as $z \to -\infty$; while for $z' \to -\infty$ with fixed z (or vice versa), P_{xc} is dominated by a sum of Friedel terms of the same form as equ. (51). P_c also has a similar, Friedel-rippled derivative discontinuity at $z = z'$. To establish the behaviour of P_c as $z' \to -\infty$ for fixed z, we note from (56) that it obeys $(\partial^2/\partial z'^2 - q_\|^2)P = \chi_0(z,z')$, and hence from (51)

$$P_c(z,z') \underset{z'\to-\infty}{\sim} -\sum_\mu \frac{C_\mu(z)}{Q_\mu^2 + q_\|^2} \frac{\exp[iQ_\mu z']}{z'^2} \tag{58}$$

where $Q_{\mu 2} + q_\mu^2 = k_\mu^2$ from (50). Thus both P_c and P_{xc} have the same simple form for z' deep in the metal. (For the case $\omega = 0$ where $\int \chi_0 dz' \neq 0$, for small $q_\|$ we also need to include an $\exp(q_\| z)$ term in P_c).

We now make use of (55) to see whether the dynamic density perturbation $\delta n(z)$ can be adequately represented for $z \to -\infty$ by the same Friedel form (51) as sufficed for χ_0 at fixed z' (with $C_\mu(z')$ replaced by a constant H_μ). We assume that, as $z \to -\infty$, $\delta n(z)$ has at least those Friedel terms present in δn_0, so that a $2k_F$ Friedel term and an $\exp(q_\| z)$ term must be added to the Q_μ terms, for $q_\| \approx 0$. It remains to investigate whether the integral (call it I) on the right-hand side of (55) can introduce any further asymptotic terms into $\delta n(z)$. In view of the qualitative behaviour discussed above for P_c and P_{xc}, for large negative z deep in the metal the z' integration on (55) can be broken into three parts:

(i) the contribution from the "main" density response $\delta n(z')$ which occurs in a region lying near to the surface $z'=0$ but distant at least (say) $z/2$ from the peak in P at $z'=z$:

$$I^{surf}(z) = \int_{z/2}^{\infty} P(z,z') \delta n(z') \, dz'$$

$$\approx \sum_\mu \frac{\exp(iQ_\mu z)}{z^2} \int_{z/2}^{\infty} H_\mu(z') \delta n(z') \, dz' \tag{59}$$

assuming only that H is bounded, we note that the integral on the second line of (59) is convergent as $z \to -\infty$ and so I^{surf} is a sum of Friedel terms involving Q_μ, as for χ_0.

(ii) the next contribution to I is due to the peak and derivative discontinuity at $z' = z$ where χ_0 is not small;

$$I^{peak}(z) = \int_{3z/2}^{z/2} P(z,z') \delta n(z') dz'$$

$$\approx \int_{3z/2}^{z/2} [P^{bulk}(|z-z'|) + O(\frac{\exp(i2k'_{Fz})}{z^2})] \sum_\mu \frac{c_\mu \exp(iQ_u z')}{(z')^2} dz'$$

$$\approx \sum_\mu T_\mu \frac{\exp(iQ_\mu z)}{z^2} + O(z^{-3}),$$

$$T_\mu = c_\mu \int_{z/2}^{-z/2} P_{bulk}(|u|) \exp(iQ_\mu u) du \to \text{constant as } z \to -\infty.$$

(60)

This contribution merely reproduces any $O(z^{-2})$ Friedel terms already present in $\delta n(z)$.

(iii) the final contribution is the integral over $-\infty < z' < 3z/2$ where both $P(z,z')$ and δn are of $O[(z')^{-2}]$ and so the integral is of higher order than z^{-2}.

The above highly simplified treatment suggests that no extra oscillatory $O(z^{-2})$ terms are introduced by the screening kernel $P \equiv 1-\varepsilon$, beyond those already found to occur in the bare response or source term $\delta n_0 = \chi_0 * \delta V_{ext}$ (though no conclusions emerge about terms decaying more rapidly than z^{-2}). This is not in accord with the work of Feibelman[27], who investigated a similar integral to I(z) for the case $q_\parallel \approx 0$. (M_{00} in Ref [27] is essentially $\int \chi_0(z,z') \delta n(z') dz'$ with the z axis reversed from our notation). His conclusion, in present notation, is that integration of a Friedel function $\delta n(z)$ behaving asymptotically as $\exp(iqz)/z^2$ could lead to new Friedel forms $\exp(iq'z)/z^2$ with $q' \neq q$. Our conclusion, from the argument above as well as others not given here, is that any such terms appear to be of higher order in inverse powers of z. The new Friedel wavenumbers introduced into the interacting δn in Feibelman's analysis are $\pm 2k_F$ and $\pm m\omega/(2k_F)$, and these were found to beat with the $\{k_\mu:\mu=1,...,4\}$ to produce further Friedel terms. If the above wavevector-changing effect is valid then Feibelman's analysis will produce further wavenumbers q' from the interaction of the screening kernel with the "photoemission" Friedel wavenumber k_5, k_6: Feibelman did not consider k_5 and k_6 in his analysis. Schaich and co-workers[31,34] have also obtained such wavenumber-changing behaviour in a k-space analysis but found[35] the $\omega/2k_F$ terms to be unimportant for jellium sufaces. Song et al[33] reported experimental second harmonic generation results as a function of admetal layer thickness, showing very pronounced oscillations which were attributed to the $\omega/2k_F$ Friedel oscillations put forward by Feibelman. At about the same time, however, several workers[29-31] commented that photoemission Friedel oscillations (see equ. (44)) could produce spatial periods similar to $\omega/2k_F$. At the present, therefore, the author believes that the issue of the spectrum and strength of Friedel oscillations occuring at low $q_\parallel$ in jellium surface problems, as well as their physical origin, remains a somewhat cloudy one. At finite $q_\parallel$ the author's analysis (not given here) suggests that, if present to $O(z^{-2})$ at $q_\parallel=0$, the $\omega/2k_F$ oscillations, like the photoemission Friedel terms considered above, will be removed by phase-averaging, leaving only the set Q_μ to $O(z^{-2})$, as in equations (50).

VIII. NUMERICAL TECHNIQUES FOR TREATING FRIEDEL OSCILLATIONS

The solution of the screening equations ((55), (11) or equivalent) for planar jellium problems is important in a variety of physical situations. Examples include metal-vacuum interfaces[24,26,27,28,32], metal-metal interfaces[36] and metal-vacuum-metal[37] situations.

Similar formalism is relevant to wide semiconductor quantum wells[13]. The current state of the art (for real-space methods of solution) was largely pioneered by Feibelman[26,27], and furthered by Liebsch[32]. Methods based on k space have also been developed[29-31,34,38], but here we will only discuss real-space methods.

As will be apparent from the preceding sections, when solving the screening equations for $\delta n(z)$, one must allow for slowly dying oscillations not only in the given functions χ_0 and P, but also in the unknown solution δn. This leads to numerical difficulties which are the subject of the remainder of this paper. These difficulties are especially acute in cases such as the calculation of the Feibelman d-coefficients[26] which determine the now-famous negative low-$q_\|$ dispersion[15] of surface plasmons on simple metals:

$$\omega = \frac{\omega_p}{\sqrt{2}} [1 - d(\omega_p/\sqrt{2}) q_\| + O(q_\|^2)] \quad , \quad (61)$$

$$d(\omega) = [\int_{-\infty}^{\infty} z \delta n(z) dz] / [\int_{-\infty}^{\infty} \delta n(z) dz]$$

Here $\delta n(z)$ is the charge density perturbation resulting from a spatially uniform, temporally oscillating $q_\|=0$ electric field perpendicular to a metal surface. Clearly, the dipole moment spatial integral in (61) is only slowly convergent at the deep-metal limit because of Friedel terms $\exp(iQz)/z^2$ occurring in $\delta n(z)$. Liebsch[32] has used a dynamic sum-rule of Sorbello[39] for $q_\|=0$, to avoid the deep-metal parts of the integration in the special case of a jellium halfspace. One would like, however, to be able to handle more general geometries such as sandwiches, and also to obtain reliable results at $q_\| \neq 0$. Therefore we discuss here methods which do not rely on special sum rules.

In these methods, the groundstate, left and right eigenfunctions ψ are obtained on a discrete set of z points z_i (i = 1,....,N) by numerical integration of the one-dimensional Schrodinger equation (see Figure 4).

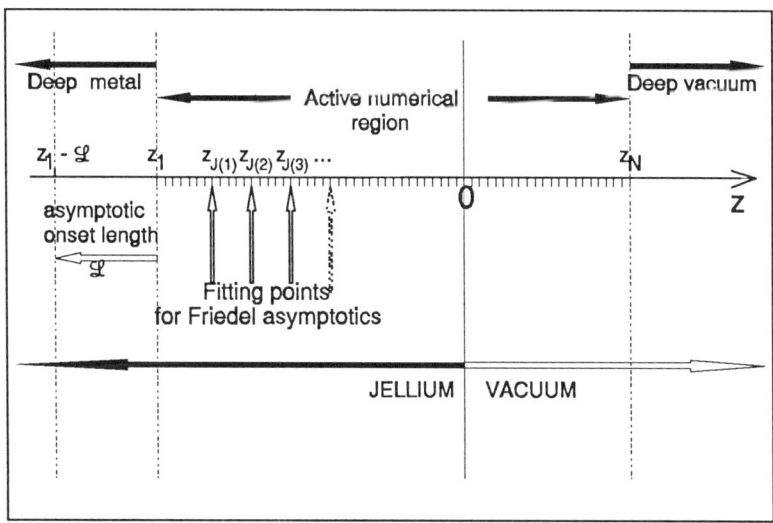

Figure 4. Coordinates for numerical solution of the screening problem

The following procedures are carried out at each desired fixed pair of values of $q_\parallel$ and ω: for simplicity these two quantities will be suppressed in the notation. We now demonstrate a number of specific numerical problem and discuss ways to treat them.

Calculating ε

The bare susceptibilty $\chi_0(z=z_i, z'=z_j)$ is calculated numerically from (32) on the same set of z_i values used for the wavefunctions. The first step in which the Friedel oscillations are a problem is the calculation of the effective nonlocal screening kernel P = I-ε where ε is the generalised dielectric function (see equs. (11) and (56)). The z'' integration in (56) must be carried out numerically by transforming it approximately into a summation over the discrete set of points z_k in the "active" region (see Figure 4), using a set of weights w_k. The weights need to account for the derivative discontinuity in χ_0 at z'' = z, and in the Coulomb potential at z''=z': this is most simply arranged by using one of the standard Simpson methods, and splitting into two regions of numerical integration joined at z and z': this method is quite satisfactory at the relatively low $q_\parallel$ values for which Friedel oscillations are still present (see (50)). Since χ_0 dies off slowly but has been computed only on a finite set of z values, we need either to make this set proceed very far into the metal (i.e., make N extremely large), or more sensibly to use the asymptotic form (51) to extend the integration over z'' = z_k deep into the metal. The simplest effective way to do this varies according to the case being treated.

Calculating P = I-ε; The Case $\omega = 0$ (single Friedel wavenumber)

This case does not present serious problems. The integral (56) over z'' yielding P=1-ε can be split into three regions of z'' as in Fig 4, giving P = P_{dv} + P_{num} + P_{dm}. Here P_{dv} represents the z'' integration over the deep vacuum region $z > z_N$ and can be neglected because χ_0 is exponentially small there. P_{num} is an explicit sum over the "active" region, which we retain. P_{dm} is the "deep metal" part of the integration, and for z not too deep in the metal is given by

$$P_{dm}(z,z') = \int_{-\infty}^{z_1} \frac{2\pi e^2 e^{q_\parallel(z''-z')}}{q_\parallel} \left\{ \frac{A(z) e^{iQz} + B(z) e^{-iQz}}{z''^2} \right\} dz''$$

$$= \frac{2\pi e^2 e^{q_\parallel(z_1-z')}}{q_\parallel z_1^2} \left[\frac{A(z)}{q_\parallel + iQ} e^{iQz_1} + \frac{B(z)}{q_\parallel - iQ} e^{-iQz_1} \right] + O(z^{-3}) \quad (62)$$

$$\sim -\frac{2\pi e^2 e^{q_\parallel(2z_1-z')}}{q_\parallel (q_\parallel^2 + Q^2)} \frac{d}{dz_1}\left[e^{-q_\parallel z_1}\left\{ \frac{A(z) e^{iQz_1} + B(z) e^{-iQz_1}}{z_1^2} \right\} \right] .$$

where $Q = (4k_F^2 - q_\parallel^2)^{1/2}$ is the single Friedel wavenumber from (50). In (62), the expression in braces is the Friedel asymptotic form of χ_0. The second line follows from an integration by parts, while the third line is verified, up to terms of $O(z^{-2})$, by direct differentiation. Thus to order z_1^{-2} we have shown

$$P_{dm}(z,z') = \frac{e^{q_\parallel(2z_1-z')}}{4k_F^2 q_\parallel} \frac{d}{dz_1}\left[e^{-q_\parallel z_1} \chi_0(z, z_1) \right] \quad (63)$$

where the relation $Q^2 + q_\parallel^2 = 4k_F^2$ has been used (see (50)). The derivative indicated in

(63) can be done numerically using the last few values values of χ_0 to the right of z_1, so that P on the active region is expressed entirely in terms of the N^2 tabulated values of χ_0 in the active region. The exception is that equation (63) is not justified if z lies deep in the metal (too close to z_1, so that the asymptotic form of χ_0 used in (62) is invalid). In this case, however, z will be deep enough that χ_0 is approximately bulk-like and so translationally invariant, and tabulated values from the active region nearer the surface may be used to integrate into the deep metal region of the variable z" until the Friedel-asymptotic region is reached. This is justified to leading order in z_1.

Calculating P = I-ε; the Case ω ≠ 0, Several Friedel Wavenumbers

The above method relied on the fact that for a function with only one Friedel wavenumber, a deep-metal integration over the function is approximately equal to the derivative divided by a wavenumber squared. This simple relation no longer holds with more than one Friedel wavenumber is present. In the case where there are M different real Friedel wavenumbers Q, since the wavenumbers themselves are given in terms of $q_\parallel$ and ω by equations (50), (44) and (16), one could in principle use 2M values of χ_0 near the edge to pin down the unknown amplitude and phase of the Friedel oscillations in χ_0. Then the method outlined above (see middle line of (62)) could then be used to extend the z" integration into the deep-metal region to obtain P. In practice this numerical fit to the Friedel oscillations uses only the minimum amount of data and is likely to be susceptible to numerical noise. We thefore outline a formalism adapted from ref. 27 which is quite flexible, involving highly computable linear algebra, and promises to be more stable. In the well-known least squares method one is given a set of functions $f_a(z)$, a= 1,...,A and a set of B data points (x_b, y_b), b = 1,...,B. The method seeks the "best-in-mean-square" coefficients c_a, a = 1,..,A in a function $F(x) = c_1 f_1(x) + ... + c_A f_A(x)$ designed to fit the given data points; i.e. it seeks the $\{c_a\}$ which minimize the sum $[F(x_1)-y_1]^2 + ... + [F(x_B)-y_B]^2$. The solution to this elementary calculus problem can be written

$$c_a = \sum_{b=1}^{B} E_{ab} y_b , \quad i.e. \quad \mathbf{a} = \mathbf{E}\ \mathbf{y} \tag{64}$$

where

$$E_{ab} = \sum_{a'=1}^{A} (\mathbf{M}^{-1})_{aa'} f_{a'}(x_b)$$
$$M_{aa'} = \sum_{b=1}^{B} f_a(x_b) f_{a'}(x_b) \tag{65}$$

We consider the case of fixed ω, $q_\parallel$ and z', and apply the above least-squares method to the case that the "data" are the values of the bare susceptibility at a subset of J tabulation points $x_b = z_{j(b)}$ lying deep in the metal: $y_b = \chi_0(z=x_b, z')$, b = 1,...,B. The expansion functions are $f_b(z) = z^{-2}\exp(iP_b z)$ where the $\{P_b\}$ are the 2, 4 or 6 generalized Friedel wavenumbers Q_μ given by equs. (50), (44) and (16). The integral $\int_{-\infty}^{z_1} \exp(iq_\parallel[z'-z''])f_b(z'')\,dz''$ can be done asymptotically as in the middle line of equation (62). Thus

$$P_{dm}(z,z') \sim \frac{1}{z_1^2} \sum_{a,b} E_{ab} \frac{2\pi e^{q_\parallel(z_1-z')}}{q_\parallel(q_\parallel + iP_a)} \chi(z,x_b) \tag{66}$$

Thus once more the contribution to P from the deep-metal integration has been expressed as a linear function of the tabulated values of χ_0 in the active region. Note that the coefficient matrix E depends only on the known Friedel wavenumbers $Q_{\alpha\beta}$ and on the chosen fitting points $x_{J(b)}$, and is independent of the values of χ.

Solving the Screening Equation

The selfconsistent screening equation (11) is a linear integral equation for the density perturbation δn, which can be written as

$$\delta n(z) - \int_{-\infty}^{\infty} P(z,z') \delta n(z') \, dz' = \delta n_0(z) \tag{67}$$

where the terms are given in (56) and (57). The aim is to convert (67) to a finite matrix problem involving N values, δn_i, of the unknown function at tabulation points x_i restricted to the "active" region (see Fig 4). The portion of the integrals where z' lies in the "active" region can be approximated by discrete sums, provided that care is taken, as described above, with the derivative discontinuities of χ_0 and P at z' = z. Just as in the previous section concerning calculation of P, the problem is to express the integrals from $-\infty$ to z_1 using only values of χ_0, P and δn on the "active" region. Here there is the added problem that the function δn is unknown. For z well to the right of z_1 we can, to lowest order, ignore the deep-metal parts of the integrals because both δn and χ_0 (or P) are of $O(z^{-2})$. The dominant deep-metal correction arises when z is near to z_1 so that $P(z,z')$ and $\chi_0(z,z')$ are not small for values of z' to the near left of z_1. Thus values of these functions are required outside their Friedel asymptotic region but also for z' outside the active region where they are tabulated. This is handled either

(i) by using bulk values of P and χ_0 calculated by inverse Fourier transformation of their analytic uniform-gas k-space forms derived from equ. (13) ; or

(ii) by noting that, for z and z' both deep in the metal, P and χ_0 are appproximately translationally invariant (i.e. dependent on z-z'): thus, for values for z' near to but to the left of z_1, we may shift both z and z' back into the active region where these two functions are already tabulated.

For z' deep to the left of z_1, all of δn, χ_0 and P are of $O(z^{-2})$ so the contribution may be ignored.

The part of the integral over z' from $-\infty$ to z_1 is then significant if $z_i < z_1 + \mathcal{L}$ (see Fig. 4) and equals

$$\int_{-\infty}^{z_1} P(z_i, z') \delta n(z') \, dz' \sim \sum_b \delta n_{I(b)} \sum_a E_{ab} \int_{z_1 - \mathcal{L}}^{z_1} P(z_i, z') \frac{\exp(iQ_a z')}{z'^2} dz'$$

$$= \sum_j \Delta M_{ij} \delta n_j \tag{68}$$

where the coefficients ΔM_{ij} are independent of the unknown $\{\delta n_i\}$:

$$\Delta M_{ij} = \theta(\lambda - i) \sum_b \delta_{j, I(b)} \sum_{k=1}^{1+\lambda-i} w_{j'}(i) P(z_{1+\lambda}, z_k) \sum_{a=1}^{4} E_{ab} \frac{\exp[iQ_a(z_k - \mathcal{L} + z_1)]}{(z_k - \mathcal{L} + z_1)^2} \tag{69}$$

Here we chose the deep metal length $\mathcal{L}$ to be a multiple of the tabulation spacing, so there is an integer λ such that $z_\lambda - z_1 = \mathcal{L}$. The screening integral equation (67) can now be written

$$\sum_{j=1}^{N} M_{ij} \delta n_j = S_i, \quad i=1,\ldots,N \tag{70}$$

$$M_{ij} = \delta_{ij} - w_j(i)(P_{ij} + \chi_{0ij} f_{xc}(n_0(z_j))) + \Delta M_{ij} .$$

which clearly involves only the N unknown values of δn on the "active" region, yet it accounts for the extension of the problem deep into the metal to lowest negative power of the Friedel asymptotic length $\mathcal{L}$. It is hoped that, once the remaining questions about the correct spectrum of Friedel wavenumbers is clarified, this method will allow stable treatment of a variety of outstanding problems in planar geometry.

IX. SUMMARY

Friedel oscillations are spatial modulations of the electron density of a metal due to a localised disturbance, and occur at low temperatures as a consequence of the sharpness of the Fermi surface. For static point perturbations in a uniform electron gas they take the well-known form $\delta n(r) \sim \sin(2k_F r - \Phi)/r^3$. Experimental observation of static (or near-static) Friedel oscillations are provided quite directly by magnetic measurements in magnet-non-magnet layer systems; and less directly by Kohn anomalies in the phonon spectra of metals. In this paper we have considered generalisations due to finite frequency and to broken spatial symmetry as at a metal surface. Both of these generalisations introduce new physics. In a uniform bulk system at finite frequency the single Friedel wavenumber $2k_F$ is replaced by up to four wavenumbers given in equation (16). They can be understood as representing the onset wavenumbers for absorption or emission of energy $\hbar\omega$ by single-particle excitations with conservation of three-dimensional momentum. The presence of a planar surface introduces further effects; the $1/r^3$ falloff typical of spherical geometry is replaced by a $1/z^2$ decay, and two new Friedel wavenumbers are added for the case of a neutral metal adjacent to vacuum. These relate to the threshold for "photoemission" of single particles from the metal into unbound states in the vacuum. Further Friedel terms have been postulated for the dynamic surface problem, related to beat phenomena in the self-consistent screening equations. The present author believes their status is still uncertain. Experimental observations of second harmonic generation in metal adlayers as a function of layer thickness have produced clear evidence of spatial periods different from $2k_F$, but it seems unclear whether these are the "photoemission" or "beat" phenomena described above. The dynamic surface Friedel oscillations produce substantial difficulties in microscopic numerical predictions of dynamic electronic response in bounded metals, and a method is outlined here for overcoming these. New results described here relate to the effect on the Friedel phenomena of excitation at a finite surface-parallel wavenumber $q_\parallel$. As $q_\parallel$ is increased the photoemission oscillations (as well as the beat phenomena, if they exist) are expected to be blurred. The bulk-like oscillations remain sharp, however, up to a finite $q_\parallel$ value (see equation (50).)

ACKNOWLEDGEMENTS

I have benefited from discussions with G.H. Harris, W. Kohn, A. Liebsch and W.L.

Schaich. Part of this work was supported by an ARC Large Grant. I acknowledge the hospitality of the Physics Department and support from the Gordon Godfrey Bequest of the University of New South Wales, as well as the hospitality of the Physics Department and the Quantum Institute of the University of California at Santa Barbara, where some of this was written.

REFERENCES

1. J. Friedel, Phil. Mag. **43**, 153 (1952)
2. N.D. Lang and W. Kohn, Phys. Rev. B **1**, 4555 (1970): N.D. Lang, Adv. Sol. State Phys. **28**, 225 (1973)
3. W.A. Harrison, "Solid State Theory", McGraw Hill, New York 1970, p. 299
4. E. Bruno & B. L. Gyorffy PRL 71, 181 (1993): L.M. Roth, H.J. Zeigler and T.A. Kaplan, Phys. Rev. **149**, 519 (1966)
5. W. Kohn, Phys. Rev. Lett. **2**, 393 (1959)
6. J.M. Ziman, "Principles of the Theory of Solids", Cambridge University Press, Cambridge, 1972.
7. W.R. Bennet, W. Schwarzacher and W.F. Egelhoff, Phys. Rev. Lett. **65**, 3169 (1990)
8. J. Unguris, R.J. Celotta and D. T. Pierce , Phys. Rev. Lett. **67**, 140 (1991)
9. P. Bruno and C. Chappert, Phys. Rev. Lett. **67**, 1602 (1991): Phys. Rev. B **46**, 261 (1992)
10. V. Grolier, D. Renard, B. Bartenlian, P. Beauvillain, C. Chappert, C. Dupas, J. Ferré, M. Galtier, E. Kolb, M. Mulloy, J.P. Renard and P. Veillet, Phys. Rev. Lett **71**, 3023 (1993)
11. N.E. Bickers, D.J. Scalapino and R.T. Scalettar, Internat. J. Mod. Phys. B **1**, 687 (1987): N.E. Bickers, D.J. Scalapino and S.R. White, Phys. Rev. Lett. **62**, 961 (1989)
12. F.K. Schulte, Surf. Sci. **55**, 427 (1976)
13. J.F. Dobson, Phys. Rev. B **46**, 10163 (1992)
14. W. Kohn and L.J. Sham, Phys. Rev. **137**, A1697 (1965)
15. K.-D. Tsuei, E.W. Plummer and P.J. Feibelman, Phys. Rev. Lett. **63**, 2256 (1989)
16. K.-D. Tsuei, E.W. Plummer, A. Liebsch, K. Kempa and P. Bakshi, Phys. Rev. Lett. **64**, 44 (1990)
17. P.R. Pinsukanjana, E.G. Gwinn, J.F. Dobson, E.L. Yuh, N.G. Asmar, M. Sundaram, and A.C. Gossard, Phys. Rev. B **46**, 7284 (1992)
18. W. Kohn and L.J. Sham, Phys. Rev. **140**, A1133 (1965)
19. A.L. Fetter and J.D. Walecka, "Quantum Theory of Many-Particle Systems", McGraw-Hill, New York 1971
20. T. Ando, Z. Phys. B **26**, 263 (1977)
21. A. Zangwill and P. Soven, Phys. Rev. A **21**, 1561 (1980)
22. E. Runge and E.K.U. Gross, Phys. Rev. Lett. **52**, 997 (1984)
23. E.K.U. Gross and W. Kohn, Phys. Rev. Lett. **55**, 2850 (1985): erratum *ibid.*, **57**, 923 (1986): N. Iwamoto and E.K.U. Gross, Phys. Rev. B **35**, 3003 (1987)
24. J.F. Dobson and G. H. Harris, J. Phys. C: Solid St. Phys. **21**, L729 (1988)
25. B.N.J. Persson, J. Phys. C: Solid St. Phys. **13**, 435 (1980)
26. P.J. Feibelman, Prog. Surf. Sci. **12**, 287 (1982)
27. P.J. Feibelman, Phys. Rev. B **9**, 5077 (1974): Phys. Rev. B **12**, 1319 (1975)
28. J.F. Dobson and G.H. Harris, J. Phys. C: Solid St. Phys **19**, 3971 (1986)
29. K. Kempa and W.L. Schaich, Phys. Rev. B **34**, 547 (1986)
30. K. Kempa and W.L. Schaich, Phys. Rev. B **37**, 6711 (1988)
31. K. Kempa, A. Liebsch and W.L. Schaich, Phys. Rev. B **38**, 12645 (1988)
32. A. Liebsch, Phys. Rev. B **36**, 7378 (1987)
33. K.J. Song, D. Heskett, H.L. Dai, A. Liebsch, and E.W. Plummer, Phys. Rev. Lett. **61**, 1380 (1988)
34. K. Burke and W.L. Schaich, Phys. Rev. B **48**, 14599 (1993)
35. W.L. Schaich, private communication (1993)
36. A. Liebsch, Phys. Rev. Lett. **67**, 2858 (1991)
37. G.H. Harris "Electronic properties of juxtaposed metal surfaces", Ph.D. thesis, Griffith University, Nathan, Queensland Australia 1992
38. A.G. Eguiluz, Phys. Rev. Lett. **51**, 1907 (1983)
39. R.S. Sorbello, Solid State Commun. **56**, 821 (1985): *ibid* **48**, 989 (1983)

COLLECTIVE ELECTRONIC OSCILLATIONS ON C_{60}

Marek T. Michalewicz[1] and Mukunda P. Das[2]

[1] Supercomputing Support Group, Division of Information Technology
Commonwealth Scientific and Industrial Research Organisation
723 Swanston Street, Carlton, Victoria 3053, Australia

[2] Department of Theoretical Physics
Research School of Physical Sciences and Engineering
The Australian National University, ACT 2601, Australia

INTRODUCTION

The C_{60} molecule, known also as buckyball, has a very interesting history. During 1980-85 mass spectroscopic studies of carbon by laser vaporisation of graphite revealed magic numbers of carbon clusters [1,2]. A special stablity was found for C_{60} and C_{70} in the joint Sussex-Rice experiment [2] and at the same time the correct structure was identified. The structure is analogous to the geodesic domes invented by Buckminster Fuller, hence the name Buckminsterfullerene or in short fullerene. In 1990 fullerene was isolated and a major breakthrough took place after Krätchmer et. al. [3] were successful in producing sizable quantities of the new compound. Apart from C_{60} and C_{70} other stable 3D cage-shaped molecules, for example C_{28}, C_{32}, C_{50}, C_{84}, C_{240}, C_{540} were identified. During the past 4-5 years the study of the fullerenes has been fascinating because of several of its remarkable properties, such as the interesting chemistry, crystalline structure and structural transitions, electronic structure, collective electronic and vibrational properties, optical properties, phase transition and ferromagnetism and superconductivity upon doping. [4-10].

The C_{60} is a cluster of 60 carbon atoms forming a soccer ball-like structure with the symmetry of a truncated icosahedron, having 60 vertices, 20 hexagonal faces, 12 pentagonal faces with six five-fold axes and ten 3-fold axes running through their centres, 30 double and 60 single bonds. All carbon sites on the surface of the spherical fullerine are equivalent. The pentagons are regular whereas the hexagons have alternate single and double bonds. The ball is hollow inside.

Experimental determination of the structure of C_{60} and C_{70} have been carried out by a variety of techniques, such as x-ray, Raman, IR, NMR, SIM, neutron and electron diffraction methods. Results from these experiments confirm the expected molecular cage structure. One can construct a C_{60} molecule by wrapping a graphite layer with 12 pentagonal *defects*. It is now known that the sphere of C_{60} has a diameter of 7.1 Å with 2 bond lengths (6-6 ring) 1.391Å and (6-5 ring) 1.455Å. In comparison, the graphite has intralayer bond length 1.42 Å and interlayer bond length 3.35 Å. The curvature of the network of the C_{60} mixes states of some sp^3 character to the mostly sp^2 hybridization. Out of 4 valence electrons on each carbon atom, 3 electrons participate in the 90 σ bond (180 electrons) along the edges of the truncated icosahedron. The remaining 60 electrons are in the π orbitals. The short bond (6-6 ring) has more charge compared to the other bond (6-5 ring). The bonding in C_{60} has a closer relationship to graphite than to diamond. Because of the large diameter of the C_{60} sphere, the electron density is localized near the surface of the sphere with zero density in the central region. The *nonresonant* nature of the single and double bonds (unlike on graphite) makes the charge distribution along the bonds spatially inhomogeneous.

The C_{60} molecules form a crystalline solid having an fcc structure at room temperature. The unit cell constant $a_0 = 14.1$ Å, corresponding to a nearest neighbour separation of 10 Å. The distance between the nearest approach in the crystalline solid is 3.1 Å. From the x-ray and NMR studies it is well-known that C_{60} molecules possess high degree of orientational disorder at the room temperature. When the temperature is reduced below 250 K, the C_{60} solid undergoes a first order freezing transiton forming a simple cubic phase with 4 bucky balls per unit cell. Molecular dynamical studies of vibrational densty of states are discussed elsewhere in this volume.

The electronic structure of the fullerenes is a fascinating topic which is extensively studied both experimentally [11–27] and theoretically [28–41] by numerous available methods. We will very briefly review this topic in the next section. In this paper we limit our discussion to the collective electronic structure of C_{60} molecule. We analyse the collective electronic excitations using a hydrodynamic model. The results are discussed in relation to the recent experiments of electron-energy-loss and photo-emission spectroscopies.

ELECTRONIC STRUCTURE AND COLLECTIVE OSCILLATIONS

Study of the electronic structure of the molecular unit of C_{60} is basic to the understanding of the equilibrium and transport properties of its crystals. The photo-emission and the inverse photo-emission spectra presents the occupied and the unoccupied density of states of the solid respectively [41]. In contrast to the density of states in diamond and graphite, the electronic spectra of C_{60} is rich in structures. There are now empirical and ab-initio electronic structure calculations of the C_{60} [45–47]. An extensive discussion of the electronic structure of both molecular and solid C_{60} obtained by wide variety of methods, as well as references to the original literature can be found in a number of review papers [4,6,7,26,41]. The main observation is that the valence band states and the leading conduction band states have well defined angular momentum and molecular σ or π character [45,6].

An important technique in the study of the electronic excitations is the electron energy loss spectroscopy (EELS). Recently the EELS study has been very actively pursued.

The excitation spectra of fullerene was obtained in high-resolution electron-energy-loss spectroscopy experiments both in transmission [12–14,20] and reflection [15,18,19] geometry for fullerite films and for C_{60} in gas phase [16]. It shows highly structured excitations of collective nature related to the high degree of molecular symmetry. Even on a free molecule the delocalized valence electrons exhibit collective oscillations at high frequency. The first experimental report on the EELS by Saito *et. al.* showed two loss peaks at 6.5 eV and at 26 eV and was thought to be due to weakly bound π electrons and all the valence (σ and π) electrons. These peaks are observed around 7 eV and 27 eV in the case of graphite. In a number of papers the high energy peak (in solids) has been shown to occur in the range of 25 to 30 eV. It is now accepted view that these two broad features around 6.5 eV and 25-28 eV can be attributed to molecular collective $\pi \rightarrow \pi^*$ electronic transitions and to plasma oscillations of all valence electrons, respectively. It is argued that the shift in the observed plasmon peak between 21 ev and 28 eV is essentially of instrumental origin [32]. The EELS data for C_{60} gas phase [16] indicates possible dispersion and excitation of higher multipole ($l > 1$) plasmons depending on the scattering angle. The palsmon peak shifts from 22 eV for low-angle inelastic (dipole) scattering to 25 eV at larger angles. This is very reminiscent of the dispersion of higher multipole surface plasmons on small metallic clusters reported in Ref. [48,49].

Similarily, depending on the primary-electron energies the broad plasmon peak centered at 28 eV for a ~ 50 eV beam [15,18] shifts to around 23 eV for primary-electron energy ~ 2 keV [21]. It was suggested "that reducing the incident energy E_p of the electrons increases the momentum transfers with the plasmon excitations and enhances thereby the cross section for inelastic multipolar scattering" [32].

Although the ab-initio local density theory underestimates the correlation effects, it is found that the valence electrons are delocalized on the surface of the sphere. The results conform to the photoemission and inverse photoemission, optical and UV absorption data reasonably well. The inhomogeneous distribution of electrons makes the C_{60} molecule highly electronically polarizable. The high polarizability giving rise to dispersion forces is responsible for the dominant part of the cohesion in the solid phase [45,36,42].

We are interested in a model of the C_{60} molecule in which all valence electrons fill the spherical shell of finite thickness. Figure 1 depicts the hollow nuclear cage of C_{60} molecule and the 2D cross section of the total electronic density at the mid-section of the molecule, parallel to the pentagonal face. The electronic density was obtained from MNDO90 semi-empirical program [43]. MNDO Hamiltonian was used for geometry optimization and calculations of electronic density and molecular orbitals. The structure obtained this way is in a very good agreement (within 0.1%) with the experimental results quoted above. It is also seen that the distribution of charge is nonuniform along (6-5 ring) and (6-6 ring) bonds. Very similar results for the electron density distribution on the C_{60} were obtained by number of other methods [45,6,7]. The spread of electron gas normal to the sphere surface, (the thickness of the electron gas), clearly seen on the Figure 1, will serve as the only parameter of our model. We will return to this point shortly.

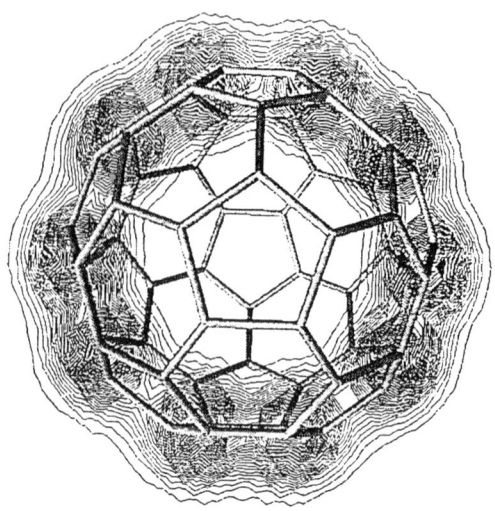

Figure 1. Two dimensional cross-section along the plane parallel to the pentagonal face of the C_{60} molecule showing the total density of valence electrons. The thickness parameter α is estimated to be $0.4 \leq \alpha \leq 0.7$. The density was calculated using MNDO method [43]. The contour lines are plotted every 0.02 $e\text{Å}^{-3}$. The highest electron density in the vicinity of carbon cores is 2.14 $e\text{Å}^{-3}$

HYDRODYNAMIC DESCRIPTION OF A SPHERICAL SHELL MODEL OF C_{60} MOLECULE

We have presented a preliminary analysis of the multipole surface plasmons on C_{60} elsewhere [31]. Although the hydrodynamic model may be considered oversimplifying, it highlights some very interesting qualitative features of the system which were not elucidated in other treatments, and which are observed in experiments. For example the experimental data from EELS experiments reveal trends in favour of an excitation energy of the plasmons that increases with their multipolar order [32].

In Ref. [48] we disscussed relative merits and shortcomings of hydrodynamic model compared with time-dependent local density approximation. The discussion of applicability of simple hydrodynamic model in *non-dispersive* limit of infinitely thin shell can be found in Ref. [33]. Here we use the *dispersive* hydrodynamic description [49].

The linearized hydrodynamic equation of motion for the time dependent first order electron density fluctuation n_1 is written as

$$[\omega^2 - \omega_P^2 + \beta^2 \nabla^2] n_1(\mathbf{r}) = 0 \tag{1}$$

where $\omega_P^2 = \frac{4\pi e^2 n_0}{m}$ (in dimensional units), n_0 is the ground state electron density of valence electrons, ω_P is the plasmon frequency and β is the dispersion parameter, which at high frequencies is $\sqrt{\frac{3}{5}} v_F$.

The electron gas is modelled by a spherical jellium shell of finite thickness. Jellium is defined here as a uniform positive charge density that confines electrons within spherical shell volume. The electrons on the C_{60} cage are confined to a spherical shell of certain thickness characterised by a parameter α given by the ratio of the inner and outer radii, $\alpha = R_1/R_2$.

The equilibrium electron density is assumed to be uniform in the shell. Equation (1) was solved in the spherical coordinates for the spherical shell geometry of the C_{60} molecule.

The general solutions for the surface plasmons ($\omega_P > \omega$, $K^2 = \kappa^2 = \frac{\omega_P^2 - \omega^2}{\beta^2}$) are

$$n_1(\mathbf{r}) = \left\{ S_l m_l(\kappa r) + P_l g_l(\kappa r) \right\} Y_l^m(\theta, \phi). \tag{2}$$

Here, m_l is the modified spherical Bessel function of the first kind and g_l the modified spherical Bessel function of the third kind [44]. S_l and P_l are the normalization coefficients.

The electrostatic potential induced by the plasmons is solved from the Poisson equation subject to the boundary conditions of vanishing radial component of displacement at $r = R_1$ and $r = R_2$.

The general dispersion relation for the surface plasmons on C_{60} shell model is given by

$$\left\{ \left[\frac{x^2}{l} - \frac{1}{2l+1} \right] m_{l+1}(\alpha xy) + \frac{\alpha^{l-1}}{2l+1} m_{l+1}(xy) + [x^2 - 1] \frac{m_l(\alpha xy)}{\alpha xy} + \frac{\alpha^{l-1} m_l(xy)}{xy} \right\}$$
$$\times \left\{ \left[x^2 \frac{2l+1}{l+1} - 1 \right] g_{l+1}(xy) - \left[x^2 \frac{l(2l+1)}{l+1} \right] \frac{g_l(xy)}{xy} + \alpha^{l+2} g_{l+1}(\alpha xy) \right\}$$
$$- \left\{ \left[x^2 \frac{2l+1}{l+1} - 1 \right] m_{l+1}(xy) + \left[x^2 \frac{l(2l+1)}{l+1} \right] \frac{m_l(xy)}{xy} + \alpha^{l+2} m_{l+1}(\alpha xy) \right\}$$
$$\times \left\{ \left[\frac{x^2}{l} - \frac{1}{2l+1} \right] g_{l+1}(\alpha xy) + \frac{\alpha^{l-1}}{2l+1} g_{l+1}(xy) - [x^2 - 1] \frac{g_l(\alpha xy)}{\alpha xy} - \frac{\alpha^{l-1} g_l(xy)}{xy} \right\} = 0 \tag{3}$$

where $x^2 = \frac{\beta^2 \kappa^2}{\omega_P^2} = 1 - \frac{\omega^2}{\omega_P^2}$, $xy = \kappa R_2$, $\kappa^2 = \frac{\omega_P^2 - \omega^2}{\beta^2}$, $\alpha = \frac{R_1}{R_2}$ and $y = \frac{R_2}{\beta} \omega_P$.

It is easily seen that the dispersion relation (3) in the limit $m_l(R_2) = 0, R_1 = 0, \alpha = 0$, reduces to the dispersion relation for the surface plasmons on a jellium sphere in vacuum [49].

$$m_l(xy) + \frac{xy}{l} m_{l+1}(xy) \left[1 - \frac{l+1}{2l+1} x^{-2} \right] = 0 \tag{4}$$

Similarily, for the limit $g_l(R_1) = 0, R_2 = \infty$, we obtain the dispersion relation for the spherical void:

$$g_l(xy') - \frac{xy'}{x^2 - 1} \frac{g_{l+1}(xy')}{l} \left[x^2 - \frac{l}{2l+1} \right] = 0 \tag{5}$$

where $y' = \frac{R_1}{\beta} \omega_P^2$.

The only parameter of the model, for the fixed number of valence electrons (both π and σ) of 240, and for the diameter of the nuclear cage of C_{60} molecule being $2R_0 = 7.1$ Å, is $\alpha = \frac{R_1}{R_2}$. The external radius is expressed in terms of the radius of the nuclear cage as $R_2 = R_0 [\frac{2}{1+\alpha}]$. The bulk plasmon energy according to the Drude formula is $\omega_P^2 = 27.2 \frac{720 a_0}{R_2^3 (1-\alpha^3)}$ eV. The radius of a sphere available for one electron, r_s, is given as $r_s = R_2 \sqrt[3]{\frac{1-\alpha^3}{240}}$ Å. The estimate for the value of the parameter α is $0.4 \leq \alpha \leq 0.7$ [31].

The dispersion relation (3) was solved numerically. The calculated dispersion curves for surface plasmons ω vs. α are presented on Figure 2. Precisely four surface plasmons can be supported on a shell of thickness $\alpha \leq 0.7$.

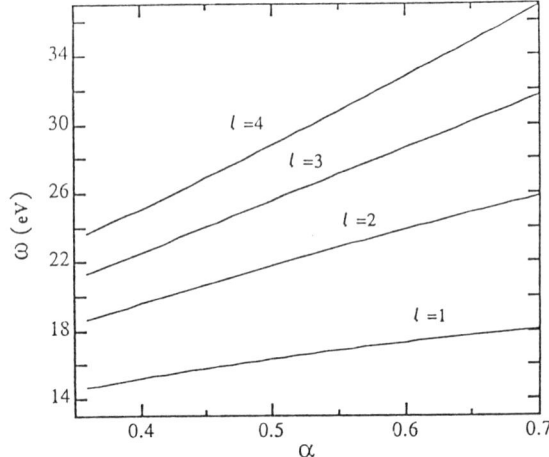

Figure 2. The dispersion curves ω_l vs. α for the four surface plasmons obtained from Eq. (3). The values of the thickness parameter α correspond to the estimates based on Figure 1 (From Ref. [31].)

The formal solution of Eq. (3) in the limit of a very thin shell gives up to 12 surface plasmons, with the plasmons of low "azimuthal number" $l = 1, 2$ and 3 disapearing for $\alpha > 0.9$. This corresponds to unphysical situation of a system of very small volume and high electron density.

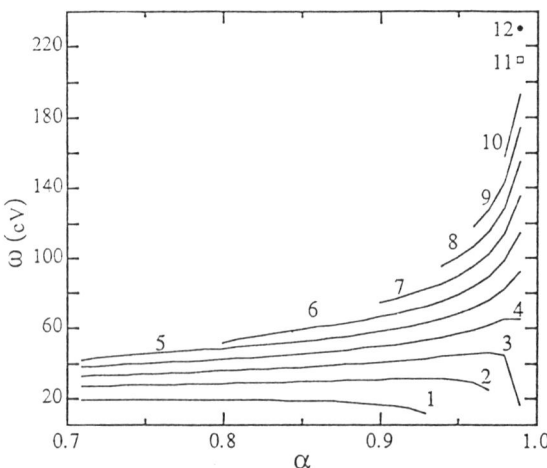

Figure 3. The dispersion curves ω_l vs. α for the surface plasmons obtained from Eq. (3) for higher values of the thickness paramenter α.

EXPERIMENTS vs. THEORY: COMPARISON AND CONCLUSIONS

It is now desirable to have an estimate of the thickness of the electron gas shell that supports surface plasmons. There are two ways of estimating this quantity. The first is to consider the diffraction data on the crystalline structure of C_{60}. The lattice constant of the fcc latice is 14.2 Å. This gives the closest approach of 3.1 Å. Assuming there is no overlap of electronic shells, the outer radius is $R_2 = 5.1$ and $\alpha \leq 0.39$ *in the solid*. By inspection of contour map of the valence electron density of fcc C_{60} crystal obtained on the basis of density functional theory in local density approximation [45], we get $\alpha \sim 0.33$. The second estimate is from our MNDO quantum chemistry calculations. From Figure 1, and similar results for cross-section along different planes we see that the parameter $0.4 \leq \alpha \leq 0.7$. The outer contour of constant electronic density was at 0.02 $e\text{Å}^{-3}$ for the lower limit and up to 0.2 $e\text{Å}^{-3}$ for the upper limit of α. The separation between the contour lines was 0.02 $e\text{Å}^{-3}$ and the maximum valence density in the vicinity of carbon cores was 2.14 $e\text{Å}^{-3}$.

Finally, we compare the eigenenergies of the four plasmons with the electron-energy-loss spectroscopy data. Recently EELS study [12–27] has been persued actively. The sample is usually a solid C_{60}. EELS probes both single and collective electronic excitations. It shows highly structured excitations of collective nature related to the high degree of molecular symmetry. Even on a free molecule the delocalized valence electrons exhibit collective oscillations at high frequency.

In a number of papers [3,12–24,11] the high energy peak (in solids) has been shown to occur in the range of 25 to 30 eV. The detailed spectra was presented by Kuzuo et al [14] and Sohmen et al [20]. The experimental data from the electron-energy-loss spectroscopy from the work of Sohmen et al [20] are presented on Figure 4. A very similar results were obtained by Kuzuo et al [14].

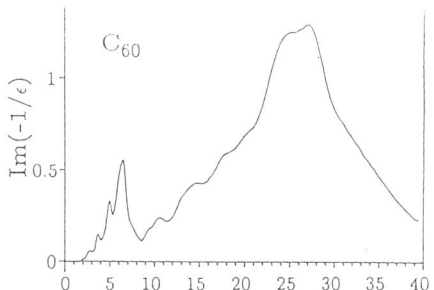

Figure 4. The experimental electron energy-loss spectroscopy loss function $Im(-1/\epsilon)$ of crystalline C_{60} for momentum transfer $q = 0.1\text{Å}^{-1}$ (From Ref. [20].)

In both cases the spectra exhibits broad peak with maximum between 25 and 27 eV. There are less pronounced shoulders at about 15, 18 and 20 eV. Those features could perhaps be attributed to surface plasmons, which in our model for $\alpha = 0.44$ appear at 15.6, 20.4, 23.7 and 26.5 eV. This is to certain extent speculative, Kuzuo et al claim that those features originate from interband transitions. But it is not yet clear if these could be signals from $l = 1, 2, 3$ and 4 surface plasmons. Saito et al [12] observed the double plasmon loss at 52 eV but the peak is broadly smeared and it is difficult to see if 30, 36, 40 eV features due to harmonic plasmon effects are present. Sohmen et al [20] removed contributions due to multiple scattering from their results. It is interesting to note that the outer radius corresponding to $\alpha = 0.44$ is $R_2 = 4.93$ Å. The volume of the shell is then 440 Å and $r_s = 0.77$. The Drúde "bulk" plasmon is then $\omega_P = 26.8 eV$. This is in contrast with other estimates [20], where the volume of the valence electron was estimated on the basis of diffraction data to be 710 Å^3. If that was the volume of the valence electron gas, the Drude plasmon would have energy $\omega_P = 21.2 eV$ and the thickness of the shell would have to be parametrized by $\alpha = 0.26$.

Although the number of EELS experiments reported in the literature is reasonably large, there is still lack of systematic series of measurements at high resolution and for the variable conditions of detection angle and primary-electron energy. Another interesting feature could be the second derivative of the EELS spectra with respect to the energy loss. The only one such measurement reported by Cohen et. al. [21] is presented on Figure 5.

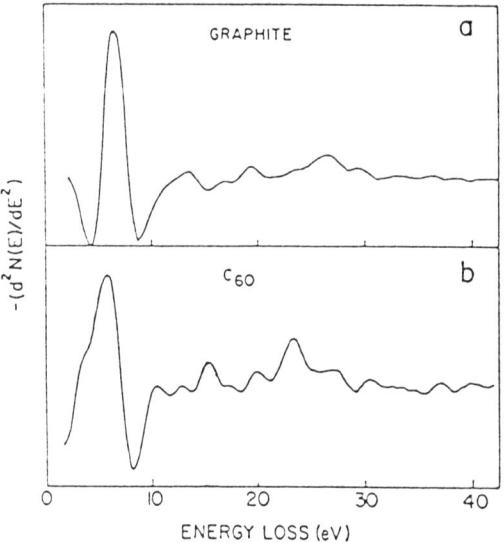

Figure 5. The second derivative of the experimental EELS loss function for the fullerite (From Ref. [21]).

There are number of interesting features in the $[10 - 30]$ eV interval which could perhaps be attributed to multipole plasmons of all valence electrons.

The giant plasmon resonace have been reported in the photoionization spectra around ~ 20 eV [25] for which microscopic calculations based on the linear-response theory have been done [39,38]. Lambin et al [36] have performed calculations of a spherical dielectric shell of finite thickness. They have shown that multipolar surface plasmons arise in the energy range $20 - 30$ eV. However their model had an adjustable resonance energy parameter fixed at the value which would reproduce the strong VUV absorbance of C_{60} around 20 eV. An empirical infinitely thin spherical shell model has been discussed by Barton and Eberlein [33] within the framework of two fluid hydro-dynamic description. There are few important differences in our treatment and that of Ref. [33]. Barton and Eberlein neglected the hydrostatic pressure, this is equvalent to the *nondispersive* model. The dynamics of π and σ electrons was separated, hence the density of electron gas was lower, and consequently the Drude plasmon energy was lower as well. Our present model is similar in spirit to that of Lambin *et al.*. They solved the problem of collective modes through the dielectric function structure. High energy peak around 28 eV was interpreted as the molecular property [36] in contrast to that due to solid state effects [11,16].

Our model does not account for the 6.0 − 6.5 eV plasmon found in EELS experiments. It could be related to the tangential character of this mode [33,15] which is not taken into account by the type of bounadry conditions used. Van Giai and Lipparini [30] studied the 6.5 eV π plasmon excitation using the sum rule approach applied to 60 π electrons. They also studied the diffuse jellium density on the basis of Local Density Approximation for a spherical jellium. Östling *et al* [37] performed similar calculations to the ones described in this work. However those workers used the *non-dispersive* hydrodynamics for a finite thickness shell. This naturally led to purely geometrical dispersion dependance and unbounded number of plasmon modes. Finally, Bulgac and Ju [38] performed the RPA calculations of collective excitations on the C_{60} molecule. Their model leads to multipole plasmons, however it does not show any variation of the excitation energies with the order l. Consequently it would be hard to de-convolute higher multipole plasmon signal from the loss spectra. However the experiments indicate that multipole plasmons show dispersion [16].

Our model can be treated as a guiding tool in studies of collective electronic excitations on C_{60} molecule. There is no complication of the single particle excitations, which makes the experimental structure so rich. The only parameter is the thickness of the electron gas shell. The model gives exactly four surface plasmons on the shell (coupled on both inner and outer surface), this is in good agreement with Barton and Eberlein estimates on the basis of dimensional analysis. In Ref. [33] they quoted $L < 6$, but the more conservative value of $L = 3$ was emphasized.

ACKNOWLEDGEMENTS

The authors wish to thank Dr J. Mahanty for many fruitful discussions, help and encouragement. MTM thanks Dr A. Vassallo for the discussions and for providing illustrative materials for this lecture. We also express our thanks to J.H. Weaver, W.E. Pickett, R.C. Haddon, A.A. Lucas, Y.M. Shulga, K. Yabana, E. Lipparini, A.Bulgac and S. Satpathy for sending us the pre- and re-prints of their work.

REFERENCES

[1] E.A. Rohlfing, D. Cox and A. Kaldor, *J.Chem. Phys.*, **81** (1984) 3322

[2] H.W. Kroto, J.R. Heath, S.C. O'Brien, R.F. Curl and R.E. Smalley, *Nature*, **318**, (1985) 162

[3] W. Krätchmer, L.D. Lamb, K. Fostiropoulos and D.R. Huffman, *Nature* , **347** (1990) 354

[4] H.W. Kroto, A.W. Allaf and S.P. Balm, *Chem. Rev.*, **91** (1991) 1213. This review contains extensive references to the original papers on the discovery of C_{60}.

[5] R. Taylor and R.M. Walton, *Nature*, **363** (1993) 685

[6] J.H. Weaver and D.M. Poirier, "Solid state properties of fullerenes and fullerene-based materials" preprint, to appear in: *Solid State Physics* Vol. 48, Academic Press, New York, 1994

[7] W.E. Pickett "Electrons and phonons in C_{60}-based materials" preprint, to appear in: *Solid State Physics* Vol. 48, Academic Press, New York, 1994

[8] R.C. Haddon, *Accounts of Chemical Research* **25** (1992) 127

[9] R.C. Haddon, A.F. Hebard, M.J. Rosseinsky, D.W. Murphy, S.H. Glarum, T.T.M. Palstra, A.P. Ramirez, S.J. Duclos, R.M. Fleming, T. Siegrist and R. Tycko, in: *"Fullerenes: Synthesis, properties, and chemistry of large carbon clusters"*, G.S. Hammond and V.J. Kuck, Eds., American Chemical Society, 1992, Chapter 5, pp.71-89

[10] The current bibliography on fullerenes can be found by sending a single line message: "BIBLIO" to bucky@sol1.lrsm.upenn.edu. To learn about information services on fullerene physics and chemistry on Internet send one line message: "INTRO" to the same electronic address.

[11] J.H. Weaver, J.L. Martins, T. Komeda, Y. Chen, T.R. Ohno, G.H. Kroll, N. Troullier, R.E. Haufner and R.E. Smalley, *Phys. Rev. Lett.* **66** (1991) 1741

[12] Y. Saito, H. Shinohara and A. Ohshita, *Jap. J. Appl. Phys.* **30**, (1991) L1068

[13] P.L. Hansen, P.S Fallon and W. Kratchmer, *Chem. Phys. Lett.*, **181** (1991) 367

[14] R. Kuzuo, M. Terauchi, M. Tanaka, Y. Saito and H. Shinohara, *Jap. J. Appl. Phys.*, **30**, part B, (1991) L1817

[15] G. Gensterblum, J.J. Pireaux, P.A. Thiry, R. Caudano, J.P. Vigneron, Ph. Lambin, A.A. Lucas and W. Krätschmer, *Phys. Rev. Lett.*, **67** (1991) 2171

[16] J.W. Keller and M.A. Coplan, *Chem. Phys. Lett.*, **193** (1992) 89

[17] A.A. Lucas, *J. Phys. Chem. Solids* **53** (1992) 1415

[18] A. Lucas, G. Gensterblum, J.J. Pireaux, P.A. Thiry, R. Caudano, J.P. Vigneron and Ph. Lambin, it *Phys. Rev. B* **45** (1992) 13694

[19] Y.M. Shulga, V.I. Rubtsov, A.P. Moravskii and A.S. Lobach, *Doklady Acad. Nauk* (in Russian) preprint April 1992

[20] E. Sohmen. J. Fink and W. Krätschmer, *Z. Phys. B - Condensed Matter*, **86**, (1992) 87

[21] H. Cohen, E. Kolodney, T. Maniv and M. Folman, *Solid State Comm.* **81** (1992) 183

[22] Y. Saito, N. Suzuki, H. Shinohara, T. Hayashi and M. Tomita, *Ultramicroscopy*, **41** (1992) 1

[23] V.I. Rubtsov and Y.M. Shulga, *Zh. Eksp. Teor. Fiz.* **103** (1993) 1026

[24] Y.M. Shulga, V.I. Rubtsov and A.S. Lobach, *Z. Phys. B* (1993) preprint

[25] I.V. Hertel, H. Steger, J. de Vries, B. Weisser, C. Manzel, B. Kamke and W. Kamke, *Phys. Rev. Lett.* **68** (1992) 784

[26] J.H. Weaver, in *Handbook of Surface Imaging and Visualisation*, A.T. Hubbard, Ed., CRC Press, Boca Raton, 1993

[27] Y.M. Shulga, A.P. Moravskii, A.S. Lobach and V.I. Rubtsov, *JETP Letters* **55** (1992) 132

[28] K. Yabana and G.F. Bertsch, *Physica Scripta* **48** (1993) 633

[29] K. Yabana and G.F. Bertsch, "Inelastic electron scattering on C_{60} clusters" preprint

[30] N. Van Giai and E. Lipparini, *Z. Phys. D* **27** (1993) 193

[31] M.T. Michalewicz and M.P. Das, *Solid State Comm.*, **84** (1992) 1121

[32] Ph. Lambin and A.A. Lucas, *Fullerene Science and Technology* **1** (1993) 159

[33] G. Barton and C. Eberlein, *J. Chem. Phys.*, **95**, (1991) 1512

[34] B. Carazza, "On the plasmon-like excitations of atomic clusters with application to fullerene" (1994) preprint

[35] P. Senet, L. Henrard, Ph. Lambin and A.A. Lucas, "A one parameter model of the uv spectra of carbon" Proceedings of The International Winter School on Progress in Fullerene Research, Kirchberg, preprint

[36] Ph. Lambin, A.A. Lucas and J.-P. Vigneron, *Phys. Rev. B* **46** (1992) 1796

[37] D. Östling, P. Apell and A. Rosen, *Europhys. Lett.* **21** (1993) 539

[38] A. Bulgac and N. Ju, *Phys. Rev. B*, **46** (1992) 4297

[39] G.F. Bertsch, A. Bulgac, D. Tomanek and Y. Wang, *Phys. Rev. Lett.* **67** (1991) 2690

[40] S.G. Louie and E. L. Shirley, *J. Phys. Chem. Solids* **54** (1993) 1767

[41] A.A. Lucas and J.H. Weaver, "Electron spectroscopies of C_{60} and C_{70} fullerites and fullerides" (1992) preprint

[42] N.W. Ashcroft, Europhys. Lett. **16** 355 (1991)

[43] MNDO90 written by M.J.S Dewar and co-workers is a part of $UniChem^{TM}$ quantum chemistry software environment from $CrayResearch^{TM}$.

[44] M. Abramowitz and I.A. Stegun, *Handbook of Mathematical Functions*, New York: Dover, 1972, Ch.10

[45] S. Saito and A. Oshiyama, Phys. Rev. Lett. **66** 2637 (1991)

[46] S.C. Erwin and W.E. Pickett, Science 254,842(1991)

[47] S. Satpathy, V.P. Antropov, O.K. Andersen, O. Jepson, O. Gunnarsson and A.I. Liechtenstein, Phys. Rev. B (to be published)

[48] M. T. Michalewicz, *Phys. Rev. B*, **45** (1992) 13664

[49] M. T. Michalewicz, in: *Topics in Condensed Matter Physics*, M.P. Das, Ed., Nova Science Publishers, New York, 1994

THEORETICAL STUDIES OF SEMICONDUCTOR SURFACES WITH PARTICULAR REFERENCE TO FLUORINE AND CHLORINE CHEMISORPTION ON SI(001)

P.V. Smith, M.W. Radny and A.J. Dyson

Physics Department
University of Newcastle
Callaghan, New South Wales
Australia 2308

INTRODUCTION

The study of semiconductor surfaces is of both considerable fundamental and technological importance. Unlike metal surfaces which maintain their bulk geometry and simply undergo changes in the surface interlayer spacings, semiconductor surfaces usually reconstruct to form completely new geometrical structures.[1] Perhaps the best example of this is the Si(111) surface. When cleaved along a (111) plane at room temperature silicon exhibits a (2×1) reconstruction.[2] Annealing this surface at high temperature for a short time produces a $\sqrt{3}\times\sqrt{3}R30°$ LEED pattern believed to be associated with a vacancy model.[3] Further annealing yields diffraction spots characteristic of the now famous Si(111)7×7 adatom-dimer-stacking-fault structure of Takayanagi et al.[4] It is this diversity of surface reconstructions and their associated properties which makes semiconductor surfaces such an interesting and fascinating area of study.

The main technological importance of semiconductor surfaces arises from the significant role played by semiconductors and semiconductor interfaces in most modern day electronic devices. This is probably best illustrated by the use of silicon microchips in computers, and the importance of the SiO_2-Si interface in MOSFET devices. Understanding semiconductor surfaces, and the chemisorption and growth processes that take place upon them, is obviously important in helping us to determine the extent to which these surfaces can be modified by processes such as etching and ion implantation. The presence of adsorbates and the growth of thin films on semiconducting surfaces can also greatly modify their wear characteristics, susceptibilty to corrosion, reaction rates and usefulness as catalytic agents.

It is thus not surprising that the last two decades have seen many different theoretical and experimental techniques developed for studying semiconductor surfaces. The main experimental methods are described in reference 5 and include Auger electron specroscopy (AES), low energy electron diffraction (LEED), core and valence photo-electron spectroscopy (XPS, UPS, ARUPS), low-energy ion scattering (LEIS), scanning tunnelling microscopy (STM), scanning tunnelling spectroscopy (STS), low energy electron loss spectroscopy (LEELS), low energy electron microscopy (LEEM), electron spin ion assisted desorption (ESDIAD) and X-ray adsorption fine structure measurements (SEXAFS, NEXAFS). The main theoretical techniques that are currently available for investigating semiconductor surfaces are listed in Table 1.

Table 1. Main theoretical techniques for determining surface structure

semi-empirical	ab initio
ASED-MO [6]	Hartree-Fock calculations [11]
MINDO/3 [7]	Density Functional Theory [12]
AM1 [8], PM3 [9]	Pseudopotential calculations [13]
Classical molecular dynamics [10]	Green function techniques [14]
	Quantum molecular dynamics [15]
	Quantum Monte Carlo [16]

Broadly speaking, these can be divided into semi-empirical (requiring parameter input) and ab initio (first-principles) techniques, and whether they are cluster and/or periodic formalisms. While some of these theoretical methods are very sophisticated, none of them provide a completely definitive approach for studying semiconductor surfaces. The ab initio calculations are more rigorous but are generally restricted to relatively small systems. Semi-empirical calculations, on the other hand, are less accurate but are much faster. They thus allow for a far greater sampling of geometry space and a more ready treatment of many of the more physically interesting systems such as the Si(111)7×7 surface. Periodic calculations are not suited to the early stages of chemisorption whilst cluster calculations, which are ideal for this purpose, have a problem with the boundaries, and so on. A further choice one needs to make is that between performing static or dynamic calculations. All of the former techniques determine the optimum structure via a multi-variate optimisation of the total energy whilst the latter methods (such as the molecular dynamics and Monte Carlo techniques) enable one to study the evolution of the surface topology as a function of both time and temperature, and in so doing, determine the thermodynamic properties of the system. All of these calculations should be carried out self-consistently to allow for any charge transfer which may occur within the surface region.

Experience has shown that in order to gain an adequate understanding of the geometrical and electronic properties of semiconducting surfaces, and the chemisorption of different species onto such surfaces, it is necessary to employ a number of different and complementary techniques. We shall now endeavour to demonstrate this important principle, and the application of some of the above theoretical techniques, by looking in detail at the clean silicon (001) surface and the chemisorption of fluorine and chlorine onto this surface.

THE RECONSTRUCTED CLEAN SILICON (001) SURFACE

Apart from perhaps the (111) surface of silicon, the Si(001) surface has been the most widely studied of all semiconductor surfaces. While higher-order reconstructions such as c(4×2) and p(2×2) have been shown to occur on this surface[17], most work has been devoted to the (2×1) dimer reconstruction. This 2×1 topology is now universally considered to arise from alternate rows of surface atoms moving together to form rows of dimers as shown in figure 1. The dimer bondlength has been experimentally determined to lie in the range 2.35-2.54 Å and the surface reconstruction to extend down to about the sixth layer. The surface bandstructure is found to be semiconducting[18]. Considerable controversy still exists, however, over whether the dimers are parallel to the surface (symmetrical), or tilted (asymmetrical). Buckling of the surface dimers is supported by photoemission[19], ion-scattering[20], core electron spectroscopy[21] and some LEED[22] experiments, whilst other LEED work[23] and polarisation dependent ARUPS measurements[24] indicate a symmetric dimer. Room temperature scanning tunnelling microscopy studies of the Si(001) surface have revealed the co-existence of both symmetric and asymmetric dimers with the symmetric dimers constituting the majority species and the asymmetric dimers appearing to be concentrated near defect sites[25]. It has been argued, however, that asymmetric dimers would appear to be symmetric in the STM topographs as a result of time-averageing over the thermal vibrations.

Theoretical studies of the silicon (001) surface also predict both symmetrical and asymmetrical dimer configurations. Whilst the original self-consistent pseudopotential calculations[13] favoured buckling of the dimers, subsequent calculations[26] have yielded symmetric dimers. Semi-empirical tight-binding calculations[27] have predicted tilting of the dimers whilst SLAB-MINDO calculations[28] have found that buckling increases the energy. More recent work[29,30] employing the fairly sophisticated local density functional formalism has also been almost equally divided between symmetric and asymmetric dimers. Virtually all of these calculations, however, have found that the difference in energy between these two configurations is quite small, suggesting that both might co-exist at room temperature. STM topographs of the Si(001)2×1 reconstructed surface taken over a substantial range of temperature have shown that the proportion of buckled to symmetric dimers increases with decreasing temperature[31]. This has been interpreted as strong evidence for asymmetric dimers constituting the minimum energy configuration. Theoretical calculations just published[32] have also indicated that the interaction between the surface and tip in a typical STM experiment is sufficient to cause asymmetrical dimers to produce a "symmetric" dimer STM image. Finally, first-principles theoretical calculations employing a symmetric dimer topology predict metallic behaviour whereas buckling of the dimers is found to open up an energy gap to produce a semiconducting surface bandstructure in agreement with experiment[18]. It has now been shown, however, that the semiconducting character of the Si(001)2×1 surface can also be obtained through the inclusion of magnetic effects without invoking any asymmetry of the dimer reconstruction.[30,33]

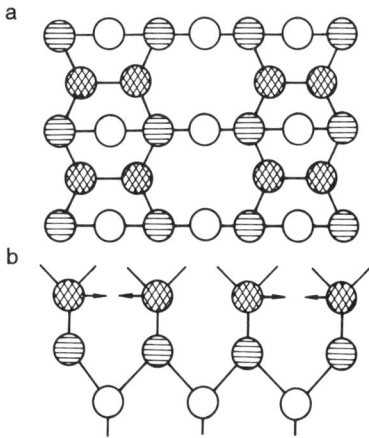

Figure 1. The 2×1 dimer reconstruction of the Si(001) surface. (a) top view (b) side view

CHEMISORPTION OF ATOMIC FLUORINE AND CHLORINE ONTO THE SILICON (001)2×1 SURFACE

The study of the chemisorption of fluorine and chlorine on the Si(001) surface is important because of the widespread use of plasma and reactive ion etching using halogen containing radicals in the etching of silicon surfaces. Such processes are of direct application in the fabrication of semiconductor devices and the manufacture of patterned silicon substrates for the very-large scale integrated (VLSI) circuits which underlie so much of modern day electronics. Experiments have shown that it is the action of atomic fluorine and chlorine in forming volatile silicon fluorides and chlorides that is actually responsible for the etching of silicon surfaces[34] and that the use of these halogens can significantly enhance the etching rate compared with that produced by physical sputtering.[35] Layer by layer etching techniques employing these reactive ions have also been proposed for the future fabrication of electronic devices at an atomic level.[36] In addition, experiments have revealed that chlorine and fluorine exhibit quite different reactivities on silicon, that etching rates increase significantly in the presence of UV irradiation, and that the relative abundance of the various desorbed species depends strongly on the surface structure.[34]

Despite this large amount of experimental work, however, little is known about the actual etching mechanisms and associated surface chemistry at a fundamental level. This is because most etching technology to date has been driven by manufacturing needs. A more fundamental understanding is now necessary for the future development and optimisation of these etching processes. The chemisorption studies of atomic fluorine and chlorine onto the (001) surface of silicon presented in this paper represent an initial step in the acquisition of such an understanding.

In order to investigate the adsorption of atomic fluorine and chlorine onto the silicon (001)2×1 reconstructed surface at all concentrations up to one monolayer, we have performed both cluster and periodic calculations. Our approach has been to first delineate possible minimum energy configurations using the semi-empirical SLAB-MINDO molecular orbital method[37] and then employ the ab initio GAMESS '92 [38], GAUSSIAN '90 [39] and CRYSTAL '92 [40] programs to investigate these structures in greater detail. The SLAB-MINDO method is based on the well-known MINDO/3 molecular orbital method developed by Bingham et al.[7] and has been shown through a variety of applications to constitute a powerfully predictive and computationally viable surface analytical technique.[41] The GAMESS '92, GAUSSIAN '90 and CRYSTAL '92 codes collectively allow one to perform both cluster and periodic first-principles Hartree-Fock calculations using a wide range of different basis sets. GAMESS '92 and GAUSSIAN '90 also enable the inclusion of correlation effects. An overview of these calculational schemes is given in table 2.

Half Monolayer Fluorine Coverage

To determine the various equilibrium configurations appropriate to a 0.5 ML coverage of fluorine on the Si(001)2×1 surface we have employed a 2×2 surface unit cell with the fluorines initially at the positions denoted 1 through 5 in figure 2. The optimum topology has then been determined by minimising the total energy with respect to the co-ordinates of all of the atoms in the topmost surface layers. For the initial starting geometry of the reconstructed surface we have used the symmetric dimer topology predicted by the SLAB-MINDO method.[28]

Table 2. Summary of the MINDO/3, GAUSSIAN '90, GAMESS '92 and CRYSTAL '92 codes.

	Type of approach	Geometry	Properties
MINDO/3	semi-empirical HF	cluster, slab, bulk	Geometry optimisation Populations Electronic structure Local density of states
GAUSSIAN '90	ab initio HF, variable basis sets, freezing of variables and core orbitals Includes CNDO, MINDO/3, MNDO & AM1.	cluster	Molecular energy levels Geometry optimisation Populations Force constants Vibrational frequencies Potential energy surfaces Electrostatic properties Electronic charge densities Electric field & gradient Electric polarizabilities
GAMESS '92	ab initio HF, variable basis sets. Includes MNDO, AM1 & PM3.	cluster	As for GAUSSIAN '88, plus graphics software: Structure plots Equipotential surfaces Molecular orbital contours Electron density maps
CRYSTAL '92	ab initio Restricted HF (RHF), variable basis sets, core pseudopotentials.	cluster, slab, bulk	Geometry optimisation Band structure Electron density of states Charge density maps Populations X-ray structure factors Electron momentum distribution Compton profiles

Fluorine chemisorbed at the open positions 1 and 5 is found to penetrate the silicon surface without any activation to form the configuration shown in figure 3. The surface Si-Si dimer bond is essentially the same as that before chemisorption whilst the chemisorbed fluorine occupies a highly co-ordinated position just below the second layer and acquires a nett charge of 0.13e (see table 3).

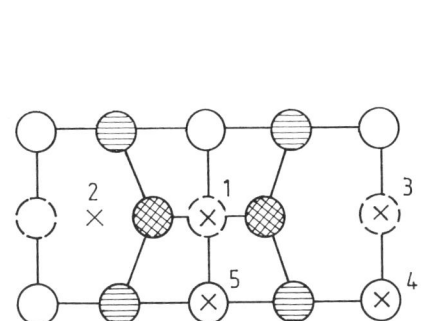

Figure 2. Schematic of the Si(001) 2×1 reconstructed surface indicating various possible chemisorption sites.

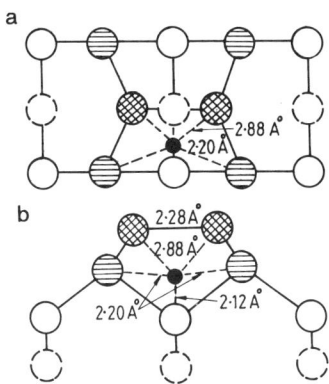

Figure 3. 0.5 ML subsurface configuration for the Si(001) 2×1:F system (a) top view (b) side view.

Table 3. The electronic charge on the fluorine and top layer silicons, and the total energy values, for the 0.5 ML and 1.0 ML minimum energy configurations shown in figs 3-6 and 8. The single asterisks denote silicon atoms directly bonded to a fluorine whilst the double asterisks label the upper fluorine atom in each of the 1.0 ML configurations. The energy values are in eV per 2×2 unit cell (0.5 ML) and eV per 2×1 unit cell (1.0 ML).

	Si (001) 2×1:F - 0.5ML			Si (001) 2×1:F - 1.0ML	
	Dangling Bond	Shared Dimer	Subsurface	Molecular-like	Shared Dimer: Dangling bond
F	7.43	7.16	7.13	7.52** 7.24	7.39** 7.16
Si (1st layer)	3.23* 4.35	3.71* 4.33	4.22* 4.22*	2.62* 4.22*	3.32* 3.81*
Si (2nd layer)		3.99	3.99* 3.99*		
Si (3rd layer)			3.85*		
Energy	-3470.91	-3474.72	-3476.62	-2943.03	-2943.58

At the non dimer bridge site (position 3 in figure 2), a small activation energy is needed to move the F atom below the surface. In the resulting process the neighbouring Si-Si dimer bonds are broken and then reconstituted above each subsurface fluorine to yield the same identical 2x1 topology as that shown in figure 3. The precursor to this process, however, is an on-top bridge site geometry with a relatively high total energy and hence may not occur.

Previous work has suggested two possible subsurface configurations for fluorine chemisorption on silicon. In one of these[42], it is postulated that the fluorine might exist as an F⁻ ion isoelectric with neon at an interstitial position between the second and third layers, whilst in the second model[43], the fluorine breaks a Si-Si backbond to form a Si-F-Si bonding configuration between the two topmost surface layers. This latter picture is supported by the semi-empirical molecular dynamics calculations of Weakliem et al.[44] which reveal some breaking of the Si-Si dimer bonds at 0.5 ML coverage and fluorine atoms interacting with the second layer. While our subsurface configuration is clearly consistent with the first model, we find no evidence for the formation of Si-F-Si backbonds, nor for the depletion of the Si-Si dimer bonds at 0.5 ML coverage. Rather, we find that F can diffuse below the surface without significantly changing the surface reconstruction.

Two stable on-top geometries have also been determined for half monolayer coverage. In the first of these, the F atom near position 2 of figure 2 reacts spontaneously with its neighbouring dimer atom to saturate one of the surface dangling bonds without breaking the Si-Si dimer bond. This yields the dangling-bond topology shown in figure 4 with Si-F and Si-Si bondlengths of 1.69 Å and 2.34 Å, and a F-Si-Si bondangle of 118°. The corresponding charge transfer to the chemisorbed fluorine is 0.43e. These values are in very good agreement with the corresponding values of 1.646 Å, 2.426 Å, 113.4° and 0.38e obtained from the very sophisticated first-principles cluster calculations of Wu and Carter.[45] Both sets of theoretical results for this dangling-bond structure are also consistent with the ESDIAD experiments which determine F⁻ exit angles of 29°±3°[46], 36°±5°[47] and 30°±5°.[48]

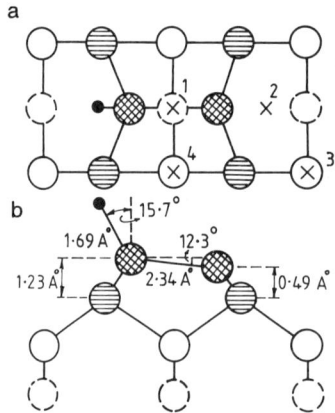

Figure 4. 0.5 ML dangling-bond topology for fluorine on Si(001):2×1 (a) top view (b) side view.

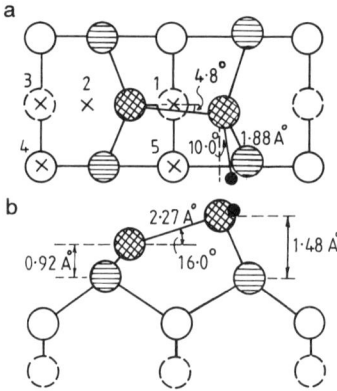

Figure 5. 0.5 ML shared-dimer configuration for the Si(001)2×1:F system (a) top view (b) side view.

The second on-top configuration, shown in figure 5, is obtained from initial position 4 of figure 2. The fluorine atom is now bound between two adjacent dimers, has a charge of 0.16e and lies in almost the same plane as that of the bonded dimer atoms. This topology is 1.9 eV per fluorine more stable than that of figure 4, but 0.95 eV per fluorine less stable than the subsurface configuration of figure 3 (see table 3). Whilst this shared-dimer configuration has not been previously reported in the literature, recent STM experiments have provided clear evidence for the adsorption of Cs and K at such chemisorption sites.[49] Experimental studies of the chemisorption of boron onto the Si(001)2×1 surface have also been interpreted in terms of possible adsorption at these sites, but in a subsurface rather than slightly on-top position.[50] Periodic RHF calculations which we have just completed using CRYSTAL '92 also show this shared dimer configuration to constitute a stable structure for the Si(001)2×1:F system at 0.5 ML coverage.

Monolayer Fluorine Coverage

In order to study the adsorption of fluorine at monolayer coverage we have started from the 0.5 ML on-top configurations shown in figures 4 and 5 and added an additional F atom per 2×1 surface unit cell at the various positions denoted by the crosses.

Starting from initial positions 1 and 2 in figure 4 for the second F atom (per 2×1 surface unit cell) of the dangling-bond configuration, we obtain the "molecular-like" geometry shown in figure 6. The two fluorines are separated by 1.62 Å with one fluorine being essentially in an on-top position and the other at a dimer bridge site. The additional charge acquired by the upper and lower fluorines is 0.52e and 0.24e respectively, gained predominantly from the immediately adjacent silicons (see table 3). The corresponding charge distribution, which has been calculated using the GAMESS '92 program, is shown in figure 7. The occurrence of a well-defined F-F bond is clearly evident.

No evidence is found from these SLAB-MINDO calculations for a monolayer configuration in which both dangling bonds of the surface dimers are saturated with fluorine. Wu and Carter[45,51], on the basis of their ab initio electronic structure calculations using

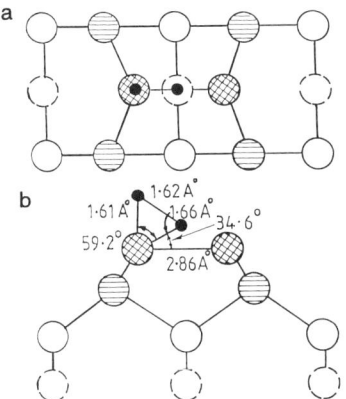

Figure 6. Monolayer SLAB-MINDO "molecular-like" geometry for fluorine on the Si(001)2×1 surface. (a) top view (b) side view.

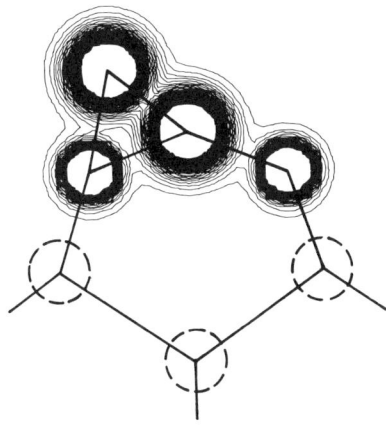

Figure 7. Electronic charge distribution for the topology of figure 6.

relatively small clusters, have claimed that fluorine will preferentially adsorb onto the surface dangling bonds to form a uniform dangling bond configuration at monolayer coverage. This is supported by their semi-empirical molecular dynamics calculations which also show evidence of F-Si-Si-F dangling bond configurations.[44] CRYSTAL '92 calculations with an STO-2G basis set also determine the dangling bond topology to be stable at monolayer coverage. Starting our periodic SLAB-MINDO calculations from such a dangling bond configuration, however, always results in one of the Si-F bonds bending over to form the "molecular-like" topology of figure 6. MINDO cluster calculations which we have performed using exactly the same parameters as for our periodic calculations, on the other hand, predict (in addition to the "molecular-like" topology) a stable dangling bond configuration in good agreement with that of Wu and Carter. This is a very significant result and shows that considerable care must be exercised in extrapolating from cluster calculations to the semi-infinite periodic system. Structures which are minimum energy configurations for a given cluster will not necessarily be stable structures for the extended periodic system.

Starting from initial positions 3 and 4 in figure 4 we find that the second fluorine is finally bound between two adjacent dimers giving rise to the monolayer structure shown in figure 8. This is, in essence, a combination of the two 0.5 ML on-top geometries found for 0.5 ML coverage, with one F atom saturating one of the surface dangling bonds whilst the other is bound between two adjacent surface dimers. The dangling bond fluorine is found to have a net charge of 0.39e whilst the shared dimer fluorine has acquired an extra charge of 0.16e. This on-top configuration is also obtained by starting from the 0.5 ML minimum energy topology of figure 5 and adding the second fluorine per 2×1 surface unit cell at any of the initial positions denoted 1, 2 and 3. This shared dimer:dangling bond geometry is very stable, being 0.55 eV per 2×1 surface unit cell lower in energy than the above "molecular-like" geometry. Moreover, we believe that this configuration, which has not been previously reported in the literature, will be still quite reactive to fluorine and may play a very significant role in any subsequent fluorine chemisorption and etching of the silicon surface by fluorine.

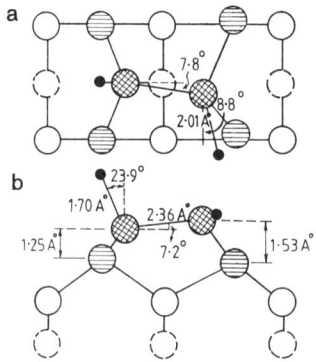 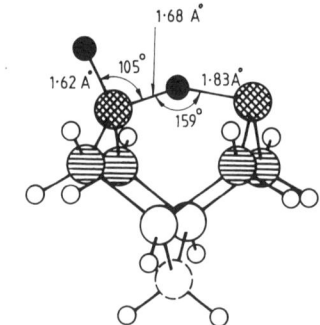

Figure 8. Shared dimer:dangling bond monolayer configuration for the Si(001)2×1:F system, (a) top view (b) side view.

Figure 9. Molecular-like geometry obtained from RHF STO-3G cluster calculations.

AM1 and ab initio Hartree-Fock cluster calculations which we have performed using a variety of different basis sets within the GAMESS '92 and GAUSSIAN '90 software packages also predict both a dangling bond and a "molecular-like" topology. While the dangling bond topology is very similar to that obtained via the MINDO method, the "molecular-like" geometry is quite different. In figure 9 we have plotted the "molecular-like" topology obtained from RHF STO-3G cluster calculations. We observe that one of the fluorines now occupies a dangling bond site and that the F-F interatomic distance is substantially greater than before at 2.51 Å. The corresponding charge density (which has again been derived from the GAMESS '92 program) is shown in figure 10. It is clear that in this case there is no F-F molecular bond. Periodic calculations which we have performed using the CRYSTAL '92 code with an STO-2G basis set also find a stable "molecular-like" geometry with a F-F separation of 2.48 Å. Despite the fact that employing increasingly sophisticated basis sets within the ab initio HF calculations results in the predicted F-F separation varying over the range 2.51 to 2.65 Å, it would appear that the MINDO formalism significantly overestimates the F-F bonding interactions. It should be remembered, however, that it was this formalism which first predicted this novel structure.[52] We thus believe that this semi-empirical technique, which requires relatively modest computer resources, still has a very valuable role to play in indicating possible minimum energy structures for employment as starting geometries within the more rigorous first-principles techniques.

In addition to the "molecular-like" and shared dimer:dangling bond topologies, our SLAB-MINDO calculations have also found the bridge site geometry shown in figure 11 to constitute a stable configuration at monolayer coverage with an energy almost identical to that of the "molecular-like" geometry[52]. While this bridge site configuration is not obtained from either of the 0.5 ML on-top configurations of figures 4 and 5, we have found that it can arise from the 0.5 ML on-top bridge site configuration which results from chemisorption onto the non-dimer bridge sites (position 3 in figure 2). As stated earlier, however, this on-top bridge site geometry has a relatively high total energy and hence both it, and the associated monolayer bridge site topology, are unlikely to occur.

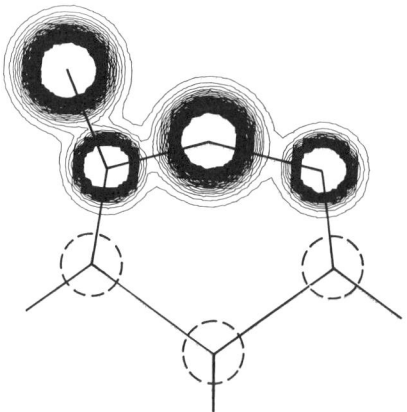

Figure 10. Electronic charge distribution for the topology of figure 9.

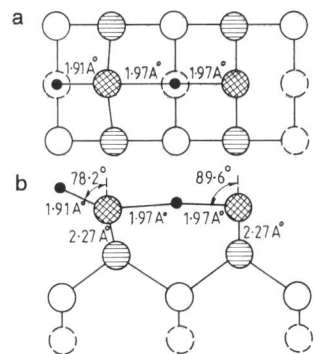

Figure 11. Monolayer bridge-site configuration for the Si(001)2×1:F system.

Half Monolayer Chlorine Coverage

To investigate the chemisorption of chlorine on the Si(001) 2×1 reconstructed surface we have followed exactly the same procedure as in the case of fluorine. The resulting behaviour is very different, however, as we shall see.

Starting with a chlorine atom at the dimer bridge sites of each 2×2 surface unit cell (position No. 1 in figure 2), and optimising the energy within the periodic SLAB-MINDO formalism, results in the chlorines moving to the adjacent dangling bond sites to produce the type 2s dangling bond configuration shown in figure 12. The Si-Si dimer bonds are buckled with the chlorine chemisorbed silicon being raised compared to its dimer counterpart. The dimer bondlengths are very close to the bulk nearest neighbour distance of 2.35 Å. The Si-Cl bonds are approximately 2.1 Å in length and are oriented approximately tetrahedrally with respect to the Si-Si dimer bonds. Each chlorine has gained ~0.46e, derived mainly from its associated silicon dimer atom (see table 4).

Commencing with chlorine at the same No.2 site of each 2×1 surface unit cell, or

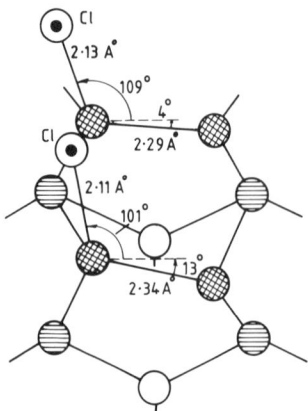
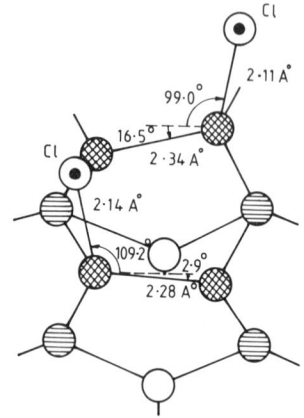

Figure 12. 0.5 ML type 2s dangling bond configuration for Cl on Si(001).

Figure 13. 0.5 ML type 2a dangling bond configuration for Cl on Si(001).

Table 4. The electronic charge on the chlorine and top layer silicons, and the total energy values, for the 0.5 ML and 1.0 ML structures shown in figs 12-16. The single asterisks denote silicon atoms directly bonded to a chlorine whilst the double asterisk labels the upper chlorine atom in the 1.0 ML bridge site configuration. The energy values are in eV per 2×2 unit cell (0.5 ML) and eV per 2×1 unit cell (1.0 ML).

	Si (001) 2×1:Cl - 0.5 ML			Si (001) 2×1:Cl - 1.0 ML	
	Type 1	Type 2s	Type 2a	Dangl. bond	Bridge site
Cl	7.42 7.42	7.48 7.44	7.52 7.44	7.57 7.57	7.30 7.34**
Si (1st layer)	3.61* 3.97	3.42* 3.52*	3.44* 3.70*	3.48* 3.48*	3.61* 3.61*
	3.61* 3.97	4.57 3.97	4.56 3.69		
Si (2nd layer)	4.13 4.13	4.23 3.91	4.17 3.95	4.05 4.05	4.20 4.20
Energy	-3208.39	-3208.28	-3208.17	-3145.83	-3142.87

chemisorbing from the No.3 sites, also produces the type 2s dangling-bond configuration of figure 12. Starting our geometry optimisation with the two chlorines on alternate No.2 sites, on the other hand, leads to the type 2a dangling bond configuration shown in figure 13. This configuration is 0.05 eV per chemisorbed chlorine higher in energy than the type 2s configuration presented in figure 12 (see table 4). The chemisorbed chlorines now have charges of 0.52e and 0.44e.

Placing the chlorines initially at the No.4 sites of the 2×2 surface unit cell yields a structure in which one of the chlorine atoms with a charge of 0.47e lies 2.10 Å almost directly above one of the silicon dimer atoms whilst the other has a charge of 0.52e and is sited 1.07 Å above the ideal surface, close to an adjacent No.4 site. This configuration, which is not shown, is unlikely to occur, however, as it is roughly 0.4 eV per chemisorbed chlorine less stable than the above dangling-bond configurations.

Commencing our optimisation process with the chlorines sited above the reconstructed

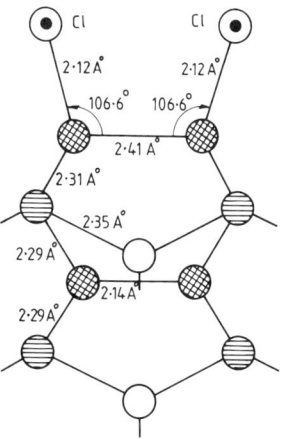

Figure 14. 0.5 ML type 1 dangling bond configuration for Cl on Si(001).

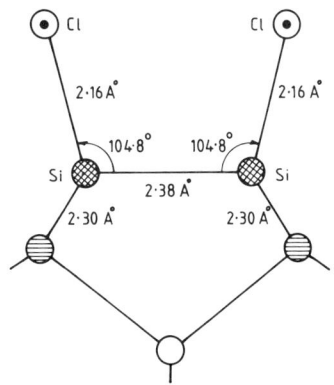

Figure 15. Monolayer dangling bond configuration for the Si(001)2×1:Cl system.

Si(001) surface at the No.5 sites yields the symmetric type 1 dangling bond topology shown in figure 14. Each Si-Cl bond is of length 2.12 Å and makes an angle of 106.6° with its associated Si-Si dimer bond. This structure is 0.05 eV per chemisorbed chlorine more stable than the type 2s dangling bond configuration of figure 12 and represents the lowest energy 0.5 ML structure that we have found for the Si(001) surface. In this configuration each chlorine has a nett charge of 0.42e derived primarily from the corresponding silicon dimer atoms (see table 4).

A recent STM study by John Boland[53] on a sub-monolayer chlorine-dosed Si(001)2×1 surface has shown that type 1 configurations are symmetric, that type 2s dangling bond configurations also occur, but not type 2a, and that the chlorines bond preferentially onto the same side of immediately adjacent dimers. Type 3 configurations in which only a single isolated dangling bond is saturated were also observed and shown to be strongly buckled. The above MINDO results are consistent with these observations in that the type 1 configuration is indeed symmetric and the type 2s topology is predicted to be lower in energy than the type 2a geometry.

Monolayer Chlorine Coverage

From all of the above calculations it is clear that the preferred chemisorption site for chlorine chemisorption on the silicon (001) surface is the dangling bond site. In order to study monolayer chemisorption we thus start with one chlorine per 2×1 surface unit cell occupying a dangling bond (No. 2) site and determine the resultant topologies when the second chlorine atom is placed at the positions 1, 3, 4 and 5 of figure 2.

Commencing our geometry optimisation with the second chlorine sited at either the No. 1, 3 or 5 sites results in the additional chlorine moving directly to the unoccupied dangling bond site to form the symmetric dangling bond topology shown in figure 15. This structure is very stable against both buckling and twisting of the dimers and is the lowest energy structure which we have found for monolayer chemisorption of chlorine onto the silicon (001) surface. This result is consistent with both experiment[54-60] and the first-principles calculations of Kruger and Pollmann.[61] The Si-Si dimer bondlength of 2.38 Å is marginally greater than the bulk nearest neighbour distance of 2.35 Å whilst the Si-Si backbonds are slightly shorter at 2.30 Å. The Si-Cl bonds are of length 2.16 Å and are oriented at 104.8° to the Si-Si dimer bonds. These structural parameters are in excellent agreement with the values derived by Kruger and Pollmann (2.40 Å, 2.33 Å, 2.05 Å and 105°)[61] and with those we have obtained independently from RHF STO-3G cluster calculations using GAMESS '92 (2.40 Å, 2.36 Å, 2.13 Å and 113.7°). Not suprisingly, they are also very similar to the half-monolayer values shown in figure 14. The charge on each chemisorbed chlorine in this case is 0.57e.

It is interesting to note that whilst all three of the theoretically determined Si-Cl bondlengths referred to above are significantly larger than the value of 1.95±0.04 Å derived from the NEXAFS experiments of Thornton et al.[56], all of the calculated bond angles lie within their determination of a tilt angle of somewhat less than 25°. Both the SLAB-MINDO value of 14.8° and the Kruger and Pollmann value of 15°, however, are substantially smaller than the recent ESDIAD determinations of 25°±4°[58] and ~30°[60]. The local density of states distribution corresponding to our SLAB-MINDO dangling bond configuration has been shown[62] to be in good agreement with both the LEELS spectra of Aoto et al.[55] and the polarisation-dependent angle-resolved photemission studies of Johansson et al.[57]

A different on-top geometry to the dangling bond configuration is obtained when the additional chlorine is initially positioned over the No. 4 sites. In this case, each dangling bond chlorine is found to move to a position 2.17 Å almost directly above its associated silicon dimer atom while the other chlorine atom per 2×1 surface unit cell remains close to its No. 4 site, 1.16 Å above the ideal surface. This is virtually identical to the behaviour observed for chemisorption from the No. 4 sites in the 0.5 ML case. This monolayer topology is unlikely to occur, however, as its energy is about 0.45 eV per chemisorbed chlorine higher than the above symmetric dangling bond configuration.

In previous work[62] we have also found a stable bridge site configuration for monolayer coverage of Cl on the Si(001) 2×1 surface. This structure is shown in figure 16. The chlorines lie directly on top of the bridge sites of the ideal surface, 0.98 and 1.20 Å above the silicon surface layer. The silicon substrate is essentially that of the ideal undistorted Si(001) 1×1 surface apart from an inward relaxation of the surface layer of 0.17 Å. The Si-Cl bond lengths are 2.17 and 2.25 Å. The upper and lower chlorines have aquired a charge of 0.34e and 0.30e respectively, with the surface silicons giving up 0.39e and the second layer silicons gaining 0.20e. This bridge site configuration is determined to be 1.50 eV per surface dimer less stable than the symmetric dangling bond configuration of figure 15. A similar result has been obtained by Kruger & Pollmann. It is interesting to note, however, that evidence for both the dangling bond and bridge site topologies has been observed in recent ESDIAD experiments.[48,58]

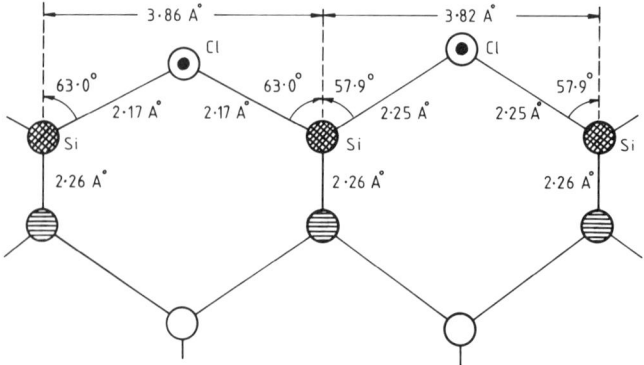

Figure 16. Monolayer bridge site configuration for the Si(001)2×1:Cl system.

DISCUSSION AND CONCLUSIONS

It is obvious from all of the above theoretical calculations that fluorine and chlorine should exhibit quite different chemisorption behaviour on the silicon (001)2×1 dimer reconstructed surface in agreement with experiment. Whilst the chemisorption of fluorine from above the surface can give rise to subsurface fluorine, the chemisorption of chlorine always results in an on-top configuration. Moreover, while fluorine will happily occupy either a dangling bond or a shared-dimer site above the surface, chlorine shows a strong preference for the dangling bond sites. The minimum energy configuration for chlorine at monolayer coverage is predicted to be a symmetric tetrahedral dangling bond configuration whereas the SLAB-MINDO formalism finds this structure to be unstable in the case of fluorine chemisorption. Starting our SLAB-MINDO geometry optimisation for fluorine from the dangling bond topology always results in one of the dangling bond fluorines moving to the dimer bridge site to form the "molecular-like" geometry. This behaviour, however, may simply be the result of an inadequate MINDO parameterisation of the F-F interactions and we are currently investigating this using CRYSTAL '92.

In addition, no evidence has been found in the case of chlorine for either the "molecular-like" or shared dimer:dangling-bond configurations which have been determined for monolayer coverage of fluorine on the silicon (001) surface. As we have seen, starting with one of the dangling-bonds per surface 2×1 unit cell saturated with fluorine, and chemisorbing a further fluorine atom from above the No.1 or No.2 positions, produces a "molecular-like" topology, whereas performing the same calculation with chlorine gives rise to the symmetric dangling-bond configuration. Similarly, saturating one dangling bond per 2×1 surface unit cell and chemisorbing onto the No.5 sites (of figure 2) gives rise to the shared dimer:dangling-bond configuration for fluorine, whilst chlorine again yields the symmetric dangling-bond configuration. Starting our chlorine optimisation from the "molecular-like" topology found for F_2 chemisorption on the Si(001)2×1 surface results in the dimer bridge site chlorine atom moving back to the unoccupied dangling bond site to form the fully saturated (symmetric) dangling bond configuration. Analogous calculations which we have performed using the GAMESS '92 RHF molecular orbital method with STO-3G basis functions also give the same result.

As a result of all of the calculations reported in this paper we now have a fairly good understanding of the chemisorption of atomic fluorine and chlorine onto the silicon (001) 2×1 reconstructed surface for all concentrations up to one monolayer. Our plan now is to use this knowledge to extend these calculations to higher exposures in an endeavour to understand the processes whereby the interaction of fluorine and chlorine with the silicon (001) surface leads to actual etching of the surface. To achieve this end, we propose to augment our current theoretical techniques with molecular dynamics calculations. This will enable us to study these processes dynamically as a function of both time and temperature. The required Si-F and Si-Cl interaction potentials will be determined to fitting to potential energy curves derived from first-principles calculations along the lines suggested in references 44 and 45. The resulting calculations will be similar to those already performed on the Si(001)-F system by other authors[44,63] but based on the novel structures which have been reported here. Such calculations should provide valuable insights into these very complex and fascinating chemisorption systems.

ACKNOWLEDGEMENTS

We would like to thank the Australian Research Council (ARC) and our own university Research Management Committee for financial support during the course of this work. We also wish to acknowledge the Quantum Chemistry Group at North Dakota State University for providing us with the GAMESS '92 software.

REFERENCES

1. C. B. Duke, *J. Vac. Sci. Technol.* B14:1336 (1993).
2. K. C. Pandey, *Phys. Rev. Lett.* 47:1913 (1981).

3. W. C. Fan, A. Ignatiev, H. Huang and S. Y. Tong, *Phys. Rev. Lett.* 62:1516 (1989).
4. V. Takayanagi, Y. Tanishino, M. Takahashi and S. Takahashi, *J. Vac. Sci. Technol.* A3:1502 (1985).
5. A. Zangwill, "Physics at Surfaces," Cambridge University Press, (1988).
6. J. A. Pople and D. L. Beveridge, "Approximate Molecular Orbital Theory," McGraw-Hill, New York, (1970).
7. R. C. Bingham, M. J. S. Dewar and D. H. Lo, *J. Amer. Chem. Soc.* 97:1285 (1975).
8. M. J. S. Dewar, E. G. Zoebisch, E. F. Healy and J. J. P. Stewart, *J. Amer. Chem. Soc.* 107:3902. (1985).
9. J. J. P. Stewart, *J. Comp. Chem.* 10:209 (1989).
10. M. P. Allen and D. J. Tildesley, "Computer Simulations in Liquids," Oxford University Press, Oxford, (1992).
11. C. Pisani, R. Dovesi and C. Roetti, "Lecture Notes in Chemistry," Vol. 48, Springer-Verlag, Heidelberg, (1988).
12. R. M. Dreizler and E. K. U. Gross, "Density Functional Theory," Springer-Verlag, Berlin, (1990).
13. J. Ihm, M. L. Cohen, and D. J. Chadi, *Phys. Rev. B* 21:4592 (1980).
    M. T. Yin and M. L. Cohen, *Phys. Rev. B* 24:2303 (1981).
14. J. Pollman and S. K. Pantelides, *Phys. Rev. B* 18:5524 (1978).
    S. Crampin, D. D. Vvedensky, J. M. MacLaren, and M. E. Eberhardt, *Phys. Rev. B* 40:3413 (1989).
15. G. Galli and M. Parinello, *in*: "Computer Simulation in Material Science," NATO ASI Series E: Applied Sciences, Vol. 205, M. Meyer and V. Pontikis, eds., Kluwer Academic, Dordrecht, (1991).
16. D. M. Ceperly and M. H. Kalos, "Quantum Monte Carlo Methods," M. Suzuki, ed., Springer-Verlag, New York, (1987).
17. Y. Enta, S. Suzuki and S. Kono, *Phys. Rev. Lett.* 65:2704 (1990).
18. See the review articles by:
    D. Haneman, *Rep. Prog. Phys.* 50:1045 (1987).
    J. E. Griffith and G. P. Kochanski, *Crit. Rev. Solid State Mater. Sci.* 16:255 (1990).
19. F. J. Himpsel and D. E. Eastman, *J. Vac. Sci. Technol.* 16:1297 (1979).
    H. A. Van Hoof, G. V. Hansson, J. M. Nicholls and S. A. Flodstrom, *Phys. Rev. B* 24:4684 (1981).
20. R. M. Tromp, R. G. Smeek and F. W. Saris, *Phys. Rev. Lett* 46:9392 (1981).
    R. M. Tromp, R. G. Smeek, F. W. Saris and D. J. Chadi, *Surf. Sci.* 133:137 (1983).
21. E. Landemark, C. J. Karlsson, Y. C. Chao and R. I. G. Uhrgerg, *Surf. Sci.* 287/288:529 (1993).
22. Y. S. Shu, W. S. Yang, F. Jona and P. M. Marcus, *in*: "The Structure of Surfaces," M. A. Van Hove and S. Y. Tong, eds., Springer-Verlag, Berlin, (1985).
    B. W. Holland, C. B. Duke and A. Paton, *Surf. Sci.* 140:L269 (1984).
23. S. Y. Tong and A. L. Maldonado, *Surf. Sci.* 34:90 (1973).
24. L. S. O. Johansson, R. I. G. Uhrberg, P. Martensson and G. V. Hansson, *Phys. Rev. B* 42:1303 (1990)
25. R. J. Hamers, R. M. Tromp and J. E. Demuth, *Phys. Rev. B* 34:5343 (1986).
    P. Avouris, *J. Phys. Chem.* 94:2246 (1990)
26. K. C. Pandey, *in*: "Proceedings of the 17th International Conference on the Physics of Semiconductors," D. J. Chadi and W. A. Harrison, eds., Springer-Verlag, Berlin, (1984)
27. D. J. Chadi, *Phys. Rev. Lett.* 43:43 (1979).
28. B. I. Craig and P. V. Smith, *Surf. Sci.* 218:569 (1989).
29. Z. Zhu, N. Shima and M. Tsukada, *Phys. Rev. B* 40:11868 (1989).
    J. Dabrowski and M. Scheffler, *Appl. Surf. Sci.* 56:15 (1992).
    K. Kobayashi, Y. Morikawa, K. Therakura and S. Blägel, *Phys. Rev. B* 45:3469 (1992).
    M. C. Payne, N. Roberts, R. J. Needs, M. Needels and J. D. Joannopoulos, *Surf. Sci.* 211/212:1 (1989).
    I. P. Batra, *Phys. Rev. B* 41:5048 (1990).
    S. Tang, A. J. Freeman and B. Delley, *Phys. Rev. B* 45:1776 (1992).
30. E. Artacho and F. Yndurain, *Phys. Rev. Lett.* 62:2491 (1984).
31. R. A. Wolkow, *Phys. Rev. Lett.* 68:2636 (1992).
32. K. Cho and J. D. Joannopoulos, *Phys. Rev. Lett.* 71:1387 (1993).
33. T. Vichon, A. M. Oles, D. Spanjaard and M. C. Desjonquères, *Surf. Sci.* 287/288:534 (1993).
34. W. Frotzheim, *in*: "The Chemical Physics of Solid and Heterogeneous Catalysis," Vol. 5, B. A. King and D. W. Woodruff, eds., Elsevier, Amsterdam (1988), p215.
    T. M. Mayer, M. S. Ameen and D. J. Vitkavage, *ibid*, p427.
    H. S. Winters and J. W. Coburn, *Surf. Sci. Rep.* 14:161 (1992).
35. K. Asakawa and S. Sugata, *J. Vac. Sci. Technol.* B3:402 (1985).
    S. J. Pearton, U. K. Chakrabarti, W. S. Hobson and A. P. Kinsella, *J. Vac. Sci. Technol.* B8:607 (1990).
36. P. A. Moka and D. J. Ehrlich, *Appl. Phys. Lett.* 55:91 (1989).
37. B. I. Craig and P. V. Smith, *Surf. Sci.* 210:468 (1989).

38. M. Dupuis, D. Spangler, and J. J. Wendoloski, National Resource for Computations in Chemistry Software Catalog, University of California: Berkeley, CA (1980), Program QG01.
    M. W. Schmidt et al., *QCPE Bulletin*, 10:52 (1990).
39. M. J. Frisch et al., "Gaussian '90," Gaussian Inc., Pittsburgh, (1990).
40. R. Dovesi et al., "Crystal '92," University of Turin, (1992).
41. See, for example, X. M. Zheng and P. V. Smith, *Surf. Sci.* 279:127 (1992) and references therein.
42. C. G. Van de Walle, F. R. McFeely and S. T. Pantelides, *Phys. Rev. Lett.* 61:1867 (1988).
43. P. J. Van der Hoek, W. Ravenek and E. J. Baerend, *Phys. Rev. B* 38:1208 (1988).
44. P. C. Weakliem, C. J. Wu, and E. A. Carter, *Phys. Rev. Letters* 69:200 (1992).
45. C. J. Wu and E. A. Carter, *Phys. Rev. B* 45:9065 (1992).
46. A. L. Johnson, M. M. Walczak and T. E. Madey, *Langmuir* 4:277 (1988).
47. M. J. Bozack, M. J. Dresser, W. J. Choyke, P. A. Taylor and J. T. Yates Jr., *Surf. Sci.* 184:L332 (1987).
48. S. L. Bennett, C. L. Greenwood and E. M. Williams, *Surf. Sci.* 290:267 (1993).
49. T. Hashizume, K. Motai, Y. Hasegawa, L. Sumita, H. Tanaka, S. Amano, S. Hydo and T. Sakurai, *J. Vac. Sci. Technol.* B9:745 (1991).
50. R. Cao, X. Yang and P. Pianetta, *J. Vac. Sci. Technol.* B11:1455 (1993).
51. C. J. Wu and E. A. Carter, *J. Amer. Chem. Soc.* 113:9061 (1991).
52. B. I. Craig and P. V. Smith, *Surf. Sci.* 262:235 (1992).
53. J. J. Boland, private communication.
54. J. E. Rowe, G. Margaritondo and S. B. Christman, *Phys. Rev. B* 16:1581 (1977).
55. N. Aoto, E. Ihawa and Y. Kuragi, *Surf. Sci.* 199:408 (1988).
56. G. Thornton, P. L. Wincott, R. McGrath, I. T. McGovern, F. M. Quinn, D. Norman and D. D. Vvedensky, *Surf. Sci.* 211/212:959 (1989).
57. L. S. O. Johansson, R. I. G. Uhrberg, R. Lindsay, P. L. Wincott and G. Thornton, *Phys. Rev. B* 42:9534 (1990).
58. C. C. Cheng, Q. Gao, W. J. Choyke and J. T. Yates, Jr., *Phys. Rev. B* 46:12810 (1992).
59. D. Purdie, C. A. Muryn, N. S. Prakash, K. G. Purcell, P. L. Wincott, G. Thornton and D. S. L. Law, *J. Phys. Cond. Matt.* 3:7751 (1991).
60. J. T. Yates, Jr., M. D. Alvey, M. J. Dresser, M. A. Henderson, M. Kiskinova, R. D. Ramsier and A. Szabó, *Science* 255:1397 (1992).
61. P. Kruger and J. Pollman, *Phys. Rev. B* 47:1898 (1993).
62. B. I. Craig and P. V. Smith, *Surf. Sci. Lett.* 290:L662 (1993).
63. F. H. Stillinger and T. A. Weber, *Phys. Rev. Lett.* 62:2144 (1989).
    T. A. Schoolcraft and B. J. Garrison, *J. Vac. Sci. Technol.* A8:3496 (1990).

FUNCTIONAL INTEGRAL TECHNIQUES IN CONDENSED MATTER PHYSICS

Nguyen Van Hieu

Institute of Materials Science
P.O.Box 607 Hanoi
Vietnam

Path - integral formulation of quantum mechanics has been proposed long ago by Feynman (Feynman and Hibbs, 1965) and was then widely applied in quantum physics. Nowadays an unified theory based on the use of the functional integrals was firmly established and developed for studying all quantum systems, including many - body systems in statistical physics and condensed matter theory (Sakita, 1985; Wiegel, 1986; de Wit, 1981) as well as those of interacting quantized fields (Fadeev, 1976; Itzykson and Zuber, 1980; Lee, 1976; Ramond, 1981). This review is an introduction to the functional integral techniques in statistical physics and condensed matter theory. It is written for the readers who had no experience to work with the functional integrals. Therefore together with the elementary proof of the basic formulae of the functional integral techniques we explain the meaning of the functional integrals and show how can they be expressed in terms of the conventional integrals. We suppose that the readers had some basic understandings of the traditional methods of the quantum field theory in statistical physics (Abrikosov et al., 1965). The basic formulae of the functional integral techniques are derived in Sec. I for the bosonic systems and in Sec. II for the fermionic ones. They were applied widely in studying many problems of quantum theory of condensed matters. Some examples of these applications are presented in the concluding Sec. III.

I. FUNCTIONAL INTEGRALS FOR BOSONIC SYSTEMS

We begin the review by considering the quantum system with one bosonic degree of freedom. In this simplest example we explain what does the functional integral mean, and verify the basic formulae of the functional integral technique for boson fields. In the second quantization formalism the physical quantities are expressed in terms of the particle destruction and creation operators $\hat{c}$ and $\hat{c}^+$ satisfying the commutation relation

$$[\hat{c}, \hat{c}^+] = \hat{c}\hat{c}^+ - \hat{c}^+\hat{c} = 1. \tag{I.1}$$

The Hamiltonian of the free particle system and the particle number operator equal

$$\hat{H} = E_0 \hat{c}^+ \hat{c}, \tag{I.2}$$

$$\hat{N} = \hat{c}^+ \hat{c}, \tag{I.3}$$

where E_0 is the energy of each particle. The partition funtion involves directly the operator

$$\tilde{H}(\hat{c}^+, \hat{c}) = \hat{H} - \mu \hat{N} = E \hat{c}^+ \hat{c}, \tag{I.4}$$

$$E = E_0 - \mu, \tag{I.5}$$

where μ is the chemical potential. $\tilde{H}(\hat{c}^+, \hat{c})$ can be also considered as the Hamiltonian of the system with some energy shift.

Instead of the operators $\hat{c}$ and $\hat{c}^+$ the functional integral contains commuting functions $c(\tau)$ and $c^*(\tau)$ of the imaginary time τ. These functions are periodical

$$c(\tau + \beta) = c(\tau), \quad c^*(\tau + \beta) = c^*(\tau) \tag{I.6}$$

with the period

$$\beta = \frac{1}{kT},$$

where k is the Boltzman constant and T is the temperature. Replacing $\hat{c}$ and $\hat{c}^+$ by $c(\tau)$ and $c^*(\tau)$ in the Hamiltonian (I.4) we obtain the corresponding function

$$\tilde{H}(\tau) = Ec^*(\tau)c(\tau) \tag{I.7}$$

We shall show that the partition function

$$\text{Tr}\left\{e^{-\beta \tilde{H}(\hat{c}^+,\hat{c})}\right\}$$

is expressed in terms of the functional integral

$$\int [Dc][Dc^*] \exp\left\{-\int_0^\beta [c^*(\tau)\frac{dc(\tau)}{d\tau} + \tilde{H}(\tau)]d\tau\right\}$$

with some obvious definition of the later.

Before to do that we note that the Hamiltonian (I.4) is not identical to the expression

$$\tilde{H}(\hat{c},\hat{c}^+) = E\hat{c}\hat{c}^+ \tag{I.8}$$

obtained from $\tilde{H}(\hat{c}^+,\hat{c})$ by permuting $\hat{c}$ and $\hat{c}^+$. However, if we replace $\hat{c}$ and $\hat{c}^+$ by $c(\tau)$ and $c^*(\tau)$, from two different operators (I.4) and (I.8) we obtain one and the same function (I.7), because $c(\tau)$ and $c^*(\tau)$ commute one with another. In order to have the one - to - one correspondence it is natural to form the symmetrized product of two operators and define

$$\tilde{H}_S(\hat{c}^+,\hat{c}) = \tilde{H}_S(\hat{c},\hat{c}^+) = E\frac{\hat{c}^+\hat{c} + \hat{c}\hat{c}^+}{2}. \tag{I.9}$$

The correspondence between $\tilde{H}_S(\hat{c}^+,\hat{c})$ and $\tilde{H}(\tau)$ is one - to - one.

Now we show that the partition function for the system of free bosons with one degree of freedom can be calculated either by directly evaluating the matrix elements of the operators or by means of the functional integral technique

$$Z_0 = Z_0(E) = \text{Tr}\left\{e^{-\beta \tilde{H}_S(\hat{c}^+,\hat{c})}\right\} =$$

$$= \int [Dc][Dc^*]\exp\left\{-\int_0^\beta [c^*(\tau)\frac{dc(\tau)}{d\tau} + \tilde{H}(\tau)]d\tau\right\}. \tag{I.10}$$

In order to understand the meaning of the functional integral in Eq. (I.10) we express it in terms of the conventional integrals. For that purpose we perform the Fourier expansion of $c(\tau)$

$$c(\tau) = \frac{1}{\sqrt{\beta}} \sum_\nu e^{i\omega_\nu \tau} c_\nu, \qquad \omega_\nu = 2\nu \frac{\pi}{\beta}, \tag{I.11}$$

ν being integers. The periodicity condition (I.6) is satisfied automatically. Then we have

$$\int_0^\beta \left[c^*(\tau) \frac{dc(\tau)}{d\tau} + \tilde{H}(\tau) \right] d\tau = \sum_\nu (i\omega_\nu + E) c_\nu^* c_\nu. \tag{I.12}$$

We define the functional integral in the following manner

$$\int [Dc][Dc^*] \cdots \cdots = \text{const} \prod_\nu \int \frac{dc_\nu dc_\nu^*}{2\pi i} \cdots \cdots \tag{I.13}$$

and obtain

$$\int [Dc][Dc^*] \exp\left\{ -\int_0^\beta \left[c^*(\tau) \frac{dc(\tau)}{d\tau} + \tilde{H}(\tau) \right] d\tau \right\} =$$

$$= \text{const} \prod_\nu \left(\int e^{-(i\omega_\nu + E) c_\nu^* c_\nu} \frac{dc_\nu dc_\nu^*}{2\pi i} \right), \tag{I.14}$$

with some unspecified constant. By means of the analytical continuation of the formula

$$\int e^{-\alpha z^* z} \frac{dz dz^*}{2\pi i} = \int e^{-\alpha(x^2 + y^2)} \frac{dx dy}{\pi} = \frac{1}{\alpha}$$

in the variable α we get

$$\int e^{-(i\omega_\nu + E) c_\nu^* c_\nu} \frac{dc_\nu dc_\nu^*}{2\pi i} = \frac{1}{i\omega_\nu + E}.$$

Therefore

$$Z_0(E) = \int [Dc][Dc^*] \exp\left\{ -\int_0^\beta \left[c^*(\tau) \frac{dc(\tau)}{d\tau} + \tilde{H}(\tau) \right] d\tau \right\} = \text{const} \prod_\nu \frac{1}{i\omega_\nu + E}.$$

The infinite product can be evaluated formally in the following manner. We write

$$\ln Z_0(E) = \text{const} - \sum_\nu \ln(i\omega_\nu + E)$$

and calculate the derivative in the variable E

$$\frac{d\ln Z_0(E)}{dE} = -\sum_\nu \frac{1}{i\omega_\nu + E}.$$

Because

$$\sum_\nu \frac{1}{i\omega_\nu + E} = \frac{\beta}{2}\frac{1+e^{-\beta E}}{1-e^{-\beta E}}$$

we have then

$$\frac{d\ln Z_0(E)}{dE} = -\frac{\beta}{2} - \frac{d\ln(1-e^{-\beta E})}{dE}$$

and therefore

$$Z_0(E) = \text{const}\,\frac{e^{-\frac{\beta E}{2}}}{1-e^{-\beta E}}. \tag{I.15}$$

The same result can be also derived directly from the definition

$$Z_0(E) = \text{tr}\left\{e^{-\beta \tilde{H}_S(\hat{c}^+,\hat{c})}\right\}.$$

Indeed, denote $|n\rangle$ the eigenvector with n particles

$$\hat{c}^+\hat{c}|n\rangle = n|n\rangle,$$

we have

$$\text{Tr}\left\{e^{-\beta E \frac{\hat{c}^+\hat{c}+\hat{c}\hat{c}^+}{2}}\right\} = e^{-\frac{\beta E}{2}}\text{Tr}\left\{e^{-\beta E \hat{c}^+\hat{c}}\right\} = e^{-\frac{\beta E}{2}}\sum_n \langle n|e^{-\beta E \hat{c}^+\hat{c}}|n\rangle =$$

$$= e^{-\frac{\beta E}{2}}\sum_n (e^{-\beta E})^n = \frac{e^{-\frac{\beta E}{2}}}{1-e^{-\beta E}}.$$

Thus the basic formula (I.10) has been verified, and the meaning of the functional integral was also explained.

Considering the expression under the functional integral in the r.h.s. of Eq. (I.10) as the statistical weight we define the expectation values of the products of the functions $c(\tau)$ and $c^*(\tau)$ at different values of τ in the following manner

$$\langle c(\tau_1)\ldots c(\tau_n)c^*(\sigma_1)\ldots c^*(\sigma_n)\rangle = \frac{1}{Z_0}\int [Dc][Dc^*]c(\tau_1)\ldots c(\tau_n)c^*(\sigma_1)\ldots c^*(\sigma_n).$$

$$\exp\left\{-\int_0^\beta c^*(\tau)\left(\frac{d}{d\tau}+E\right)c(\tau)d\tau\right\}. \qquad (I.16)$$

In order to calculate these expectation values we introduce the generating functional

$$Z_0[\eta,\eta^*] = \int [Dc][Dc^*]\exp\left\{-\int_0^\beta \left[c^*(\tau)\left(\frac{d}{d\tau}+E\right)c(\tau)\right.\right.$$
$$\left.\left.+\eta^*(\tau)c(\tau)+c^*(\tau)\eta(\tau)\right]d\tau\right\} \qquad (I.17)$$

depending on two functions $\eta(\tau)$ and $\eta^*(\tau)$. These functions are also periodical in the variable τ with the period β

$$\eta(\tau+\beta) = \eta(\tau), \quad \eta^*(\tau+\beta) = \eta^*(\tau). \qquad (I.18)$$

It is obvious that

$$\langle c(\tau_1)\ldots c(\tau_n)c^*(\sigma_1)\ldots c^*(\sigma_n)\rangle =$$
$$= \frac{1}{Z_0}\frac{\delta^{2n} Z_0[\eta,\eta^*]}{\delta\eta^*(\tau_1)\ldots\delta\eta^*(\tau_n)\delta\eta(\sigma_1)\ldots\delta\eta(\sigma_n)}\bigg|_{\eta=\eta^*=0}. \qquad (I.19)$$

Now we rewrite the generating functional (I.17) in a new form convenient for the calculation of the functional derivatives in the r.h.s. of Eq. (I.19). Denote $D_E(\tau)$ the periodical solution of the differential equation

$$\left(\frac{d}{d\tau}+E\right)D_E(\tau) = \delta(\tau) \qquad (I.20)$$

with the period β

$$D_E(\tau+\beta) = D_E(\tau). \qquad (I.21)$$

We can verify that

$$D_E(\tau) = \frac{\theta(\tau)e^{-E\tau} + \theta(-\tau)e^{-E(\tau+\beta)}}{1 - e^{-\beta E}}. \tag{I.22}$$

In the r.h.s. of the definition (I.10) of the partition function Z_0 let us perform the shift of the integration variables

$$c(\tau) \to c(\tau) + \int d\sigma D_E(\tau - \sigma)\eta(\sigma),$$

$$c^*(\tau) \to c^*(\tau) + \int d\sigma \eta^*(\sigma) D_E(\sigma - \tau).$$

Using the differential equation (I.20) and the periodicity conditions (I.6), (I.18) and (I.21) we can verify the relation

$$\int_0^\beta \left[c^*(\tau) + \int d\sigma \eta^*(\sigma) D_E(\sigma - \tau)\right]\left(\frac{d}{d\tau} + E\right)\left[c(\tau) + \int d\sigma D_E(\tau - \sigma)\eta(\sigma)\right] d\tau$$

$$= \int_0^\beta \left[c^*(\tau)\left(\frac{d}{d\tau} + E\right)c(\tau) + \eta^*(\tau)c(\tau) + c^*(\tau)\eta(\tau)\right] d\tau$$

$$+ \int_0^\beta d\tau \int_0^\beta d\sigma \eta^*(\tau) D_E(\tau - \sigma)\eta(\sigma).$$

Because of the invariance of the functional integral under any shift of the integration variables we have now

$$Z_0 = \int [Dc][Dc^*]\exp\left\{-\int_0^\beta \left[c^*(\tau)\left(\frac{d}{d\tau} + E\right)c(\tau) + \eta^*(\tau)c(\tau) + c^*(\tau)\eta(\tau)\right] d\tau - \int_0^\beta d\tau \int_0^\beta d\sigma \eta^*(\tau) D_E(\tau - \sigma)\eta(\sigma)\right\}. \tag{I.23}$$

From Eqs. (I.17) and (I.23) it follows that

$$Z_0[\eta, \eta^*] = Z_0 \exp\left\{\int_0^\beta d\tau \int_0^\beta d\sigma \eta^*(\tau) D_E(\tau - \sigma)\eta(\sigma)\right\}. \tag{I.24}$$

Inserting this expression into the r.h.s. of Eq. (I.19) we obtain immediately the expectation values of the products of the functions $c(\tau)$ and $c^*(\tau)$:

$$\langle c(\tau_1)\ldots c(\tau_n)c^*(\sigma_1)\ldots c^*(\sigma_n)\rangle = \sum_{P(\sigma_1,\ldots,\sigma_n)} D_E(\tau_1-\sigma_1)\ldots D_E(\tau_n-\sigma_n) \quad (I.25)$$

where

$$\sum_{P(\sigma_1,\ldots,\sigma_n)}$$

denotes the sum over all permutations of the variables $\sigma_1, \sigma_2, \ldots, \sigma_n$. In particular

$$\langle c(\tau)c^*(\sigma)\rangle = \frac{1}{Z_0}\int [Dc][Dc^*]c(\tau)c^*(\sigma)\exp\left\{-\int_0^\beta c^*(\tau)\left(\frac{d}{d\tau}+E\right)c(\tau)d\tau\right\}$$

$$= \frac{1}{Z_0}\frac{\delta^2 Z_0[\eta,\eta^*]}{\delta\eta^*(\tau)\delta\eta(\sigma)}\bigg|_{\eta=\eta^*=0} = D_E(\tau-\sigma). \quad (I.26)$$

Now we show that the expectation values (I.25) of the products of the functions $c(\tau)$ and $c^*(\tau)$ coincide with the Matsubara imaginary time Green functions in the second quantization method. Denote $\hat{c}(\tau)$ and $\hat{\tilde{c}}(\tau)$ the imaginary time - dependent quantum operators

$$\hat{c}(\tau) = e^{\tau\tilde{H}(\hat{c}^+,\hat{c})}\hat{c}\,e^{-\tau\tilde{H}(\hat{c}^+,\hat{c})},$$
$$\hat{\tilde{c}}(\tau) = e^{\tau\tilde{H}(\hat{c}^+,\hat{c})}\hat{c}^+ e^{-\tau\tilde{H}(\hat{c}^+,\hat{c})}. \quad (I.27)$$

The two-point Green function is the mean value of the imaginary time - ordered product of the operators $\hat{c}(\tau)$ and $\hat{\tilde{c}}(\sigma)$ in the canonical ensemble with the Hamiltonian $\tilde{H}(\hat{c}^+,\hat{c})$ at the temperature T

$$G_0^B(\tau-\sigma) = \langle T(\hat{c}(\tau)\,\hat{\tilde{c}}(\sigma))\rangle, \quad (I.28)$$

$$\langle\ldots\ldots\rangle = \frac{\text{Tr}\left\{\ldots e^{-\beta\tilde{H}(\hat{c}^+,\hat{c})}\right\}}{\text{Tr}\left\{e^{-\beta\tilde{H}(\hat{c}^+,\hat{c})}\right\}}. \quad (I.29)$$

By the definition of the time - ordered product

$$T(\hat{c}(\tau)\hat{\bar{c}}(\sigma)) = \theta(\tau - \sigma)\hat{c}(\tau)\hat{\bar{c}}(\sigma) + \theta(\sigma - \tau)\hat{\bar{c}}(\sigma)\hat{c}(\tau). \tag{I.30}$$

In the case of the Hamiltonian (I.4) we have

$$\hat{c}(\tau) = e^{-\tau E}\hat{c}, \quad \hat{\bar{c}}(\sigma) = e^{\sigma E}\hat{c}^+.$$

It is easy to verify that

$$\mathrm{Tr}\{\hat{c}^+\hat{c}e^{-\beta E\hat{c}^+\hat{c}}\} = \sum_n \langle n|\hat{c}^+\hat{c}e^{-\beta E\hat{c}^+\hat{c}}|n\rangle = \frac{e^{-\beta E}}{(1 - e^{-\beta E})^2},$$

$$\mathrm{Tr}\{\hat{c}\hat{c}^+e^{-\beta E\hat{c}^+\hat{c}}\} = \sum_n \langle n|\hat{c}\hat{c}^+e^{-\beta E\hat{c}^+\hat{c}}|n\rangle = \frac{1}{(1 - e^{-\beta E})^2}.$$

Therefore

$$G_0^B(\tau - \sigma) = e^{-E(\tau-\sigma)}\frac{\theta(\tau - \sigma) + \theta(\sigma - \tau)e^{-\beta E}}{1 - e^{-\beta E}} = D_E(\tau - \sigma). \tag{I.31}$$

Thus we obtain the relation

$$\langle T(\hat{c}(\tau)\hat{\bar{c}}(\sigma))\rangle = \langle c(\tau)c^*(\sigma)\rangle. \tag{I.32}$$

Consider the $2n$ - point Green function

$$G_0^B(\tau_1,\ldots,\tau_n;\sigma_1,\ldots,\sigma_n) = \langle T(\hat{c}(\tau_1)\ldots\hat{c}(\tau_n)\hat{\bar{c}}(\sigma_1)\ldots\hat{\bar{c}}(\sigma_n))\rangle \tag{I.33}$$

By means of the straightforward calculations we can express it in terms of the two - point Green functions in the following manner (Wick theorem)

$$G_0^B(\tau_1,\ldots,\tau_n;\sigma_1,\ldots,\sigma_n) = \sum_{P(\sigma_1,\ldots,\sigma_n)} G_0^B(\tau_1 - \sigma_1)\ldots G_0^B(\tau_n - \sigma_n). \tag{I.34}$$

Comparing Eq. (I.34) with Eq. (I.25) and using the relation (I.31) we conclude that the $2n$- point Green functions are expressed in terms of the functional integral (I.16) of the products of the functions $c(\tau_i)$ and $c^*(\tau_i)$

$$\langle T(\hat{c}(\tau_1)\ldots\hat{c}(\tau_n)\hat{\bar{c}}(\sigma_1)\ldots\hat{\bar{c}}(\sigma_n))\rangle =$$

$$= \frac{1}{Z_0}\int [Dc][Dc^*]c(\tau_1)\ldots c(\tau_n)c^*(\sigma_1)\ldots c^*(\sigma_n)\exp\left\{-\int_0^\beta c^*(\tau)\left(\frac{d}{d\tau}\right.\right.$$

$$\left.\left.+E\right)c(\tau)d\tau\right\} = \frac{1}{Z_0}\cdot\frac{\delta^{2n}Z_0[\eta,\eta^*]}{\delta\eta^*(\tau_1)\ldots\delta\eta^*(\tau_n)\delta\eta(\sigma_1)\ldots\delta\eta(\sigma_n)}\bigg|_{\eta=\eta^*=0}. \tag{I.35}$$

This is the basic formula of the functionnal integral technique for a free bosonic system with one degree of freedom.

Above results can be easily generalized to the case of a free bosonic system with many degrees of freedom. For example, consider the system of bosons with momenta $\mathbf{p}$ and energies $E_0(\mathbf{p})$. Denote $\hat{c}(\mathbf{p})$ and $\hat{c}^+(\mathbf{p})$ their destruction and creation operators, $\hat{c}(\mathbf{p},\tau)$ and $\hat{\bar{c}}(\mathbf{p},\tau)$ the corresponding τ - dependent quantum operators, and introduce the commuting functions $c(\mathbf{p},\tau)$ and $c^*(\mathbf{p},\tau)$ as the functional integration variables. The Hamiltonian and the particle number operator equal

$$\hat{H} = \sum_\mathbf{p} E_0(\mathbf{p})\hat{c}^+(\mathbf{p})\hat{c}(\mathbf{p}), \tag{I.36}$$

$$\hat{N} = \sum_\mathbf{p} \hat{c}^+(\mathbf{p})\hat{c}(\mathbf{p}). \tag{I.37}$$

The partition function

$$Z_0 = \mathrm{Tr}\left\{e^{-\beta\tilde{H}_s}\right\}$$

is expresses in terms of the functional integral

$$Z_0 = \int [Dc][Dc^*]\exp\left\{-\int_0^\beta \sum_\mathbf{p} c^*(\mathbf{p},\tau)\left[\frac{d}{d\tau}+E_0(\mathbf{p})-\mu\right]c(\mathbf{p},\tau)d\tau\right\} \tag{I.38}$$

The generating functional depends on the functions $\eta(\mathbf{p},\tau)$ and $\eta^*(\mathbf{p},\tau)$

$$Z_0[\eta,\eta^*] = \int [Dc][Dc^*]\exp\left\{-\int_0^\beta \sum_\mathbf{p}\left(c^*(\mathbf{p},\tau)\left[\frac{d}{d\tau}+E_0(\mathbf{p})-\mu\right]c(\mathbf{p},\tau)+\right.\right.$$

$$\left.\left.+c^*(\mathbf{p},\tau)\eta(\mathbf{p},\tau)+\eta^*(\mathbf{p},\tau)c(\mathbf{p},\tau)\right)d\tau\right\} \tag{I.39}$$

In this case we have following basic formula of the functional integral technique

$$
\begin{aligned}
G_0^B&(\mathbf{p}_1,\tau_1;\ldots\mathbf{p}_n,\tau_n;\mathbf{q}_1,\sigma_1;\ldots\mathbf{q}_n,\sigma_n) = \\
&= \langle T(\hat{c}(\mathbf{p}_1,\tau_1)\ldots\hat{c}(\mathbf{p}_n,\tau_n)\hat{\bar{c}}(\mathbf{q}_1,\sigma_1)\ldots\hat{\bar{c}}(\mathbf{q}_n,\sigma_n))\rangle = \\
&= \frac{1}{Z_0}\int[Dc][Dc^*]c(\mathbf{p}_1,\tau_1)\ldots c(\mathbf{p}_n,\tau_n)c^*(\mathbf{q}_1,\sigma_1)\ldots c^*(\mathbf{q}_n,\sigma_n)\cdot \\
&\exp\left\{-\int_0^\beta \sum_\mathbf{p} c^*(\mathbf{p},\tau)\left[\frac{d}{d\tau}+E_0(\mathbf{p})-\mu\right]c(\mathbf{p},\tau)d\tau\right\} = \\
&= \frac{1}{Z_0}\frac{\delta^{2n}Z_0[\eta,\eta^*]}{\delta\eta^*(\mathbf{p}_1,\tau_1)\ldots\delta\eta^*(\mathbf{p}_n,\tau_n)\delta\eta(\mathbf{q}_1,\sigma_1)\ldots\delta\eta(\mathbf{q}_n,\sigma_n)}\bigg|_{\eta=\eta^*=0}.
\end{aligned}
\quad (\text{I.40})
$$

With the same reasonings as in the derivation of the formula (I.24) we can establish a similar relation between the generating functional $Z_0[\eta,\eta^*]$ defined in Eq. (I.39) and the partition function Z_0 defined by Eq. (I.38). This relation reads

$$Z_0[\eta,\eta^*] = Z_0\exp\left\{\int_0^\beta d\tau\int_0^\beta d\sigma\sum_\mathbf{p}\eta^*(\mathbf{p},\tau)D(\mathbf{p},\tau-\sigma)\eta(\mathbf{p},\sigma)\right\} \quad (\text{I.41})$$

where $D(\mathbf{p},\tau)$ is the periodical solution of the equation

$$\left[\frac{d}{d\tau}+E_0(\mathbf{p})-\mu\right]D(\mathbf{p},\tau) = \delta(\tau) \quad (\text{I.42})$$

and equals

$$D(\mathbf{p},\tau) = \frac{\theta(\tau)e^{-[E_0(\mathbf{p})-\mu]\tau}+\theta(-\tau)e^{-[E_0(\mathbf{p})-\mu](\tau+\beta)}}{1-e^{-\beta[E_0(\mathbf{p})-\mu]}}. \quad (\text{I.43})$$

For the system of interacting bosons with the destruction and creation operators $\hat{c}(\mathbf{p})$ and $\hat{c}^+(\mathbf{p})$ the total Hamiltonian has the general form

$$H[\hat{c}^+(\mathbf{p}),\hat{c}(\mathbf{p})] = \sum_\mathbf{p} E_0(\mathbf{p})\hat{c}^+(\mathbf{p})\hat{c}(\mathbf{p}) + H_{int}[\hat{c}^+(\mathbf{p}),\hat{c}(\mathbf{p})]. \quad (\text{I.44})$$

In the Schrodinger picture the Matsubara imaginary time Green functions equal

$$
\begin{aligned}
G^B&(\mathbf{p}_1,\tau_1;\ldots\mathbf{p}_n,\tau_n;\mathbf{q}_1,\sigma_1,\ldots\mathbf{q}_n,\sigma_n) = \\
&\frac{\langle T(\hat{c}(\mathbf{p}_1,\tau_1)\ldots\hat{c}(\mathbf{p}_n,\tau_n)\hat{\bar{c}}(\mathbf{q}_1,\sigma_1)\ldots\hat{\bar{c}}(\mathbf{q}_n,\sigma_n)S)\rangle}{\langle S\rangle}
\end{aligned}
\quad (\text{I. 45})
$$

where S is the S-matrix

$$S = T e^{-\int_0^\beta H_{int}\left[\hat{\bar{c}}(\mathbf{p},\tau),\hat{c}(\mathbf{p},\tau)\right]d\tau} \tag{I.46}$$

By expanding the exponential in the r.h.s. of Eq. (I.46) and applying formula (I.40) for each product of quantum operators we can show that the Green functions (I.45) are expressed in terms of following partition function and generating functional

$$Z = \int [Dc][Dc^*] \exp\left\{-\int_0^\beta \sum_\mathbf{p} c^*(\mathbf{p},\tau)\left[\frac{d}{d\tau} + E_0(\mathbf{p}) - \mu\right]c(\mathbf{p},\tau)d\tau - \int_0^\beta H_{int}\left[c^*(\mathbf{p},\tau),c(\mathbf{p},\tau)\right]d\tau\right\}, \tag{I.47}$$

$$Z[\eta,\eta^*] = \int [Dc][Dc^*] \exp\left\{-\int_0^\beta \sum_\mathbf{p} \left(c^*(\mathbf{p},\tau)\left[\frac{d}{d\tau} + E_0(\mathbf{p}) - \mu\right]c(\mathbf{p},\tau) + \eta^*(\mathbf{p},\tau)c(\mathbf{p},\tau) + c^*(\mathbf{p},\tau)\eta(\mathbf{p},\tau)\right)d\tau - \int_0^\beta H_{int}\left[c^*(\mathbf{p},\tau),c(\mathbf{p},\tau)\right]d\tau\right\}, \tag{I.48}$$

$$G^B\left(\mathbf{p}_1,\tau_1;\ldots\mathbf{p}_n,\tau_n;\mathbf{q}_1,\sigma_1;\ldots\mathbf{q}_n,\sigma_n\right) =$$
$$\frac{1}{Z}\int [Dc][Dc^*] c(\mathbf{p}_1,\tau_1)\ldots c(\mathbf{p}_n,\tau_n) c^*(\mathbf{q}_1,\sigma_1)\ldots c^*(\mathbf{q}_n,\sigma_n) \cdot$$
$$\exp\left\{-\int_0^\beta \sum_\mathbf{p} c^*(\mathbf{p},\tau)\left[\frac{d}{d\tau} + E_0(\mathbf{p}) - \mu\right]c(\mathbf{p},\tau)d\tau - \int_0^\beta H_{int}\left[c^*(\mathbf{p},\tau),c(\mathbf{p},\tau)\right]d\tau\right\}$$
$$= \frac{1}{Z}\cdot\frac{\delta^{2n} Z[\eta,\eta^*]}{\delta\eta^*(\mathbf{p}_1,\tau_1)\ldots\delta\eta^*(\mathbf{p}_n,\tau_n)\delta\eta(\mathbf{q}_1,\sigma_1)\ldots\delta\eta(\mathbf{q}_n,\sigma_n)}\bigg|_{\eta=\eta^*=0}. \tag{I.49}$$

The formulae and reasonings presented above are often being used in the theoretical study of the Bose gas by means of the functional integral technique. In particular, for the application to the perturbation theory we expand the exponential into the power series and rewrite the functional integral (I.47) in the form

$$Z = \int [Dc][Dc^*] \sum_{n=0}^{\infty} \frac{(-1)^n}{n!} \int_0^\beta d\tau_1 \cdots \int_0^\beta d\tau_n . H_{int}\left[c^*(\boldsymbol{p},\tau_1), c(\boldsymbol{p},\tau_1)\right] \cdots$$

$$H_{int}\left[c^*(\boldsymbol{p},\tau_n), c(\boldsymbol{p},\tau_n)\right] \exp\left\{ -\int_0^\beta \sum_{\boldsymbol{p}} c^*(\boldsymbol{p},\tau)\left[\frac{d}{d\tau} + E_0(\boldsymbol{p}) - \mu\right] c(\boldsymbol{p},\tau) d\tau \right\}. \tag{I.50}$$

From the definition (I.39) of the generating functional for the system of free bosons we obtain

$$Z = \sum_{n=0}^{\infty} \frac{(-1)^n}{n!} \int_0^\beta d\tau_1 \cdots \int_0^\beta d\tau_n H_{int}\left[-\frac{\delta}{\delta\eta(\boldsymbol{p},\tau_1)}, -\frac{\delta}{\delta\eta^*(\boldsymbol{p},\tau_1)} \right]$$

$$\cdots H_{int}\left[-\frac{\delta}{\delta\eta(\boldsymbol{p},\tau_n)}, -\frac{\delta}{\delta\eta^*(\boldsymbol{p},\tau_n)} \right] Z_0[\eta,\eta^*]\bigg|_{\eta=\eta^*=0} = \tag{I.51}$$

$$= \exp\left\{ -\int_0^\beta H_{int}\left[-\frac{\delta}{\delta\eta(\boldsymbol{p},\tau)}, -\frac{\delta}{\delta\eta^*(\boldsymbol{p},\tau)} \right] d\tau \right\} Z_0[\eta,\eta^*]\bigg|_{\eta=\eta^*=0}.$$

Similarly

$$Z[\eta,\eta^*] = \sum_{n=0}^{\infty} \frac{(-1)^n}{n!} \int_0^\beta d\tau_1 \cdots \int_0^\beta d\tau_n H_{int}\left[-\frac{\delta}{\delta\eta(\boldsymbol{p},\tau_1)}, -\frac{\delta}{\delta\eta^*(\boldsymbol{p},\tau_1)} \right]$$

$$\cdots H_{int}\left[-\frac{\delta}{\delta\eta(\boldsymbol{p},\tau_n)}, -\frac{\delta}{\delta\eta^*(\boldsymbol{p},\tau_n)} \right] Z_0[\eta,\eta^*] = \tag{I.52}$$

$$= \exp\left\{ -\int_0^\beta H_{int}\left[-\frac{\delta}{\delta\eta(\boldsymbol{p},\tau)}, -\frac{\delta}{\delta\eta^*(\boldsymbol{p},\tau)} \right] d\tau \right\} Z_0[\eta,\eta^*].$$

It is very easy to generalize all above formulae to the case of a system consisting of different types of interacting bosons with the momentum - dependent energies.

In the study of physical processes we must to work also with the quantum operators and the Green functions depending on the real time t. The functional integrals in this case can be obtained from those of the imaginary time formalism by replacing

$$\tau = it, \qquad \frac{d}{d\tau} = -i\frac{d}{dt}$$

and analytically continuing from the imaginary values of the time t to the real ones $-\infty < t < \infty$. As the example consider again the system with the partition function (I.38) and the generating functional (I.39) in the imaginary time formalism. Denote $\hat{c}(\mathbf{p},t)$ and $\hat{c}^+(\mathbf{p},t)$ the destruction and creation operators in the Schrödinger picture, $c(\mathbf{p},t)$ and $c^*(\mathbf{p},t)$ the corresponding commuting functions in the functional integrals. The real time Green functions at the zero temperature are defined in the following manner

$$G^B\left(\mathbf{p}_1,t_1;\ldots \mathbf{p}_n,t_n;\mathbf{q}_1,s_1;\ldots \mathbf{q}_n,s_n\right) = \frac{1}{\langle S\rangle_0}\langle T\big(\hat{c}(\mathbf{p}_1,t_1)\ldots \hat{c}(\mathbf{p}_n,t_n)\hat{c}^+(\mathbf{q}_1,s_1)\ldots \hat{c}^+(\mathbf{q}_n,s_n)S\big)\rangle_0 \qquad (I.53)$$

where

$$\langle\ldots\rangle_0$$

denotes the vacuum expectation value and S is the S- matrix

$$S = T e^{-i\int_{-\infty}^{\infty} H_{int}\left[\hat{c}^+(\mathbf{p},t),\hat{c}(\mathbf{p},t)\right]dt}. \qquad (I.54)$$

Instead of the formulae (I.47) and (I.48) in the imaginary time formalism now we have the functional integrals

$$Z = \int [Dc][Dc^*]\exp\left\{i\int_{-\infty}^{\infty}\sum_{\mathbf{p}} c^*(\mathbf{p},t)\left[i\frac{d}{dt} - E_0(\mathbf{p}) + \mu + io\right]c(\mathbf{p},t)dt\right.$$
$$\left. - i\int_{-\infty}^{\infty} H_{int}\left[c^*(\mathbf{p},t),c(\mathbf{p},t)\right]dt\right\}, \qquad (I.55)$$

$$Z[\eta,\eta^*] = \int [Dc][Dc^*] \exp\left\{ i \int_0^\beta \sum_{\boldsymbol{p}} \left(c^*(\boldsymbol{p},t)\left[i\frac{d}{dt} - E_0(\boldsymbol{p}) + \mu + io\right] c(\boldsymbol{p},t) \right.\right.$$
$$\left.\left. + \eta^*(\boldsymbol{p},t)c(\boldsymbol{p},t) + c^*(\boldsymbol{p},t)\eta(\boldsymbol{p},t) \right) dt - i \int_{-\infty}^\infty H_{int}\left[c^*(\boldsymbol{p},t), c(\boldsymbol{p},t)\right] dt \right\} \quad (I.56)$$

In terms of them we can express the zero temperature real time Green functions (I.53)

$$G^B\left(\boldsymbol{p}_1,t_1;\ldots\boldsymbol{p}_n,t_n;\boldsymbol{q}_1,s_1;\ldots\boldsymbol{q}_n,s_n\right) =$$
$$\frac{1}{Z}\int [Dc][Dc^*] c(\boldsymbol{p}_1,t_1)\ldots c(\boldsymbol{p}_n,t_n) c^*(\boldsymbol{q}_1,s_1)\ldots c^*(\boldsymbol{q}_n,s_n).$$
$$\exp\left\{ i \int_{-\infty}^\infty \sum_{\boldsymbol{p}} c^*(\boldsymbol{p},t)\left[i\frac{d}{dt} - E_0(\boldsymbol{p}) + \mu + io\right] c(\boldsymbol{p},t) dt \right.$$
$$\left. - i \int_{-\infty}^\infty H_{int}\left[c^*(\boldsymbol{p},t), c(\boldsymbol{p},t)\right] dt \right\} = \quad (I.57)$$
$$= (-1)^n \frac{1}{Z} \cdot \frac{\delta^{2n} Z[\eta,\eta^*]}{\delta\eta^*(\boldsymbol{p}_1,t_1)\ldots\delta\eta^*(\boldsymbol{p}_n,t_n)\delta\eta(\boldsymbol{q}_1,s_1)\ldots\delta\eta(\boldsymbol{q}_n,s_n)}\bigg|_{\eta=\eta^*=0}.$$

The symbol $+io$ in above formulae means that we add the small imaginary number $i\varepsilon$ to the real quantity $\omega - E_0(\boldsymbol{p}) + \mu$ and take the limit of the whole expressions containing $\omega - E_0(\boldsymbol{p}) + \mu + i\varepsilon$ when ε tends to zero from the positive values. Due to the presence of the imaginary number $i\varepsilon$ the integrals over $c(\boldsymbol{p},t)$ and $c^*(\boldsymbol{p},t)$ are absolutely convergent.

For the system of free bosons we have

$$H_{int}\left[\hat{c}^+(\boldsymbol{p},t), \hat{c}(\boldsymbol{p},t)\right] = 0.$$

Then the functional integrals (I.55) and (I.56) are reduced to

$$Z_0 = \int [Dc][Dc^*] \exp\left\{ i \int_{-\infty}^\infty \sum_{\boldsymbol{p}} c^*(\boldsymbol{p},t)\left[i\frac{d}{dt} - E_0(\boldsymbol{p}) + \mu + io\right] c(\boldsymbol{p},t) dt \right\}, \quad (I.58)$$

$$Z_0[\eta,\eta^*] = \int [Dc][Dc^*] \exp\left\{ i \int_{-\infty}^{\infty} \sum_{\mathbf{p}} \left(c^+(\mathbf{p},t) \left[i\frac{d}{dt} - E_0(\mathbf{p}) + \mu + io \right] c(\mathbf{p},t) \right.\right.$$
$$\left.\left. + \eta^*(\mathbf{p},t) c(\mathbf{p},t) + c^*(\mathbf{p},t) \eta(\mathbf{p},t) \right) \right\}. \tag{I.59}$$

Between the functional integrals (I.58) and (I.59) there is a relation similary to Eq. (I.41), namely

$$Z_0[\eta,\eta^*] = Z_0 \exp\left\{ -i \int_{-\infty}^{\infty} dt \int_{-\infty}^{\infty} dt' \sum_{\mathbf{p}} \eta^*(\mathbf{p},t) D(\mathbf{p},t-t') \eta(\mathbf{p},t') \right\} \tag{I.60}$$

where $D(\mathbf{p},t)$ is the solution of the differential equation

$$\left[i\frac{d}{dt} - E_0(\mathbf{p}) + \mu + io \right] D(\mathbf{p},t) = \delta(t). \tag{I.61}$$

and can be written in the form

$$D(\mathbf{p},t) = \frac{1}{2\pi} \int_{-\infty}^{\infty} e^{-i\omega t} \frac{1}{\omega - E_0(\mathbf{p}) + \mu + io} d\omega. \tag{I.62}$$

The functional integrals (I.55) and (I.56) for the system of interacting bosons are expressed in terms of the generating functional (I.59) of the free boson system by means of the perturbation theory expansion

$$Z = \exp\left\{ -i \int_{-\infty}^{\infty} H_{int}\left[-i\frac{\delta}{\delta\eta(\mathbf{p},t)}, -i\frac{\delta}{\delta\eta^*(\mathbf{p},t)} \right] dt \right\} Z_0[\eta,\eta^*]\Big|_{\eta=\eta^*=0}, \tag{I.63}$$

$$Z[\eta,\eta^*] = \exp\left\{ -i \int_{-\infty}^{\infty} H_{int}\left[-i\frac{\delta}{\delta\eta(\mathbf{p},t)}, -i\frac{\delta}{\delta\eta^*(\mathbf{p},t)} \right] dt \right\} Z_0[\eta,\eta^*]. \tag{I.64}$$

If the interaction Hamiltonian is known, then by inserting the expression (I.60) into the r.h.s. of Eqs. (I.63) and (I.64) we can derive the explicite expressions of the perturbation theory series for the partition function Z and the generating functional $Z[\eta,\eta^*]$ of the interacting boson systems.

The two - point real time Green function

$$G_0^B(\mathbf{p}, t - t') = \langle T(\hat{c}(\mathbf{p},t)\hat{c}^+(\mathbf{p},t'))\rangle_0 \tag{I.65}$$

of the free boson system with the generating functional (I.59), which can be rewritten in the form (I.60), is proportional to the solution $D(\mathbf{p},t)$ of the inhomogeneous first order differential equation (I.61) in the time variable t

$$\left[i\frac{d}{dt} - E_0(\mathbf{p}) + \mu + io\right]G_0^B(\mathbf{p},t) = -i\delta(t). \tag{I.66}$$

The quantum operators $\hat{c}(\mathbf{p},t)$ and $\hat{c}^+(\mathbf{p},t)$ of this free boson system must satisfy the corresponding homogeneous first order differential equation

$$\begin{aligned}\left[i\frac{d}{dt} - E_0(\mathbf{p}) + \mu\right]\hat{c}(\mathbf{p},t) &= 0, \\ \left[i\frac{d}{dt} + E_0(\mathbf{p}) - \mu\right]\hat{c}^+(\mathbf{p},t) &= 0.\end{aligned} \tag{I.67}$$

However, in the condensed matter theory there do exist also the boson systems, for example the phonons, with the quantum operators and the two - point Green functions satisfying second order differential equations in the time variable t. Consider the system of free bosons with the homogeneous equations

$$\begin{aligned}\frac{d^2\hat{c}(\mathbf{p},t)}{dt^2} &= -[E_0(\mathbf{p}) - \mu]^2\hat{c}(\mathbf{p},t), \\ \frac{d^2\hat{c}^+(\mathbf{p},t)}{dt^2} &= -[E_0(\mathbf{p}) - \mu]^2\hat{c}^+(\mathbf{p},t)\end{aligned} \tag{I.68}$$

for the quantum operators, and the inhomogeneous one

$$\left\{\frac{d^2}{dt^2} + [E_0(\mathbf{p}) - \mu]^2 - io\right\}G_0^B(\mathbf{p},t) = i\delta(t) \tag{I.69}$$

for the Green function. Instead of the partition function (I.58) and the generating functional (I.59) we have following functional integrals

$$\begin{aligned}Z_0 &= \int [Dc][Dc^*]\exp\left\{i\int_{-\infty}^{\infty}\sum_{\mathbf{p}}\left[\frac{dc^*(\mathbf{p},t)}{dt}\cdot\frac{dc(\mathbf{p},t)}{dt} - \left([E_0(\mathbf{p}) - \mu]^2 - io\right)c^*(\mathbf{p},t)c(\mathbf{p},t)\right]dt\right\} = \\ &= \int [Dc][Dc^*]\exp\left\{-i\int_{-\infty}^{\infty}\sum_{\mathbf{p}}c^*(\mathbf{p},t)\left(\frac{d^2}{dt^2} + [E_0(\mathbf{p}) - \mu]^2 - io\right)c(\mathbf{p},t)dt\right\},\end{aligned} \tag{I.70}$$

$$Z_0[\eta,\eta^*] = \int [Dc][Dc^*] \exp\left\{-i \int_{-\infty}^{\infty} \sum_{\boldsymbol{p}} \left[c^*(\boldsymbol{p},t)\left(\frac{d^2}{dt^2} + [E_0(\boldsymbol{p}) - \mu]^2 - io\right)c(\boldsymbol{p},t)\right.\right.$$
$$\left.\left. + c^*(\boldsymbol{p},t)\eta(\boldsymbol{p},t) + \eta^*(\boldsymbol{p},t)c(\boldsymbol{p},t)\right]dt\right\}. \tag{I.71}$$

In the presence of the interaction instead of the functional integrals (I.70) and (I.71) we have

$$Z = \int [Dc][Dc^*] \exp\left\{-i \int_{-\infty}^{\infty} \sum_{\boldsymbol{p}} c^*(\boldsymbol{p},t)\left(\frac{d^2}{dt^2} + [E_0(\boldsymbol{p}) - \mu]^2 - io\right)c(\boldsymbol{p},t)\right.$$
$$\left. -i \int_{-\infty}^{\infty} H_{int}[c^*(\boldsymbol{p},t), c(\boldsymbol{p},t)]dt\right\} = \tag{I.72}$$
$$= \exp\left\{-i \int_{-\infty}^{\infty} H_{int}\left[i\frac{\delta}{\delta\eta(\boldsymbol{p},t)}, i\frac{\delta}{\delta\eta^*(\boldsymbol{p},t)}\right]dt\right\} Z_0[\eta,\eta^*]\bigg|_{\eta=\eta^*=0},$$

$$Z[\eta,\eta^*] = \int [Dc][Dc^*] \exp\left\{-i \int_{-\infty}^{\infty} \sum_{\boldsymbol{p}} \left[c^*(\boldsymbol{p},t)\left(\frac{d^2}{dt^2} + [E_0(\boldsymbol{p}) - \mu]^2 - io\right)c(\boldsymbol{p},t)\right.\right.$$
$$\left.\left. + c^*(\boldsymbol{p},t)\eta(\boldsymbol{p},t) + \eta^*(\boldsymbol{p})c(\boldsymbol{p},t)\right]dt - i\int_{-\infty}^{\infty} H_{int}[c^*(\boldsymbol{p},t), c(\boldsymbol{p},t)]dt\right\} = \tag{I.73}$$
$$= \exp\left\{-i \int_{-\infty}^{\infty} H_{int}\left[i\frac{\delta}{\delta\eta(\boldsymbol{p},t)}, i\frac{\delta}{\delta\eta^*(\boldsymbol{p},t)}\right]dt\right\} Z_0[\eta,\eta^*].$$

By means of the shift of the functional integration variables in the r.h.s. of Eq. (I.70) we can show that

$$Z_0[\eta,\eta^*] = Z_0 \exp\left\{i \int_{-\infty}^{\infty} dt \int_{-\infty}^{\infty} dt' \sum_{\boldsymbol{p}} \eta^*(\boldsymbol{p},t) D(\boldsymbol{p},t-t')\eta(\boldsymbol{p},t')\right\}, \tag{I.74}$$

where $D(\boldsymbol{p}, t-t')$ is the solution of the second order differential equation

$$\left(\frac{d^2}{dt^2} + [E_0(\boldsymbol{p}) - \mu]^2 - io\right) D(\boldsymbol{p},t) = \delta(t). \tag{I.75}$$

In condensed matter physics there exist many boson systems of this kind.

II. FUNCTIONAL INTEGRALS FOR FERMIONIC SYSTEMS

Before to extend the results of the preceeding Section to the system of fermions obeying the Fermi-Dirac statistics we discuss the matters concerning the integration in the anticommuting variables. Consider the set of N c-numbers θ_i, $i = 1, 2, \ldots, N$, satisfying the anticommutativity condition

$$\theta_i \theta_j = -\theta_j \theta_i, \tag{II.1}$$

especially

$$\theta_i^2 = 0, \tag{II.2}$$

and called the Grassmann variables. In the Grassmann algebra generated by these anticommuting variables the products of more than N variables do not exist, and there is only one product of N variables

$$\theta_1 \theta_2 \ldots \theta_{N-1} \theta_N.$$

Therefore any function $f(\theta_i)$ of the variables θ_i is a polynome and has following general form

$$f(\theta_i) = a^{(1)} + a^{(2)}_{ij} \theta_i \theta_j + a^{(3)}_{ijk} \theta_i \theta_j \theta_k + \cdots + a^{(N)}_{i_1 i_2 \ldots i_N} \theta_{i_1} \theta_{i_2} \ldots \theta_{i_N}, \tag{II.3}$$

where the coefficients $a^{(m)}_{i_1 \ldots i_m}$ are antisymmetric with respect to the permutations of the indices $i_1, i_2, \ldots, i_m$. In particular, any function of one Grassmann variable θ is a binome

$$f(\theta) = a + b\theta. \tag{II.4}$$

Consider the integral of $f(\theta)$ over the whole range of the variable θ. It must be invariant under the shift of the variable

$$\int f(\theta) d\theta = \int f(\theta + \eta) d\theta.$$

Inserting the expression (II.4) of $f(\theta)$ into both sides of this equation we obtain immediately the condition

$$\int d\theta = 0. \tag{II.5}$$

The integral

$$\int \theta d\theta$$

is still arbitrary, and we can normalize it such that

$$\int \theta d\theta = 1. \tag{II.6}$$

According to the requirement of the Fermi-Dirac statistics the functional integrals for the fermionic systems must contain only the anticommuting functions. While the commuting functions of the imaginary time considered in the preceeding Section are periodical, the anticommuting functions in the functional integrals for the fermionic system must be antiperiodical with respect to the imaginary time τ with the same period β. In order to prove this statement we start from the study of the simplest system of the spin $1/2$ fermions with only two degrees of freedom which are those connected with two spin projections $\alpha = \uparrow, \downarrow$. Denote $\hat{a}_\alpha$ and $\hat{a}_\alpha^+$ the destruction and creation operators. They satisfy the anticommutation relations

$$\{\hat{a}_\alpha^+, \hat{a}_\beta\} = \hat{a}_\alpha^+ \hat{a}_\beta + \hat{a}_\beta \hat{a}_\alpha^+ = \delta_{\alpha\beta},$$
$$\{\hat{a}_\alpha, \hat{a}_\beta\} = \{\hat{a}_\alpha^+, \hat{a}_\beta^+\} = 0. \tag{II.7}$$

The Hamiltonian for the free fermion system is

$$\hat{H} = E_0 \hat{a}_\alpha^+ \hat{a}_\alpha, \tag{II.8}$$

and the particle number operator equals

$$\hat{N} = \hat{a}_\alpha^+ \hat{a}_\alpha. \tag{II.9}$$

We set

$$\tilde{H}(\hat{a}_\alpha^+, \hat{a}_\alpha) = \hat{H} - \mu \hat{N} = E \hat{a}_\alpha^+ \hat{a}_\alpha, \tag{II.10}$$
$$E = E_0 - \mu \tag{II.11}$$

where μ is the chemical potential. For this system there exist two one-particle states with different spin projections $|1\alpha\rangle$, $\alpha = \uparrow$ and $\downarrow$, one vacuum state $|0\rangle$ and also only one two-particle state

$$|2\rangle = |\uparrow \downarrow\rangle = -|\downarrow \uparrow\rangle,$$

because if two identical fermions have one and the same orbital wave function then their spin projections always must be different. The trace of any operator $\hat{O}$ is then

$$\text{Tr}\{\hat{O}\} = \langle 0|\hat{O}|0\rangle + \sum_\alpha \langle 1\alpha|\hat{O}|1\alpha\rangle + \langle 2|\hat{O}|2\rangle. \tag{II.12}$$

In particular

$$\text{Tr}\left\{e^{-\beta \tilde{H}(\hat{a}_\alpha^+,\hat{a}_\alpha)}\right\} = 1 + 2e^{-\beta E} + e^{-2\beta E} = \left(1 + e^{-\beta E}\right)^2. \tag{II.13}$$

Similarly

$$\text{Tr}\left\{\hat{a}_\gamma^+ \hat{a}_\delta e^{-\beta \tilde{H}(\hat{a}_\alpha^+,\hat{a}_\alpha)}\right\} = \delta_{\gamma\delta}\left(e^{-\beta E} + e^{-2\beta E}\right) = \delta_{\gamma\delta} e^{-\beta E}\left(1 + e^{-\beta E}\right),$$

$$\text{Tr}\left\{\hat{a}_\gamma \hat{a}_\delta^+ e^{-\beta \tilde{H}(\hat{a}_\alpha^+,\hat{a}_\alpha)}\right\} = \delta_{\gamma\delta}\left(1 + e^{-\beta E}\right)$$

and therefore

$$\langle \hat{a}_\gamma^+ \hat{a}_\delta \rangle = \delta_{\gamma\delta} \frac{e^{-\beta E}}{1 + e^{-\beta E}},$$
$$\langle \hat{a}_\gamma \hat{a}_\delta^+ \rangle = \delta_{\gamma\delta} \frac{1}{1 + e^{-\beta E}}. \tag{II.14}$$

In order to define the two-point Green functions we introduce the imaginary time dependent quantum operators

$$\hat{a}_\gamma(\tau) = e^{\tau \tilde{H}(\hat{a}_\alpha^+,\hat{a}_\alpha)} \hat{a}_\gamma e^{-\tau \tilde{H}(\hat{a}_\alpha^+,\hat{a}_\alpha)},$$
$$\hat{\bar{a}}_\gamma(\tau) = e^{\tau \tilde{H}(\hat{a}_\alpha^+,\hat{a}_\alpha)} \hat{a}_\gamma^+ e^{-\tau \tilde{H}(\hat{a}_\alpha^+,\hat{a}_\alpha)}. \tag{II.15}$$

Due to the anticommutativity of the fermion quantum operators the time-ordered product is defined in the following manner

$$T\left(\hat{a}_\alpha(\tau)\hat{\bar{a}}_\beta(\sigma)\right) = \theta(\tau - \sigma)\hat{a}_\alpha(\tau)\hat{\bar{a}}_\beta(\sigma) - \theta(\sigma - \tau)\hat{\bar{a}}_\beta(\sigma)\hat{a}_\alpha(\tau). \tag{II.16}$$

In the case of the Hamiltonian (II.10) we have

$$\hat{a}_\alpha(\tau) = e^{-\tau E}\hat{a}_\alpha, \quad \hat{\bar{a}}_\alpha(\sigma) = e^{\sigma E}\hat{a}_\alpha^+. \tag{II.17}$$

The two-point Green function is the mean value of the time-ordered product of the destruction and creation operators in the canonical ensemble

$$G_0^F(\tau - \sigma)_{\alpha\beta} = \langle T(\hat{a}_\alpha(\tau)\hat{\bar{a}}_\beta(\sigma))\rangle. \tag{II.18}$$

Inserting the expressions (II.14), (II.16) and (II.17) into the r.h.s. of Eq. (II.18) we obtain

$$G_0^F(\tau - \sigma)_{\alpha\beta} = \delta_{\alpha\beta} S_E(\tau - \sigma), \tag{II.19}$$

where $S_E(\tau)$ is the antiperiodical solution

$$S_E(\tau + \beta) = -S_E(\tau) \tag{II.20}$$

of the differential equation

$$\left(\frac{d}{d\tau} + E\right) S_E(\tau) = \delta(\tau), \tag{II.21}$$

namely

$$S_E(\tau) = e^{-\tau E} \frac{\theta(\tau) - \theta(-\tau)e^{-\beta E}}{1 + e^{-\beta E}}. \tag{II.22}$$

In order to represent above quantities in terms of the functional integrals we introduce the anticommuting c-number functions $a_\alpha(\tau)$ and $a_\alpha^*(\tau)$ corresponding to the operators $\hat{a}_\alpha(\tau)$ and $\hat{\bar{a}}_\alpha(\tau)$. Because the Green function (II.18) is antiperiodical the anticommuting functions $a_\alpha(\tau)$ and $a_\alpha^*(\tau)$ must be also antiperiodical

$$a_\alpha(\tau + \beta) = -a_\alpha(\tau), \quad a_\alpha^*(\tau + \beta) = -a_\alpha^*(\tau). \tag{II.23}$$

Together with the Hamiltonian (II.10) let us consider also the expression obtained from it by means of the antipermutation of $\hat{a}_\alpha$ and $\hat{a}_\alpha^+$ (permuting and changing the sign)

$$\tilde{H}(\hat{a}_\alpha^+, \hat{a}_\alpha) \rightarrow -\tilde{H}(\hat{a}_\alpha, \hat{a}_\alpha^+) = -E\hat{a}_\alpha\hat{a}_\alpha^+. \tag{II.24}$$

If we replace the operators $\hat{a}_\alpha$ and $\hat{a}_\alpha^+$ by the anticommuting functions $a_\alpha(\tau)$ and $a_\alpha^*(\tau)$ then from two different expressions (II.10) and (II.24) we obtain one and the same c-number function

$$\tilde{H}(\tau) = E a_\alpha^*(\tau) a_\alpha(\tau). \tag{II.25}$$

This function changes the sign if we permute $a_\alpha(\tau)$ with $a_\alpha^*(\tau)$. Among the set of two expression (II.10), (II.24) and their linear combinations let us select the unique combination which also changes its sign when we permute $\hat{a}_\alpha$ and $\hat{a}_\alpha^+$ - the antisymmetrized Hamiltonian

$$\tilde{H}_A(\hat{a}_\alpha^+, \hat{a}_\alpha) = -\tilde{H}_A(\hat{a}_\alpha, \hat{a}_\alpha^+) = E \frac{\hat{a}_\alpha^+ \hat{a}_\alpha - \hat{a}_\alpha \hat{a}_\alpha^+}{2}. \tag{II.26}$$

The correspondence between the antisymmetrized products of the quantum operators and the c-number functions obtained when we replace the operators $\hat{a}_\alpha$, $\hat{a}_\alpha^+$ by the functions $a_\alpha(\tau)$, $a_\alpha^*(\tau)$ is the one-to-one corespondence:

$$\tilde{H}_A(\hat{a}_\alpha^+, \hat{a}_\alpha) \longleftrightarrow \tilde{H}(\tau).$$

Now we show that indeed there exists following relation for the system of free fermion with only the spin degrees of freedom

$$Z_0 = Z_0(E) = \mathrm{Tr}\left\{ e^{-\beta \tilde{H}_A(\hat{a}_\alpha^+, \hat{a}_\alpha)} \right\}$$

$$= \int [Da][Da^*] \exp\left\{ -\int_0^\beta \left[a_\alpha^*(\tau) \frac{da_\alpha(\tau)}{d\tau} + \tilde{H}(\tau) \right] d\tau \right\} \tag{II.27}$$

The meaning of the functional integral in the r.h.s. of Eq. (II.27) can be understood if we perform the Fourier transformation

$$a_\alpha(\tau) = \frac{1}{\sqrt{\beta}} \sum_\nu e^{i\varepsilon_\nu \tau} a_{\nu\alpha},$$

$$\varepsilon_\nu = (2\nu + 1)\frac{\pi}{\beta}, \tag{II.28}$$

ν being integers. Then we have

$$\int_0^\beta \left[a_\alpha^*(\tau) \frac{da_\alpha(\tau)}{d\tau} + \tilde{H}(\tau) \right] d\tau = \sum_\nu (i\varepsilon_\nu + E) a_{\nu\alpha}^* a_{\nu\alpha} \tag{II.29}$$

It is easy to verify that functions $a_\alpha(\tau)$ determined by Eq. (II.28) with above values of ε_ν satisfy the antiperiodicity condition (II.23). We define the functional integral in the manner similar to Eq. (I.13)

$$\int [Da][Da^*]\cdots = \text{const} \prod_\nu \prod_\alpha \int da_{\nu\alpha} da^*_{\nu\alpha} \cdots \quad \text{(II.30)}$$

and obtain

$$\int [Da][Da^*] \exp\left\{-\int_0^\beta \left[a^*_\alpha(\tau)\frac{da_\alpha(\tau)}{d\tau} + \tilde{H}(\tau)\right]d\tau\right\} =$$
$$= \text{const} \prod_\alpha \prod_\alpha \left(\int e^{-(i\varepsilon_\nu + E)a^*_{\nu\alpha} a_{\nu\alpha}} da_{\nu\alpha} da^*_{\nu\alpha}\right) \quad \text{(II.31)}$$

Let us compute the integrals

$$\int e^{-(i\varepsilon_\nu + E)a^*_{\nu\alpha} a_{\nu\alpha}} da_{\nu\alpha} da^*_{\nu\alpha}$$

over the Grassmann variables. In the beginning of this Section the properties of the anticommuting c-numbers and their integrals were given. By expanding the exponential and using these properties we obtain easily

$$\prod_\alpha \int e^{-(i\varepsilon_\nu + E)a^*_{\nu\alpha} a_{\nu\alpha}} da_{\nu\alpha} da^*_{\nu\alpha} = (i\varepsilon_\nu + E)^2.$$

Therefore

$$Z_0(E) = \int [Da][Da^*] \exp\left\{-\int_0^\beta \left[a^*_\alpha(\tau)\frac{da_\alpha(\tau)}{d\tau} + \tilde{H}(\tau)\right]d\tau\right\}$$
$$= \text{const} \prod_\nu (i\varepsilon_\nu + E)^2. \quad \text{(II. 32)}$$

In order to evaluate the infinite product in the r.h.s. of Eq. (II.32) we apply again the reasoning presented in calculating $Z_0(E)$ for the boson system. We have

$$\frac{d\ln Z_0(E)}{dE} = 2\sum_\nu \frac{1}{i\varepsilon_\nu + E}.$$

Because

$$\sum_\nu \frac{1}{i\varepsilon_\nu + E} = \frac{\beta}{2}\frac{1 - e^{-\beta E}}{1 + e^{-\beta E}}$$

then we obtain

$$\frac{d\ln Z_0(E)}{dE} = \beta + 2\frac{d\ln(1 + e^{-\beta E})}{dE}.$$

Therefore
$$Z_0(E) = \text{const } e^{\beta E}\left(1 + e^{-\beta E}\right)^2. \tag{II.33}$$

On the other hand from the relations (II.13) and (II.26) it follows that the same expression of $Z_0(E)$ can be also calculated by means of the definition

$$Z_0(E) = \text{Tr}\left\{e^{-\beta \tilde{H}_A(\hat{a}_\alpha^+, \hat{a}_\alpha)}\right\}.$$

Thus the basic formula (II.27) was proved.

As in the case of the boson system we introduce the generating functional

$$Z_0[\xi, \xi^*] = \int [Da][Da^*] \exp\left\{-\int_0^\beta \left[a_\alpha^*(\tau)\left(\frac{d}{d\tau} + E\right)a_\alpha(\tau) \right.\right.$$
$$\left.\left. + \xi_\alpha^*(\tau)a_\alpha(\tau) + a_\alpha^*(\tau)\xi_\alpha(\tau)\right]d\tau\right\} \tag{II.34}$$

depending on the functions $\xi_\alpha(\tau)$, $\xi_\alpha^*(\tau)$ anticommuting ones with anothers and with the Grassmann variables $a_\alpha(\tau)$, $a_\alpha^*(\tau)$. The differentiation in these anticommuting variables is defined to be the left derivation

$$\frac{\delta}{\delta \xi_\alpha(\tau)} \exp\left\{-\int_0^\beta a_\gamma^*(\rho)\xi_\gamma(\rho)d\rho\right\} = \frac{\delta}{\delta \xi_\alpha(\tau)} \exp\int_0^\beta \xi_\gamma(\rho)a_\gamma^*(\rho)d\rho =$$
$$= a_\alpha^*(\tau)\exp\int_0^\beta \xi_\gamma(\rho)a_\gamma^*(\rho)d\rho = a_\alpha^*(\tau)\exp\left\{-\int_0^\beta a_\gamma^*(\rho)\xi_\gamma(\rho)d\rho\right\},$$

$$\frac{\delta^2}{\delta \xi_\alpha(\tau)\delta \xi_\beta(\sigma)}\exp\left\{-\int_0^\beta a_\gamma^*(\rho)\xi_\gamma(\rho)d\rho\right\} = \frac{\delta}{\delta \xi_\alpha(\tau)}a_\beta^*(\sigma)\exp\left\{-\int_0^\beta a_\gamma^*(\rho)\xi_\gamma(\rho)d\rho\right\}$$
$$= -a_\beta^*(\sigma)\frac{\delta}{\delta \xi_\alpha(\tau)}\exp\left\{-\int_0^\beta a_\gamma^*(\rho)\xi_\gamma(\rho)d\rho\right\} = -a_\beta^*(\sigma)a_\alpha^*(\tau)\exp\left\{-\int_0^\beta a_\gamma^*(\rho)\xi_\gamma(\rho)d\rho\right\}$$
$$= a_\alpha^*(\tau)a_\beta^*(\sigma)\exp\left\{-\int_0^\beta a_\gamma^*(\rho)\xi_\gamma(\rho)d\rho\right\},$$

$$\frac{\delta}{\delta \xi_\alpha^*(\tau)}\exp\left\{-\int_0^\beta \xi_\gamma^*(\rho)a_\gamma(\rho)d\rho\right\} = -a_\alpha(\tau)\exp\left\{-\int_0^\beta \xi_\gamma^*(\rho)a_\gamma(\rho)d\rho\right\},$$

$$\frac{\delta^2}{\delta\xi_\alpha^*(\tau)\delta\xi_\beta^*(\sigma)} \exp\left\{-\int_0^\beta \xi_\gamma^*(\rho)a_\gamma(\rho)d\rho\right\} = -\frac{\delta}{\delta\xi_\alpha^*(\tau)} a_\beta(\sigma) \exp\left\{-\int_0^\beta \xi_\gamma^*(\rho)a_\gamma(\rho)d\rho\right\}$$

$$= a_\beta(\sigma)\frac{\delta}{\delta\xi_\alpha^*(\tau)} \exp\left\{-\int_0^\beta \xi_\gamma^*(\rho)a_\gamma(\rho)d\rho\right\} = -a_\beta(\sigma)a_\alpha(\tau) \exp\left\{-\int_0^\beta \xi_\gamma^*(\rho)a_\gamma(\rho)d\rho\right\}$$

$$= a_\alpha(\tau)a_\beta(\sigma) \exp\left\{-\int_0^\beta \xi_\gamma^*(\rho)a_\gamma(\rho)d\rho\right\},$$

$$\frac{\delta^2}{\delta\xi_\alpha^*(\tau)\delta\xi_\beta(\sigma)} \exp\left\{-\int_0^\beta \left[a_\gamma^*(\rho)\xi_\gamma(\rho) + \xi_\gamma^*(\rho)a_\gamma(\rho)\right]d\rho\right\}$$

$$= \frac{\delta}{\delta\xi_\alpha^*(\tau)} a_\beta^*(\sigma) \exp\left\{-\int_0^\beta \left[a_\gamma^*(\rho)\xi_\gamma(\rho) + \xi_\gamma^*(\rho)a_\gamma(\rho)\right]d\rho\right\}$$

$$= -a_\beta^*(\sigma)\frac{\delta}{\delta\xi_\alpha^*(\tau)} \exp\left\{-\int_0^\beta \left[a_\gamma^*(\rho)\xi_\gamma(\rho) + \xi_\gamma^*(\rho)a_\gamma(\rho)\right]d\rho\right\}$$

$$= a_\beta^*(\rho)a_\alpha(\tau) \exp\left\{-\int_0^\beta \left[a_\gamma^*(\rho)\xi_\gamma(\rho) + \xi_\gamma^*(\rho)a_\gamma(\rho)\right]d\rho\right\}$$

$$= -a_\alpha(\tau)a_\beta^*(\sigma) \exp\left\{-\int_0^\beta \left[a_\gamma^*(\rho)\xi_\gamma(\rho) + \xi_\gamma^*(\rho)a_\gamma(\rho)\right]d\rho\right\}.$$

It is straightforward to extend all above results to other more general cases of the fermion systems.

Consider the system of fermions with momentum-dependent energies $E_0(\boldsymbol{p})$ and spin $1/2$. The destruction and creation operators are denoted $\hat{a}_\alpha(\boldsymbol{p})$ and $\hat{a}_\alpha^+(\boldsymbol{p})$. The imaginary time dependent operators $\hat{a}_\alpha(\boldsymbol{p},\tau)$ and $\hat{\bar{a}}_\alpha(\boldsymbol{p},\tau)$ correspond to the anticommuting c-number functions $a_\alpha(\boldsymbol{p},\tau)$ and $a_\alpha^*(\boldsymbol{p},\tau)$ in the functional integrals. The partition function equals

$$Z_0 = \int [Da][Da^*] \exp\left\{-\int_0^\beta \sum_{\boldsymbol{p}} a_\alpha^*(\boldsymbol{p},\tau)\left[\frac{d}{d\tau} + E_0(\boldsymbol{p}) - \mu\right]a_\alpha(\boldsymbol{p},\tau)d\tau\right\}. \quad \text{(II.35)}$$

The generating functional depends on two set of anticommuting c-number functions of the imaginary time $\xi_\alpha(\mathbf{p},\tau)$ and $\xi_\alpha^*(\mathbf{p},\tau)$ which anticommute also with the functions $a_\alpha(\mathbf{p},\tau)$ and $a_\alpha^*(\mathbf{p},\tau)$. It has the form

$$Z_0[\xi,\xi^*] = \int [Da][Da^*]\exp\left\{-\int_0^\beta \sum_\mathbf{p}\left(a_\alpha^*(\mathbf{p},\tau)\left[\frac{d}{dt}+E_0(\mathbf{p})-\mu\right]a_\alpha(\mathbf{p},\tau)\right.\right.$$
$$\left.\left.+ a_\alpha^*(\mathbf{p},\tau)\xi_\alpha(\mathbf{p},\tau)+\xi_\alpha^*(\mathbf{p},\tau)a_\alpha(\mathbf{p},\tau)\right)d\tau\right\} \tag{II.36}$$

Denote $S(\mathbf{p},\tau)_{\alpha\beta}$ the antiperiodical solution of the differental equation

$$\left[\frac{d}{d\tau}+E_0(\mathbf{p})-\mu\right]S(\mathbf{p},\tau)_{\alpha\beta} = \delta_{\alpha\beta}\delta(\tau), \tag{II.37}$$

$$S(\mathbf{p},\tau+\beta)_{\alpha\beta} = -S(\mathbf{p},\tau)_{\alpha\beta}. \tag{II.38}$$

It is easy to verify that

$$S(\mathbf{p},\tau)_{\alpha\beta} = \delta_{\alpha\beta}e^{-[E_0(\mathbf{p})-\mu]\tau}\frac{\theta(\tau)-\theta(-\tau)e^{-\beta[E_0(\mathbf{p})-\mu]}}{1+e^{-\beta[E_0(\mathbf{p})-\mu]}}. \tag{II.39}$$

By applying an appropriate shift of the functional integration variables $a_\alpha(\mathbf{p},\tau)$ and $a_\alpha^*(\mathbf{p},\tau)$ in the r.h.s. of Eq. (II.35) we can show that

$$Z_0[\xi,\xi^*] = Z_0 \exp\left\{\int_0^\beta d\tau \int_0^\beta d\sigma \sum_\mathbf{p} \xi_\alpha^*(\mathbf{p},\tau)S(\mathbf{p},\tau-\sigma)_{\alpha\beta}\xi_\beta(\mathbf{p},\sigma)\right\}. \tag{II.40}$$

For the system of interacting fermions with momentum-dependent energies $E_0(\mathbf{p})$ the partition function and the generating functional are

$$Z = \int [Da][Da^*]\exp\left\{-\int_0^\beta \sum_\mathbf{p} a_\alpha^*(\mathbf{p},\tau)\left[\frac{d}{dt}+E_0(\mathbf{p})-\mu\right]a_\alpha(\mathbf{p},\tau)d\tau\right.$$
$$\left.-\int_0^\beta H_{int}\left[a_\alpha^*(\mathbf{p},\tau),a_\alpha(\mathbf{p},\tau)\right]d\tau\right\}, \tag{II.41}$$

217

$$Z[\xi, \xi^*] = \int [Da][Da^*] \exp\left\{ -\int_0^\beta \sum_{\mathbf{p}} \left(a_\alpha^*(\mathbf{p}, \tau)\left[\frac{d}{dt} + E_0(\mathbf{p}) - \mu\right] a_\alpha(\mathbf{p}, \tau) \right. \right.$$
$$\left. + \xi_\gamma^*(\mathbf{p}, \tau) a_\alpha(\mathbf{p}, \tau) + a_\alpha^*(\mathbf{p}, \tau) \xi_\alpha(\mathbf{p}, \tau) \right) d\tau \quad \text{(II.42)}$$
$$\left. - \int_0^\beta H_{int}\left[a_\alpha^*(\mathbf{p}, \tau), a_\alpha(\mathbf{p}, \tau)\right] d\tau \right\}$$

where $H_{int}\left[\hat{a}_\alpha^+(\mathbf{p}), \hat{a}_\alpha(\mathbf{p})\right]$ is the interaction Hamiltonian. Denote S the S-matrix

$$S = T\exp\left\{ -\int_0^\beta H_{int}\left[\hat{\bar{a}}_\alpha(\mathbf{p}, \tau), \hat{a}_\alpha(\mathbf{p}, \tau)\right] d\tau \right\} \quad \text{(II.43)}$$

and define the time-ordered product of the quantum operators by generalizing the definition of that for two operators, Eq. (II.16), such that it is antisymmetric under any permutation of two operators, for example

$$T\big(a_{\alpha_1}(\mathbf{p}_1, \tau_1) \ldots \bar{a}_{\beta_1}(\mathbf{q}_1, \sigma_1) \ldots\big) = -T\big(\bar{a}_{\beta_1}(\mathbf{q}_1, \sigma_1) \ldots a_{\alpha_1}(\mathbf{p}_1, \tau_1) \ldots\big).$$

The imaginary time Green functions are

$$G^F\big(\mathbf{p}_1, \tau_1; \ldots \mathbf{p}_n, \tau_n; \mathbf{q}_1, \sigma_1; \ldots \mathbf{q}_n, \sigma_n\big)_{\alpha_1 \ldots \alpha_n \beta_1 \ldots \beta_n} = \quad \text{(II.44)}$$
$$= \frac{\langle T(\hat{a}_{\alpha_1}(\mathbf{p}_1, \tau_1) \ldots \hat{a}_{\alpha_n}(\mathbf{p}_n, \tau_n) \hat{\bar{a}}_{\beta_1}(\mathbf{q}_1, \sigma_1) \ldots \hat{\bar{a}}_{\beta_n}(\mathbf{q}_n, \sigma_n) S) \rangle}{\langle S \rangle}.$$

They can be expressed in terms of the functional integrals

$$G^F\big(\mathbf{p}_1, \tau_1; \ldots \mathbf{p}_n, \tau_n; \mathbf{q}_1, \sigma_1; \ldots \mathbf{q}_n, \sigma_n\big)_{\alpha_1 \ldots \alpha_n \beta_1 \ldots \beta_n} =$$
$$= \frac{1}{Z} \int [Da][Da^*] a_{\alpha_1}(\mathbf{p}_1, \tau_1) \ldots a_{\alpha_n}(\mathbf{p}_n, \tau_n) a_{\beta_1}^*(\mathbf{q}_1, \sigma_1) \ldots a_{\beta_n}^*(\mathbf{q}_n, \sigma_n)$$
$$\exp\left\{ -\int_0^\beta \sum_{\mathbf{p}} a_\alpha^*(\mathbf{p}, \tau)\left[\frac{d}{d\tau} + E_0(\mathbf{p}) - \mu\right] a_\alpha(\mathbf{p}, \tau) d\tau \right.$$
$$\left. - \int_0^\beta H_{int}\left[a_\alpha^*(\mathbf{p}, \tau), a_\alpha(\mathbf{p}, \tau)\right] d\tau \right\} = \quad \text{(II.45)}$$
$$(-1)^n \cdot \frac{1}{Z} \frac{\delta^{2n} Z[\xi, \xi^*]}{\delta \xi_{\alpha_1}^*(\mathbf{p}_1, \tau_1) \ldots \delta \xi_{\alpha_n}^*(\mathbf{p}_n, \tau_n) \delta \xi_{\beta_1}(\mathbf{q}_1, \sigma_1) \ldots \delta \xi_{\beta_n}(\mathbf{q}_n, \sigma_n)}\bigg|_{\xi = \xi^* = 0}.$$

By means of the perturbation theory the partition function (II.41) and the generating functional (II.42) of the interacting fermion system are expressed in terms of the generating functional (II.36) of the free fermion system

$$Z = \exp\left\{-\int_0^\beta H_{int}\left[\frac{\delta}{\delta\xi_\alpha(\bm{p},\tau)}, -\frac{\delta}{\delta\xi_\alpha^*(\bm{p},\tau)}\right]d\tau\right\}Z_0[\xi,\xi^*]\bigg|_{\xi=\xi^*=0}, \quad \text{(II.46)}$$

$$Z[\xi,\xi^*] = \exp\left\{-\int_0^\beta H_{int}\left[\frac{\delta}{\delta\xi_\alpha(\bm{p},\tau)}, -\frac{\delta}{\delta\xi_\alpha^*(\bm{p},\tau)}\right]d\tau\right\}Z_0[\xi,\xi^*]. \quad \text{(II.47)}$$

By means of the analytical continuation of above formulae derived in the imaginary time formalism we can obtain all corresponding expressions for the functional integrals and the Green functions in the case of the formalism with the real time. Instead of the formulae (II.41) and (II.42) now we have

$$Z = \int [Da][Da^*]\exp\left\{i\int_{-\infty}^{\infty}\sum_{\bm{p}} a_\alpha^*(\bm{p},t)\left[i\frac{d}{dt} - E_0(\bm{p}) + \mu + io\right]a_\alpha(\bm{p},t)dt\right.$$
$$\left. - i\int_{-\infty}^{\infty} H_{int}[a_\alpha^*(\bm{p},t), a_\alpha(\bm{p},t)]dt\right\}, \quad \text{(II.48)}$$

$$Z[\xi,\xi^*] = \int [Da][Da^*]\exp\left\{i\int_{-\infty}^{\infty}\sum_{\bm{p}}\left(a_\alpha^*(\bm{p},t)\left[i\frac{d}{dt} - E_0(\bm{p}) + \mu + io\right]a_\alpha(\bm{p},t)\right.\right.$$
$$\left.\left. + \xi_\alpha^*(\bm{p},t) + a_\alpha^*(\bm{p},t)\xi_\alpha(\bm{p},t)\right)dt - i\int_{-\infty}^{\infty} H_{int}[a_\alpha^*(\bm{p},t), a_\alpha(\bm{p},t)]dt\right\}. \quad \text{(II.49)}$$

For the free fermion system these functional integrals become

$$Z_0 = \int [Da][Da^*]\exp\left\{i\int_{-\infty}^{\infty}\sum_{\bm{p}} a_\alpha^*(\bm{p},t)\left[i\frac{d}{dt} - E_0(\bm{p}) + \mu + io\right]a_\alpha(\bm{p},t)dt\right\}, \quad \text{(II.50)}$$

$$Z_0[\xi,\xi^*] = \int [Da][Da^*]\exp\left\{i\int_{-\infty}^{\infty}\sum_{\bm{p}}\left(a_\alpha^*(\bm{p},t)\left[i\frac{d}{dt} - E_0(\bm{p}) + \mu + io\right]a_\alpha(\bm{p},t)\right.\right.$$
$$\left.\left. + \xi_\alpha^*(\bm{p},t)a_\alpha(\bm{p},t) + a_\alpha^*(\bm{p},t)\xi_\alpha(\bm{p},t)\right)dt\right\}. \quad \text{(II.51)}$$

By the shift of the functional integration variable in the r.h.s. of Eq. (II.50) we can show that

$$Z_0[\xi, \xi^*] = Z_0 \cdot \exp\left\{-i \int_{-\infty}^{\infty} dt \int_{-\infty}^{\infty} dt' \sum_{\mathbf{p}} \xi_\alpha^*(\mathbf{p}, t) S(\mathbf{p}, t - t')_{\alpha\beta} \xi_\beta(\mathbf{p}, t')\right\}, \quad (\text{II.52})$$

where $S(\mathbf{p}, t)_{\alpha\beta}$ is the solution of the differential equation

$$\left[i\frac{d}{dt} - E_0(\mathbf{p}) + \mu + io\right] S(\mathbf{p}, t)_{\alpha\beta} = \delta_{\alpha\beta} \delta(t). \quad (\text{II.53})$$

It is obvious that

$$S(\mathbf{p}, t)_{\alpha\delta} = \delta_{\alpha\beta} \frac{1}{2\pi} \int \frac{e^{-i\omega t}}{\omega - E_0(\mathbf{p}) + \mu + io} d\omega. \quad (\text{II.54})$$

The functional integrals (II.48) and (II.49) for the interacting system can be expressed in terms of $Z_0[\xi, \xi^*]$ by means of the perturbation theory

$$Z = \exp\left\{i \int_{-\infty}^{\infty} H_{int}\left[i\frac{\delta}{\delta \xi_\alpha(\mathbf{p}, t)}, -i\frac{\delta}{\delta \xi_\alpha^*(\mathbf{p}, t)}\right] dt\right\} Z_0[\xi, \xi^*]\bigg|_{\xi=\xi^*=0}, \quad (\text{II.55})$$

$$Z[\xi, \xi^*] = \exp\left\{-i \int_{-\infty}^{\infty} H_{int}\left[i\frac{\delta}{\delta \xi_\alpha(\mathbf{p}, t)}, -i\frac{\delta}{\delta \xi_\alpha^*(\mathbf{p}, t)}\right] dt\right\} Z_0[\xi, \xi^*]. \quad (\text{II.56})$$

The Green functions are now defined by a formula similary to the relation (II.44) with the change from the imaginary time to the real one and the substitution of the mean value over the canonical ensemble at the temperature T

$$\langle \ldots \ldots \rangle$$

by the vacuum expectation value

$$\langle \ldots \ldots \rangle_0.$$

So we have following definition of the real time Green functions

$$G^F(\mathbf{p}_1, t_1; \ldots \mathbf{p}_n, t_n; \mathbf{q}_1, s_1; \ldots \mathbf{q}_n, s_n)_{\alpha_1 \ldots \alpha_n \beta_1 \ldots \beta_n} = \quad (\text{II.57})$$

$$\frac{1}{\langle S \rangle_0} \langle T(\hat{a}_{\alpha_1}(\mathbf{p}_1, t_1) \ldots \hat{a}_{\alpha_n}(\mathbf{p}_n, t_n) \hat{a}_{\beta_1}^+(\mathbf{q}_1, s_1) \ldots \hat{a}_{\beta_n}^+(\mathbf{q}_n, s_n) S) \rangle_0.$$

They can be expressed in terms of the functional integrals

$$G^F\left(\boldsymbol{p}_1,t_1;\ldots \boldsymbol{p}_n,t_n;\boldsymbol{q}_1,s_1;\ldots \boldsymbol{q}_n,s_n\right)_{\alpha_1\ldots\alpha_n\beta_1\ldots\beta_n} =$$

$$\frac{1}{Z}\int [Da][Da^*]a_{\alpha_1}(\boldsymbol{p}_1,t_1)\ldots a_{\alpha_n}(\boldsymbol{p}_n,t_n)a^*_{\beta_1}(\boldsymbol{q}_1,s_1)\ldots a^*_{\beta_n}(\boldsymbol{q}_n,s_n)\cdot$$

$$\exp\left\{i\int_{-\infty}^{\infty}\sum_{\boldsymbol{p}}a^*_\alpha(\boldsymbol{p},t)\left[i\frac{d}{dt}-E_0(\boldsymbol{p})+\mu+io\right]a_\alpha(\boldsymbol{p},t)\,dt\right.$$

$$\left.-i\int_{-\infty}^{\infty}H_{int}\left[a^*_\alpha(\boldsymbol{p},t),a_\alpha(\boldsymbol{p},t)\right]dt\right\} = \qquad (\text{II.58})$$

$$\frac{1}{Z}\frac{\delta^{2n}Z[\xi,\xi^*]}{\delta\xi^*_{\alpha_1}(\boldsymbol{p}_1,t_1)\ldots\delta\xi^*_{\alpha_n}(\boldsymbol{p}_n,t_n)\delta\xi_{\beta_1}(\boldsymbol{q}_1,s_1)\ldots\delta\xi_{\beta_n}(\boldsymbol{q}_n,s_n)}\bigg|_{\xi=\xi^*=0}.$$

In many problems of the condensed matter theory we have the systems with a large number of electrons which is conserved. The equilibrium state of a many-electron system at the zero temperature $T=0$ is its ground state in which all the energy levels below the Fermi energy $(E_0(\boldsymbol{p})-\mu<0)$ are fully occupied, and all those higher the Fermi one $(E_0(\boldsymbol{p})-\mu>0)$ are empty. For the convenience in the study of the systems of this kind we consider the ground state at $T=0$ as some vacuum and introduce the notion of the hole. Then for the energy levels below the Fermi energy the destruction and creation operators $\hat{a}_\alpha(\boldsymbol{p},t)$ and $\hat{a}^+_\alpha(\boldsymbol{p},t)$ of the electron become the creation and destruction operators $\hat{b}^+_{-\alpha}(-\boldsymbol{p},t)$ and $\hat{b}_{-\alpha}(-\boldsymbol{p},t)$ of the hole, respectively,

$$\hat{a}_\alpha(\boldsymbol{p},t)=\hat{b}^+_{-\alpha}(-\boldsymbol{p},t),\hat{a}^+_\alpha(\boldsymbol{p},t)=\hat{b}_{-\alpha}(-\boldsymbol{p},t), \qquad (\text{II.59})$$
$$E_0(\boldsymbol{p})-\mu<0.$$

Let us now rewrite some above formulae in the new form appropriate to the electron-hole physical picture. For this purpose we denote $b_\alpha(\boldsymbol{p},t)$ and $b^*_\alpha(\boldsymbol{p},t)$ the anticommuting c-number functions corresponding to the quantum operators of the hole:

$$b_\alpha(\boldsymbol{p},t)=a^*_{-\alpha}(-\boldsymbol{p},t),\quad b^*_\alpha(\boldsymbol{p},t)=a_{-\alpha}(-\boldsymbol{p},t), \qquad (\text{II.60})$$
$$E_0(\boldsymbol{p})-\mu<0.$$

In the expressions (II.48) and (II.50) of the partition function we split each sum over the whole range of the variable $\boldsymbol{p}$ into two parts: one with $E_0(\boldsymbol{p})-\mu>0$ and another with $E_0(\boldsymbol{p})-\mu<0$. Performing the substitution (II.60) in the second part we obtain

$$\sum_{\boldsymbol{p}} \theta[\mu - E_0(\boldsymbol{p})] \int_{-\infty}^{\infty} b_{-\alpha}(-\boldsymbol{p},t) \left[i\frac{d}{dt} - E_0(\boldsymbol{p}) + \mu + io\right] b^*_{-\alpha}(-\boldsymbol{p},t) dt =$$

$$\sum_{\boldsymbol{p}} \theta[\mu - E_0(-\boldsymbol{p})] \int_{-\infty}^{\infty} b^*_\alpha(\boldsymbol{p},t) \left[i\frac{d}{dt} + E_0(-\boldsymbol{p}) - \mu - io\right] b_\alpha(\boldsymbol{p},t) dt.$$

Therefore the partition function of the free electron-hole system, for example, can be written in the form

$$Z_0 = \int [Da][Da^*][Db][Db^*]\cdot \tag{II.61}$$

$$\exp\left\{i \int_{-\infty}^{\infty} \left(\sum_{\boldsymbol{p}} \theta[E_0(\boldsymbol{p}) - \mu] a^*_\alpha(\boldsymbol{p},t) \left[i\frac{d}{dt} - E_0(\boldsymbol{p}) + \mu + io\right] a_\alpha(\boldsymbol{p},t)\right.\right.$$

$$\left.\left. + \sum_{\boldsymbol{p}} \theta[\mu - E_0(-\boldsymbol{p})] b^*_\alpha(\boldsymbol{p},t) \left[i\frac{d}{dt} + E_0(-\boldsymbol{p}) - \mu - io\right] b_\alpha(\boldsymbol{p},t)\right) dt\right\},$$

where the functional integration variables $a(\boldsymbol{p},t)$ and $a^*(\boldsymbol{p},t)$ are the functions of $\boldsymbol{p}$ with the range determined by the condition

$$E_0(\boldsymbol{p}) > \mu,$$

$b(\boldsymbol{p},t)$ and $b^*(\boldsymbol{p},t)$ are the functions of $\boldsymbol{p}$ with the range

$$E_0(\boldsymbol{p}) < \mu.$$

The generating functional of this system is defined in the following manner

$$Z_0[\xi, \xi^*] = \int [Da][Da^*][Db][Db^*]\cdot$$

$$\exp\left\{i \int_{-\infty}^{\infty} \left(\sum_{\boldsymbol{p}} \theta[E_0(\boldsymbol{p}) - \mu] \left[a^*_\alpha(\boldsymbol{p},t)(i\frac{d}{dt} - E_0(\boldsymbol{p}) + \mu + io) a_\alpha(\boldsymbol{p},t)\right.\right.\right.$$

$$\left. + \xi_\alpha(\boldsymbol{p},t) a_\alpha(\boldsymbol{p},t) + a^*_\alpha(\boldsymbol{p},t) \xi_\alpha(\boldsymbol{p},t)\right] + \tag{II.62}$$

$$\sum_{\dot{\boldsymbol{p}}} \theta[\mu - E_0(\boldsymbol{p})] \left[b^*_\alpha(\boldsymbol{p},t)(i\frac{d}{dt} + E_0(-\boldsymbol{p}) - \mu - io) b_\alpha(\boldsymbol{p},t)\right.$$

$$\left.\left.\left. + \xi^*_\alpha(\boldsymbol{p},t) b_\alpha(\boldsymbol{p},t) + b^*_\alpha(\boldsymbol{p},t) \xi_\alpha(\boldsymbol{p},t)\right]\right) dt\right\}.$$

Note that the definition of $\xi_\alpha(\mathbf{p},t)$ and $\xi_\alpha^*(\mathbf{p},t)$ for $E_0(\mathbf{p}) - \mu < 0$ in the relation (II.62) is different from that in the formulae (II.49) and (II.51). Applying the relation (II.52) separately to the functional integrals over the sets of variables a, a^* and b, b^* in the r.h.s. of Eq. (II.62), we obtain

$$Z_0[\xi,\xi^*] = Z_0 \exp\left\{-i\int_{-\infty}^{\infty}dt\int_{-\infty}^{\infty}dt'\sum_{\mathbf{p}}\xi_\alpha^*(\mathbf{p},t)\Big[\theta(E_0(\mathbf{p})-\mu)S^e(\mathbf{p},t-t')_{\alpha\beta}\right.$$
$$\left.+\theta(\mu-E_0(-\mathbf{p}))S^h(\mathbf{p},t-t')_{\alpha\beta}\Big]\xi_\beta(\mathbf{p},t')\right\}, \tag{II.63}$$

where $S^e(\mathbf{p},t)_{\alpha\beta}$ and $S^h(\mathbf{p},t)_{\alpha\beta}$ satisfy the differential equations

$$\left[i\frac{d}{dt} - E_0(\mathbf{p}) + \mu + io\right]S^e(\mathbf{p},t)_{\alpha\beta} = \delta_{\alpha\beta}\delta(t), \tag{II.64}$$

$$\left[i\frac{d}{dt} + E_0(\mathbf{p}) - \mu - io\right]S^h(\mathbf{p},t)_{\alpha\beta} = \delta_{\alpha\beta}\delta(t), \tag{II.65}$$

and equal

$$S^e(\mathbf{p},t)_{\alpha\beta} = \delta_{\alpha\beta}\frac{1}{2\pi}\int\frac{e^{-i\omega t}}{\omega - E_0(\mathbf{p}) + \mu + io}d\omega, \tag{II.66}$$

$$S^h(\mathbf{p},t)_{\alpha\beta} = \delta_{\alpha\beta}\frac{1}{2\pi}\int\frac{e^{-i\omega t}}{\omega + E_0(-\mathbf{p}) - \mu - io}d\omega. \tag{II.67}$$

If we set

$$S(\mathbf{p},t)_{\alpha\beta} = \theta(E_0(\mathbf{p})-\mu)S^e(\mathbf{p},t)_{\alpha\beta} + \theta(\mu - E_0(-\mathbf{p}))S^h(\mathbf{p},t)_{\alpha\beta}$$
$$= \frac{1}{2\pi}\int e^{-i\omega t}S(\mathbf{p},\omega)_{\alpha\beta}d\omega, \tag{II.68}$$

$$S(\mathbf{p},\omega)_{\alpha\beta} = \delta_{\alpha\beta}\left\{\frac{\theta(E_0(\mathbf{p})-\mu)}{\omega - E_0(\mathbf{p}) + \mu + io} + \frac{\theta(\mu - E_0(-\mathbf{p}))}{\omega + E_0(-\mathbf{p}) - \mu - io}\right\}, \tag{II.69}$$

then we rewrite the r.h.s. of Eq. (II.63) in the same form as Eq. (II.52). The expression (II.69) of the Fourier transform of the two-point Green function of the free fermion in the electron-hole formalism is used very often in the condensed matter theory.

III. HUBBARD - STRATONOVICH TRANSFORMATION

The functional integral technique has been widely applied in the condensed matter theory. In concluding our review we consider one aspect of these applications: the study of fermion systems with the quadrilinear direct four-fermion couplings. In the framework of the functional integral technique we show that each system of this kind is equivalent to the system of fermions and bosons with some trilinear interaction Hamiltonian bilinear in fermionic operators and linear in bosonic ones. The new bosonic variables in the functional integrals describe the composite quasiparticles whose constituents are the given fermions with the quadrilinear interaction. This equivalence is established by means of the Hubbard - Stratonovich transformation.

In order to understand the essence of this transformation we consider following integral in the real variable x

$$\int_{-\infty}^{\infty} e^{iax^2} dx = (1 + i\,\mathrm{sign}\,a)\sqrt{\frac{\pi}{2|a|}} \,.$$

It is invariant under any shift $x \to x + y$ of the integration variable

$$\int_{-\infty}^{\infty} e^{iax^2} dx = \int_{-\infty}^{\infty} e^{ia(x+y)^2} dx = e^{iay^2} \int_{-\infty}^{\infty} e^{ia(x^2+2xy)} dx$$

Therefore we have identity

$$e^{-iay^2} = \frac{\int_{-\infty}^{\infty} e^{ia(x^2+2xy)} dx}{\int_{-\infty}^{\infty} e^{iax^2} dx} \tag{III.1}$$

with any real number y. Similarly in the n-dimensional space we consider the multiple integral

$$\int_{-\infty}^{\infty} dx_1 \cdots \int_{-\infty}^{\infty} dx_n \exp\left\{i \sum_{k=1}^{n} a_k x_k^2\right\} =$$

$$= \int_{-\infty}^{\infty} dx_1 \cdots \int_{-\infty}^{\infty} dx_n \exp\left\{i \sum_{k=1}^{n} a_k (x_k + y_k)^2\right\} =$$

$$= \exp\left\{i \sum_{k=1}^{n} a_k y_k^2\right\} \int_{-\infty}^{\infty} dx_1 \cdots \int_{-\infty}^{\infty} dx_n \exp\left\{i \sum_{k=1}^{n} a_k (x_k^2 + 2x_k y_k)\right\}$$

and obtain the identity

$$\exp\left\{-i\sum_{k=1}^{n}a_k y_k^2\right\} = \frac{\int_{-\infty}^{\infty} dx_1 \cdots \int_{-\infty}^{\infty} dx_n \exp\left\{i\sum_{k=1}^{n} a_k(x_k^2 + 2x_k y_k)\right\}}{\int_{-\infty}^{\infty} dx_1 \cdots \int_{-\infty}^{\infty} dx_n \exp\left\{i\sum_{k=1}^{n} a_k x_k^2\right\}}. \quad \text{(III.2)}$$

It is straightforword to extend this formula to the case of the functional integral of the form

$$I_\varphi = \int [D\varphi] \exp\left\{i\int \varphi(\boldsymbol{x},t)\hat{M}\varphi(\boldsymbol{x},t)d\boldsymbol{x}dt\right\} \quad \text{(III.3)}$$

in which the functional integration variables $\varphi(\boldsymbol{x},t)$ are the functions of the n-dimensional spatial coordinate $\boldsymbol{x}$ and the real time t, $\hat{M}$ is any operator. Because this functional integral is invariant under any shift of the integration variable

$$\varphi(\boldsymbol{x},t) \to \varphi(\boldsymbol{x}+t) + f(\boldsymbol{x},t)$$

we have

$$I_\varphi = \int [D\varphi] \exp\left\{i\int [\varphi(\boldsymbol{x},t) + f(\boldsymbol{x},t)]\hat{M}[\varphi(\boldsymbol{x},t) + f(\boldsymbol{x},t)]d\boldsymbol{x}dt\right\}$$

$$= \exp\left\{i\int f(\boldsymbol{x},t)\hat{M}f(\boldsymbol{x},t)d\boldsymbol{x}.dt\right\} \int [D\varphi] \exp\left\{i\int [\varphi(\boldsymbol{x},t)\hat{M}\varphi(\boldsymbol{x},t)\right.$$
$$\left. + \hat{\varphi}(\boldsymbol{x},t)\hat{M}f(\boldsymbol{x},t) + f(\boldsymbol{x},t)\hat{M}\varphi(\boldsymbol{x},t)]d\boldsymbol{x}dt\right\}. \quad \text{(III.4)}$$

Therefore we obtain following identity for any operator $\hat{M}$ and any function $f(\boldsymbol{x},t)$

$$\exp\left\{-i\int f(\boldsymbol{x},t)\hat{M}f(\boldsymbol{x},t)d\boldsymbol{x}dt\right\} = \quad \text{(III.5)}$$

$$\frac{\int [D\varphi] \exp\left\{i\int [\varphi(\boldsymbol{x},t)\hat{M}\varphi(\boldsymbol{x},t) + \varphi(\boldsymbol{x},t)\hat{M}f(\boldsymbol{x},t) + f(\boldsymbol{x},t)\hat{M}\varphi(\boldsymbol{x},t)]d\boldsymbol{x}dt\right\}}{\int [D\varphi] \exp\left\{i\int \varphi(\boldsymbol{x},t)\hat{M}\varphi(\boldsymbol{x},t)d\boldsymbol{x}dt\right\}}.$$

In the particular case $\hat{M} = 1$ this relation becomes

$$\exp\left\{-i\int f(\boldsymbol{x},t)^2 d\boldsymbol{x}dt\right\} = \frac{\int [D\varphi] \exp\left\{i\int [\varphi(\boldsymbol{x},t)^2 + 2\varphi(\boldsymbol{x},t)f(\boldsymbol{x},t)]d\boldsymbol{x}dt\right\}}{\int [D\varphi] \exp\left\{i\int \varphi(\boldsymbol{x},t)d\boldsymbol{x}dt\right\}}.$$

(III.6)

Performing the Fourier transformation of the functions $\varphi(\pmb{x},t)$ and $f(\pmb{x},t)$ in some normalization volume V

$$\varphi(\pmb{x},t) = \frac{1}{\sqrt{V}} \sum_{\pmb{p}} e^{i\pmb{p}\pmb{x}} \tilde{\varphi}(\pmb{p},t),$$

$$f(\pmb{x},t) = \frac{1}{\sqrt{V}} \sum_{\pmb{p}} e^{i\pmb{p}\pmb{x}} \tilde{f}(\pmb{p},t), \qquad (III.7)$$

$$\hat{M}\varphi(\pmb{x},t) = \frac{1}{\sqrt{V}} \sum_{\pmb{p}} e^{i\pmb{p}\pmb{x}} \tilde{M}(\pmb{p}) \tilde{\varphi}(\pmb{p},t),$$

instead of Eqs. (III.5) and (III.6) we obtain now

$$\exp\left\{-i \int_{-\infty}^{\infty} \sum_{\pmb{p}} \tilde{M}(\pmb{p}) f(-\pmb{p},t) f(\pmb{p},t) dt \right\} = \qquad (III.8)$$

$$\frac{\int [D\varphi] \exp\left\{i \int_{-\infty}^{\infty} \sum_{\pmb{p}} \tilde{M}(\pmb{p}) \left[\varphi(-\pmb{p},t)\varphi(\pmb{p},t) + f(-\pmb{p},t)\varphi(\pmb{p},t) + \varphi(-\pmb{p},t) f(\pmb{p},t)\right] dt\right\}}{\int [D\varphi] \exp\left\{i \int_{-\infty}^{\infty} \sum_{\pmb{p}} \tilde{M}(\pmb{p}) \varphi(-\pmb{p},t)\varphi(\pmb{p},t) dt\right\}},$$

$$\exp\left\{-i \int_{-\infty}^{\infty} \sum_{\pmb{p}} f(-\pmb{p},t) f(\pmb{p},t) dt\right\} = \qquad (III.9)$$

$$\frac{\int [D\varphi] \exp\left\{i \int_{-\infty}^{\infty} \sum_{\pmb{p}} \left[\varphi(-\pmb{p},t)\varphi(\pmb{p},t) + 2\varphi(-\pmb{p},t) f(\pmb{p},t)\right] dt\right\}}{\int [D\varphi] \exp\left\{i \int_{-\infty}^{\infty} \sum_{\pmb{p}} \varphi(-\pmb{p},t)\varphi(\pmb{p},t) dt\right\}}.$$

The Hubbard - Stratonovich transformation of the functional integrals is based on the relations of the form (III.5), (III.6) or (III.8), (III.9).

As an example we consider the system of fermions described by $\pmb{x}$ - dependent quantum operators $\hat{\psi}_\alpha(\pmb{x},t)$ and $\hat{\psi}_\alpha^+(\pmb{x},t)$ with the Fourier transforms $\hat{a}_\alpha(\pmb{p},t)$ and $\hat{a}_\alpha^+(\pmb{p},t)$ being the destruction and creation operators in the one - particle state with the momentum $\pmb{p}$

$$\hat{\psi}_\alpha(\pmb{x},t) = \frac{1}{\sqrt{V}} \sum_{\pmb{p}} e^{i\pmb{p}\pmb{x}} \hat{a}_\alpha(\pmb{p},t),$$

$$\hat{\psi}_\alpha^+(\pmb{x},t) = \frac{1}{\sqrt{V}} \sum_{\pmb{p}} e^{-i\pmb{p}\pmb{x}} \hat{a}_\alpha^+(\pmb{p},t). \qquad (III.10)$$

Suppose that their interaction is the direct four - fermion coupling of the form

$$\hat{H}_{int} = \int \hat{n}(\mathbf{x},t) \mathcal{U}(\mathbf{x}-\mathbf{y}) \hat{n}(\mathbf{y},t) d\mathbf{x}d\mathbf{y} \tag{III.11}$$

where $\hat{n}(\mathbf{x},t)$ is the density operator

$$\hat{n}(\mathbf{x},t) = \hat{\psi}_\alpha^+(\mathbf{x},t)\hat{\psi}_\alpha(\mathbf{x},t). \tag{III.12}$$

Denote $\hat{\omega}(\mathbf{k},t)$ and $u(\mathbf{k})$ the Fourier transforms of $\hat{n}(\mathbf{x},t)$ and $\mathcal{U}(\mathbf{x})$

$$\begin{aligned}\hat{n}(\mathbf{x},t) &= \frac{1}{\sqrt{V}} \sum_{\mathbf{k}} e^{i\mathbf{k}\mathbf{x}} \hat{\omega}(\mathbf{k},t), \\ \mathcal{U}(\mathbf{x}) &= \frac{1}{V} \sum_{\mathbf{k}} e^{i\mathbf{k}\mathbf{x}} u(\mathbf{k}).\end{aligned} \tag{III.13}$$

It is easy to verify that

$$\hat{\omega}(\mathbf{k},t) = \frac{1}{\sqrt{V}} \sum_{\mathbf{p}} \hat{a}_\alpha^+(\mathbf{p}-\mathbf{k},t) \hat{a}_\alpha(\mathbf{p},t). \tag{III.14}$$

Inserting the Fourier expansions (III.13) into the r.h.s. of Eq. (III.11) we obtain

$$\hat{H}_{int} = \sum_{\mathbf{k}} \hat{\omega}(-\mathbf{k},t) u(\mathbf{k}) \hat{\omega}(\mathbf{k},t). \tag{III.15}$$

Then the partition function (II.48) has following concrete form

$$Z = \int [Da][Da^*] \exp\left\{ i \int_{-\infty}^{\infty} \sum_{\mathbf{p}} a_\alpha^*(\mathbf{p},t)\left[i\frac{d}{dt} - E_0(\mathbf{p}) + \mu + io\right] a_\alpha(\mathbf{p},t) dt \right\} \cdot$$

$$\exp\left\{ -i \int_{-\infty}^{\infty} \sum_{\mathbf{k}} \omega(-\mathbf{k},t) u(\mathbf{k}) \omega(\mathbf{k},t) dt \right\}, \tag{III.16}$$

where $\omega(\mathbf{k},t)$ is expressed in terms of the anticommuting functions $a_\alpha(\mathbf{p},t)$ and $a_\alpha^*(\mathbf{p},t)$ by the same formula as Eq. (III.14)

$$\omega(\mathbf{k},t) = \frac{1}{\sqrt{V}} \sum_{\mathbf{p}} a_\alpha^*(\mathbf{p}-\mathbf{k},t) a_\alpha(\mathbf{p},t). \tag{III.17}$$

The corresponding expression of Z in the electron - hole formalism is obtained from Eqs. (III.16) and (III.17) by substituting

$$a_\alpha(\boldsymbol{p},t) = b^*_{-\alpha}(-\boldsymbol{p},t), \quad a^*_\alpha(\boldsymbol{p},t) = b_{-\alpha}(-\boldsymbol{p},t)$$

at the values of $\boldsymbol{p}$ such that

$$E_0(\boldsymbol{p}) - \mu < 0.$$

For the simplicity in writting formulae we consider the special case with

$$u(\boldsymbol{k}) = u(\boldsymbol{k}^2) \geq 0,$$

i.e. $u(\boldsymbol{k})$ depends only on $\boldsymbol{k}^2$ and is always non - negative. Applying the formula (III.6) and substituting

$$f(\boldsymbol{k},t) = \sqrt{u(\boldsymbol{k}^2)}\omega(\boldsymbol{k},t)$$

we obtain

$$\exp\left\{-i\int_{-\infty}^{\infty}\sum_{\boldsymbol{k}}\omega(-\boldsymbol{k},t)u(\boldsymbol{k}^2)\omega(\boldsymbol{k},t)dt\right\} =$$

$$\mathrm{const}\int[D\varphi]\exp\left\{i\int_{-\infty}^{\infty}\sum_{\boldsymbol{k}}\varphi(-\boldsymbol{k},t)\varphi(\boldsymbol{k},t)dt\right\}. \tag{III.18}$$

$$\exp\left\{\frac{2i}{\sqrt{V}}\int_{-\infty}^{\infty}\sum_{\boldsymbol{p},\boldsymbol{k}}\sqrt{u(\boldsymbol{k}^2)}\varphi(\boldsymbol{k},t)a^*_\alpha(\boldsymbol{p}-\boldsymbol{k},t)a_\alpha(\boldsymbol{p},t)dt\right\}.$$

Inserting into the r.h.s. of Eq. (III.16) we rewrite the partition function Z in the form of a functional integral in the original fermion fields $a_\alpha(\boldsymbol{p},t)$, $a^*_\alpha(\boldsymbol{p},t)$ and the induced boson field $\varphi(\boldsymbol{k},t)$

$$Z = \mathrm{const}\int[Da][Da^*][D\varphi]\exp\left\{iS_0[a_\alpha(\boldsymbol{p},t),a^*_\alpha(\boldsymbol{p},t)]\right.$$
$$\left. iS_0[\varphi(\boldsymbol{k},t)] + iS_{int}[a_\alpha(\boldsymbol{p},t),a^*_\alpha(\boldsymbol{p},t),\varphi(\boldsymbol{k},t)]\right\}. \tag{III.19}$$

Beside of the bilinear terms $S_0[a_\alpha(\boldsymbol{p},t),a^*_\alpha(\boldsymbol{p},t)]$ and $S_0[\varphi(\boldsymbol{k},t)]$ there are the trilinear ones representing the fermion - boson coupling

$$S_{int}[a_\alpha(\boldsymbol{p},t),a^*_\alpha(\boldsymbol{p},t),\varphi(\boldsymbol{k},t)] =$$

$$\frac{2}{\sqrt{V}}\sum_{\boldsymbol{p},\boldsymbol{k}}\sqrt{u(\boldsymbol{k}^2)}\int_{-\infty}^{\infty}\varphi(\boldsymbol{k},t)a^*_\alpha(\boldsymbol{p}-\boldsymbol{k},t)a_\alpha(\boldsymbol{p},t)dt. \tag{III.20}$$

If we perform the integration in the variables $a_\alpha(\mathbf{p},t)$ and $a^*_\alpha(\mathbf{p},t)$ in the r.h.s. of Eq. (III.19) then we obtain the partition function in the form of the functional integral in the bosonic variable only

$$Z = \int [D\varphi] \exp\left\{iS_{eff}[\varphi(\mathbf{k},t)]\right\}, \qquad (\text{III.21})$$

where $S_{eff}[\varphi(\mathbf{k},t)]$ can be calculated, for example, by means of the perturbation expansion. Thus the fermion system with the direct four - fermion interaction (III.11) is equivalent to a system of fermions and bosons with the trilinear boson - fermion coupling (III.20) but without the direct fermion - fermion interaction. The bosons can be interpreted as the composite particles of the original fermion. In the absence of the fermionic excitations of the systems the partition function for these bosonic quasiparticles is determined by Eq. (III.21). In the case when the fermion system is an electron gas the bosonic quasiparticles corresponding to the scalar functional integration variable $\varphi(\mathbf{k},t)$ in Eq. (III.18) are the collective excitations generated by the charge density fluctuations - plasmons.

Now we apply above reasonings to the study of the quantum ferromagnetism or antiferromagnetism. Consider the system of itinerant electrons in a lattice and denote $\hat{a}_{i\alpha}$, $\hat{a}^+_{i\alpha}$ the destruction and creation operators of the electron with the spin projection α at the site i. The spin operator at the site i equals

$$\hat{\mathbf{S}}_i = \hat{a}^+_{i\alpha} \frac{(\boldsymbol{\sigma})_{\alpha\beta}}{2} \hat{a}_{i\beta}. \qquad (\text{III.22})$$

Suppose that the electron - electron interaction is the ferromagnetic or antiferromagnetic Heisenberg coupling

$$\hat{H}_{int} = \frac{1}{2} \sum_{i,j} J_{ij} \hat{\mathbf{S}}_i \hat{\mathbf{S}}_j, \qquad (\text{III.23})$$

where J_{ij} are the constants depending on the distance between two sites i and j. Denote $a_{i\alpha}(t)$ and $a^*_{i\alpha}(t)$ the anticommuting functional integration variables corresponding to the quantum operators $\hat{a}_{i\alpha}$ and $\hat{a}^+_{i\alpha}$. The partition function of the system equals

$$Z = \int [Da][Da^*] \exp\left\{i \int_{-\infty}^{\infty} \sum_i a^*_{i\alpha}(t) \left[i\frac{d}{dt} - E + io\right] a_{i\alpha}(t) dt \right.$$

$$\left. - \frac{i}{2} \int_{-\infty}^{\infty} \sum_{ij} J_{ij} \mathbf{S}_i(t) \mathbf{S}_j(t) dt \right\} \qquad (\text{III.24})$$

where $\mathbf{S}_i(t)$ are the vector functions expressed in terms of the integration variables $a_{i\alpha}(t)$, $a^*_{i\alpha}(t)$ by means of the same formula as Eq. (III.22)

$$S_i(t) = a_{i\alpha}^*(t) \frac{(\sigma)_{\alpha\beta}}{2} a_{i\beta}(t). \tag{III.25}$$

Introduce the space of vector - valued functions $n_i(t)$ and consider the functional integral

$$I_n = \int [Dn] \exp\left\{\frac{i}{2} \int_{-\infty}^{\infty} \sum_{i,j} J_{ij} n_i(t) n_j(t)\right\} \tag{III.26}$$

From its invariance under the shift of the functional integration variables $n_i(t) \to n_i(t) + S_i(t)$ we derive the identity similary to Eq. (III.5)

$$\exp\left\{-\frac{i}{2} \int_{-\infty}^{\infty} \sum_{i,j} J_{ij} S_i(t) S_j(t) dt\right\} = \tag{III.27}$$

$$\frac{\int [Dn] \exp\left\{\frac{i}{2} \int_{-\infty}^{\infty} \sum_{i,j} J_{ij} \left[n_i(t) n_j(t) + 2 n_i(t) S_j(t)\right] dt\right\}}{\int [Dn] \exp\left\{\frac{i}{2} \int_{-\infty}^{\infty} \sum_{i,j} J_{ij} n_i(t) n_j(t) dt\right\}}$$

Inserting into the r.h.s. of Eq. (III.24) we rewrite the partition function Z in the form of the functional integral in the fermionic variables $a_{i\alpha}(t)$, $a_{i\alpha}^*(t)$ and the bosonic vector variables $n_i(t)$

$$Z = \text{const} \int [Da][Da^*][Dn] \exp\left\{i S_0\left[a_{i\alpha}(t), a_{i\alpha}^*(t)\right]\right.$$
$$\left. + i S_0\left[n_i(t)\right] + i S_{int}\left[a_{i\alpha}(t), a_{i\alpha}^*(t), n_i(t)\right]\right\}. \tag{III.28}$$

The trilinear terms in the r.h.s. of Eq. (III. 28)

$$S_{int}\left[a_{i\alpha}(t), a_{i\alpha}^*(t), n_i(t)\right] = \int_{-\infty}^{\infty} \sum_{i,j} J_{ij} n_i(t) a_{j\alpha}^*(t) \frac{(\sigma)_{\alpha\beta}}{2} a_{j\beta}(t) dt \tag{III.29}$$

represent the fermion - boson interaction. The vector - valued functional integration variables $n_i(t)$ correspond to some composite quasipartiles with the fermion number equal to zero and the spin equal to 1. These vector bosons are the collective excitations generated by the fluctuations of the spin density and called magnons.

At last we show how to apply the Hubbard - Stratonovich transformation to the superconducting electron systems at some finite temperature T. In this case we use the imaginary time formalism with the τ - dependent quantum operators $\hat{\psi}_\alpha(\boldsymbol{x},\tau)$ and $\hat{\bar{\psi}}_\alpha(\boldsymbol{x},\tau)$. For the simplicity in writing the equations we assume following attractive four - fermion contact coupling

$$\hat{H}_{int} = -\frac{g^2}{2}\int \hat{\bar{\psi}}_\alpha(\boldsymbol{x},\tau)\hat{\psi}_\alpha(\boldsymbol{x},\tau)\hat{\bar{\psi}}_\beta(\boldsymbol{x},\tau)\hat{\psi}_\beta(\boldsymbol{x},\tau)d\boldsymbol{x}$$
$$= -g^2\int \hat{\bar{\psi}}_\uparrow(\boldsymbol{x},\tau)\hat{\bar{\psi}}_\downarrow(\boldsymbol{x},\tau)\hat{\psi}_\downarrow(\boldsymbol{x},\tau)\hat{\psi}_\uparrow(\boldsymbol{x},\tau)d\boldsymbol{x} \qquad (III.30)$$

instead of the interaction Hamiltonian of the forms (III.11). Denote $\psi_\alpha(\boldsymbol{x},\tau)$ and $\psi_\alpha^*(\boldsymbol{x},\tau)$ the anticommuting integration variables corresponding to the operators $\hat{\psi}_\alpha(\boldsymbol{x},\tau)$ and $\hat{\bar{\psi}}_\alpha(\boldsymbol{x},\tau)$. In order to study the electron pairing we introduce the space of complex scalar functions $\pi(\boldsymbol{x},\tau)$, $\pi^*(\boldsymbol{x},\tau)$ and consider the functional integral

$$I_\pi = \int [D\pi][D\pi^*]\exp\left\{-\int_0^\beta d\tau \int d\boldsymbol{x} \pi^*(\boldsymbol{x},\tau)\pi(\boldsymbol{x},\tau)\right\}. \qquad (III.31)$$

From the invariance of this functional integral under the shift

$$\pi(\boldsymbol{x},\tau) \to \pi(\boldsymbol{x},\tau) + g\psi_\downarrow(\boldsymbol{x},\tau)\psi_\uparrow(\boldsymbol{x},\tau),$$
$$\pi^*(\boldsymbol{x},\tau) \to \pi^*(\boldsymbol{x},\tau) + g\psi_\uparrow^*(\boldsymbol{x},\tau)\psi_\downarrow^*(\boldsymbol{x},\tau)$$

of the integration variables we obtain the identity

$$\exp\left\{g^2\int_0^\beta d\tau\int d\boldsymbol{x}\psi_\uparrow^*(\boldsymbol{x},\tau)\psi_\downarrow^*(\boldsymbol{x},\tau)\psi_\downarrow(\boldsymbol{x},\tau)\psi_\uparrow(\boldsymbol{x},\tau)\right\} =$$
$$\frac{1}{I_\pi}\int [D\pi][D\pi^*]\exp\left\{-\int_0^\beta d\tau\int d\boldsymbol{x}\Big[\pi^*(\boldsymbol{x},\tau)\pi(\boldsymbol{x},\tau)+ \right. \qquad (III.32)$$
$$\left. g\pi^*(\boldsymbol{x},\tau)\psi_\downarrow(\boldsymbol{x},\tau)\psi_\uparrow(\boldsymbol{x},\tau) + g\pi(\boldsymbol{x},\tau)\psi_\uparrow^*(\boldsymbol{x},\tau)\psi_\downarrow^*(\boldsymbol{x},\tau)\Big]\right\}.$$

Because the Hamiltonian of non - interacting electron system has the form

$$\hat{H}_0 = \int \hat{\bar{\psi}}(\mathbf{x},\tau)\left(-\frac{\nabla^2}{2m}\right)\hat{\psi}(\mathbf{x},\tau)d\mathbf{x} \qquad (\text{III.33})$$

the partition function of the electron system with the contact interaction (III.30) equals

$$Z = \int [D\psi][D\psi^*]\exp\left\{-\int_0^\beta d\tau \int d\mathbf{x}\psi_\alpha^*(\mathbf{x},\tau)\left[\frac{d}{d\tau}+\left(-\frac{\nabla^2}{2m}\right)-\mu\right]\psi_\alpha(\mathbf{x},\tau)\right.$$
$$\left. + g^2\int_0^\beta d\tau \int d\mathbf{x}\psi_\uparrow^*(\mathbf{x},\tau)\psi_\downarrow^*(\mathbf{x},\tau)\psi_\downarrow(\mathbf{x},\tau)\psi_\uparrow(\mathbf{x},\tau)\right\}. \qquad (\text{III.34})$$

Performing the Hubbard - Stratonovich transformation (III.32) in the r.h.s. of Eq. (III.34) we rewrite the partition function Z in the form of a functional integral in the fermionic variables $\psi(\mathbf{x},\tau)$, $\psi^*(\mathbf{x},\tau)$ and the bosonic ones $\pi(\mathbf{x},\tau)$, $\pi^*(\mathbf{x},\tau)$:

$$Z = \text{const}\int [D\psi][D\psi^*][D\pi][D\pi^*]\exp\left\{-S_0\left[\psi_\alpha(\mathbf{x},\tau),\psi_\alpha^*(\mathbf{x},\tau)\right]\right.$$
$$\left.-S_0\left[\pi(\mathbf{x},\tau),\pi^*(\mathbf{x},\tau)\right]-S_{int}\left[\psi_\alpha(\mathbf{x},\tau),\psi_\alpha^*(\mathbf{x},\tau),\pi(\mathbf{x},\tau),\pi^*(\mathbf{x},\tau)\right]\right\}. \qquad (\text{III.35})$$

The bosonic variables $\pi(\mathbf{x},\tau)$ and $\pi^*(\mathbf{x},\tau)$ correspond to the composite quasiparticles with the electron number equal to 2 - the Cooper pairs. If we perform the functional integration in the fermionic variables $\psi_\alpha(\mathbf{x},\tau)$, $\psi_\alpha^*(\mathbf{x},t)$ then from the expression (III.35) of Z we obtain the partition function as a functional integral in the bosonic variables only

$$Z = \int [D\pi][D\pi^*]\exp\left\{-S_{eff}\left[\pi(\mathbf{x},\tau),\pi^*(\mathbf{x},\tau)\right]\right\}. \qquad (\text{III.36})$$

From this expression it is straightforward to establish the equation for the scalar quantum operators $\hat{\pi}(\mathbf{x},\tau)$, $\hat{\bar{\pi}}(\mathbf{x},\tau)$ corresponding to the variable $\pi(\mathbf{x},\tau)$, $\pi^*(\mathbf{x},\tau)$. This is the Ginzburg - Landau equation in the theory of superconductivity.

In general the Hubbard - Stratonovich transformation can be applied to any system with the quadrilinear interaction and give rise to the appearance of some composite quasiparticles whose partition functions and, therefore, wave equations can be derived from the original Hamiltonian.

REFERENCES

1. Abrikosov A. A., Gorkov L. P. and Dzyaloshinski I. Ye., Quantum Field Theoretical Methods in Statistical Physics, 2 nd ed., Pergamon, Oxford, 1965.

2. Fadeev L. D., Introduction to Funtional Methods, in "Methods in Field Theory", Les Houches Summer School 1975, Eds. Balian R. and Zinn - Justin J., North - Holland, Amsterdam, 1976.

3. Feyman R. P. and Hibbs A. R., Path Integrals and Quantum Mechanics, McGraw-Hill, New York, 1965.

4. Itzykson C. and Zuber J. B., Quantum Field Theory, McGraw - Hill, New York, 1980.

5. Lee B. W., Gauge Theories, in "Methods in Field Theory", Les Houches Summer School 1975, Eds. Balian R. and Zinn - Justin J., North - Holland, Amsterdam, 1976.

6. Ramond P., Field Theory, A Modern Primer, Benjamin Cummings, London, 1983.

7. Sakita B., Quantum Theory of Many - Variable Systems and Fields, World Scientific, Singapore, 1985.

8. Wiegel F. W. Introdution to Path - Integral Methods in Physics and Polymer Science, World Scientific, Singapore, 1986.

9. de Wit B., Functional Methods in Quantum Field Theory, in "Fundamental Interactions", Cargese Lectures 1981, Eds. Levy M., Basdevand J-L., Speiser D., Weyer J., Jacov M. and Gastmans R., Plenum Press, New York, 1981.

DISORDERED ELECTRONIC MATERIALS AND SPIN GLASSES

D. J. W. Geldart

Department of Physics
Dalhousie University
Halifax, Nova Scotia
Canada B3H 3J5

I INTRODUCTION

The study of well ordered, chemically pure, crystalline solids in thermal equilibrium has occcupied a prominent position in traditional condensed matter physics. It is less well known that systematic studies of classes of poorly ordered, chemically impure, noncrystalline materials, often not in a state of thermal equilibrium, have also led to very important developments, particularly in the past decade or so. These developments have led to new fundamental insights into the physics of materials as well as to applications[1-3]. The primary purpose of this lecture is to provide a concise pedagogical introduction to a selection of properties of disordered systems, with particular emphasis on spin glasses. Numerical results for a number of spin glass models will be surveyed briefly. Some recent results for metallic spin glasses will be given and challenging areas for further work will be indicated.

Disordered systems include glasses, amorphous materials, alloys with various distributions of site occupancy, among others. In all of the examples of disordered systems of present interest, random or quasi-random microstructural features play an essential role. This feature can be illustrated by considering the formation of rapidly solidified amorphous ribbons by melt spinning. In this procedure, a molten mixture of the desired chemical composition is sprayed through a small orifice onto a rotating copper wheel. To give a typical example, the wheel may have a diameter of 15 cm, a thickness of 2.5 cm, and the wheel's surface velocity presented to the impinging liquid is about 60 m/s. Effective cooling rates in excess of 10^6 K/s can be achieved and the liquid solidifies very rapidly, spinning off the wheel in the form of a ribbon 2 or 3 mm in width, 20 to 50 microns in thickness and tens of meters in length. Commercial melt-spun ribbons can be 20 cm wide, 25 microns thick, and hundreds of meters long. It is an essential point that the high effective heat capacity of the wheel causes energy to be extracted from the liquid so rapidly that the atoms of the liquid do not have time to sample enough configurations to locate an equilibrium configuration of minimum free energy, even if such a minimum should exist. Consequently, the disorder characteristic of the high temperature liquid state is "frozen in". The degree of disorder frozen into the solid depends on the rate of cooling and details of kinetic processes including relaxation times, viscosities and other factors. The entropy of this disordered solid can be large even at low temperature. Since the newly formed rapidly solidified material will not be in an equilibrium state, it will have a finite lifetime for stability. There can be a spectrum of relaxation times for relaxation through a sequence of metastable states. The relaxation times can be hundreds of years or hours, depending on the circumstances and the materials.

It is important to appreciate that it is possible in this way to prepare materials which do not occur naturally in nature. In effect, it is possible to design and engineer special purpose multicomponent materials for specific applications. Some examples in the area of amorphous magnets will be given. By virtue of efficient second harmonic detection, FeZr alloys are used commercially in devices for protection against theft. High performance magnets are made from NdFeB, for example. Transformer parts and magnetic coupling devices are made from SmCo. A great many other multicomponent amorphous materials with specially tailored physical properties are in use. Of course, the fact that the material is only in a metastable state is not a problem for applications since it will have been ensured that this lifetime is sufficiently long for practical use. There are many other examples of systems in metastable or "glassy" states which have practical and desirable properties. Ordinary window glass and stained glass windows are examples of such systems having long lifetimes.

Quite apart from any possible desire to engineer new materials, we would like to have a fundamental understanding of the phase diagrams of these systems and of the structure and the time scales of their metastable states. Of course, there are many different types of disordered materials with a variety of physical properties[1-3]. The theoretical study of these systems presents major challenges and a great deal of theoretical effort has been devoted to these questions, often with spectacular new results such as localization and universal conductance fluctuations in disordered electronic materials and all of the physics of mesoscopic systems. Developments outside conventional condensed matter systems have also been fertilized, including optimization problems (computer design, marketing strategy, *etc.*) and applications to biological problems (neural networks, brain activity, *etc.*) [1,4].

In coming to grips with the new features of disordered systems, it has been necessary to rethink a good deal of statistical mechanics in arriving at answers to some basic questions. Exactly what is out of equilibrium (orientation of bonds, positions of ions, directions of magnetic moments, *etc.*)? How do these variables couple to other degrees of freedom and how should statistical averages over the disorder be taken? What are the essential features that must be taken into account in developing a model for a given physical system? What are common features of different systems that can lead to a classification of universality classes? It is not practical to attack directly the most general problem of this kind in full detail without some guidance from special cases which may be rather simplified but which still retain important physical features. The study of systems known as "spin glasses" has been extremely useful in this respect. The remainder of this review is focused on metallic spin glasses. In section II, we consider some prototype physical systems which have led to the spin glass concept. Theoretical models for these systems and numerical results, obtained mainly by Monte Carlo simulations, are discussed in section III. In section IV, we reconsider some important physical aspects of the canonical metallic spin glasses in the light of recent developments in the theory of disordered electronic materials. Our conclusions are summarized in section V.

II PHYSICAL SPIN GLASS SYSTEMS

To appreciate the nature of a specific spin glass system, consider Fig. 1 which is a schematic representation of a crystalline ternary alloy of chemical composition A B C. The lattice sites are taken to be more or less randomly occupied, in accordance with their relative atomic concentration, by ions of type A, B or C. Ions of type A and B are taken to have zero net magnetic moment (in the simpler case of a binary alloy, A and B ions are the same chemical species). On the other hand, ions of type C carry a net local magnetic moment, and hence a spin. It is assumed that many details of electronic configurations can be neglected. In particular, the spin of every ion of type C is the same, irrespective of its position and electronic environment.

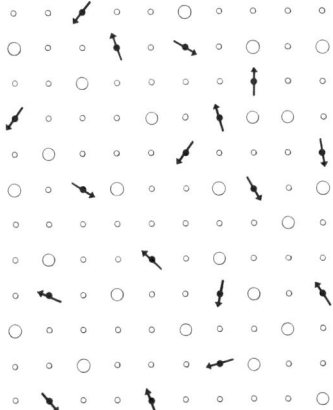

Figure 1. Schematic view of a disordered ternary alloy of composition ABC; two types of ions are nonmagnetic while the third carries a local magnetic moment.

Having localized magnetic moments, there will be exchange interactions between all pairs of type C ions. This exchange interaction may be due to direct overlap of atomic orbitals centered on different C atomic sites; this is a dominant exchange mechanism in insulating spin glasses. In metallic spin glasses, there are additional contributions to the effective interaction between a pair of ions of C type due to indirect interactions transmitted through the conduction band electrons. In all cases, the net spin dependent interaction energy depends on the relative orientation of the C ion spins and can be interpreted in terms of an effective indirect exchange interaction between the localized spins. If all spin anisotropies are neglected, the thermodynamic properties of such systems may be described by an effective Hamiltonian of Heisenberg exchange type

$$\hat{H} = -\sum_{ij} J_{ij} \vec{S}_i \cdot \vec{S}_j \qquad (1)$$

The effective exchange coupling $J_{ij} = J(R_{ij})$ is of short range in the insulating case. To be specific, consider the example of EuSrS; the nearest neighbour coupling between Eu ions is positive (ferromagnetic), the next nearest neighbour coupling is negative (antiferromagnetic) and smaller in magnitude, and the further neighbour exchange couplings are very small.

On the other hand, the metallic spin glass case is much more complex. To be specific, consider the case of a transition metal ion (such as Mn) in a noble metal host (such as Cu). The problem of the effective exchange coupling between a pair of Mn ions in an otherwise pure Cu host was examined by Ruderman and Kittel[5], by Kasuya[6] and by Yosida[7] who obtained, in a second order perturbation expansion (further discussion is given in section IV)

$$J(R) = \pi (j_{sd} N_o(0))^2 \epsilon_F W(2k_F R_{\ell\ell'}) \qquad (2)$$

where j_{sd} is the sd exchange coupling constant, $N_o(0)$ is the density of states at the Fermi energy ϵ_F and $W(x) = -(x \cos x - \sin x)/x^4$.

The long range oscillatory behaviour is the striking signature of this indirect interaction via the conduction electrons. Eq. (2) is known as the RKKY interaction. In the derivation of Eq. (2), the Mn ions were taken to be point dipoles, thus neglecting all structure. This is not strictly correct, even at very large separations. Making use of the Friedel-Anderson picture[8] of the 3d resonance in the Mn ion, Caroli[9] and Caroli and Blandin[10] gave the improved result, valid for *large* R only,

$$J(R) = (A/R^3) \cos(2k_F R + \phi) \qquad (3)$$

where the amplitude A and phase ϕ are given in terms of the 3d spin up and spin down phase shifts for conduction electron scattering. Eq. (3) reduces to Eq. (2) in the limiting case of nearly full occupancy of one of the 3d spin orbitals and nearly zero occupancy for the other.

However, Eq. (3) is still deficient in that it neglects the effect of the finite spatial extent of the magnetic form factor associated with the 3d orbitals. This effect was taken into account first by Geldart[11] and Jena and Geldart[12] in the closely related problem of the spin polarization induced by a 3d transition metal ion (Mn) in the conduction band of a noble metal (Cu) host. The effects of a finite conduction electron band width and of orthogonalization of the conduction electron wave functions (OPW's) to the Mn ion core wave functions was also considered. Malmstrom, Geldart and Blomberg[13,14] took the finite spatial extent of the Mn ion form factor into account in a detailed calculation of the the effective spin-spin pair interaction in a self-consistent Hartree-Fock approximation. It was found that the finite extent of the Mn ion form factor was essential even to obtain the correct sign of $J(R)$ at near neighbour distances. It also gave rise to very important "preasymptotic corrections" at small and intermediate Mn ion separations. The asymptotic Caroli-Blandin limit becomes an accurate representation only at separations of several nanometers. These conclusions were also obtained later by Price[15]. Both Malmstrom et. al.[13,14] and Price[15] applied their results to high concentration alloys of Heusler type. Applications to low concentration alloys were made by Levy and Zhang[16]. Although the final results for the effective interactions were given numerically, analytical interpolation forms were provided for use when semiquantitative approximations are warranted[12-14].

We can conclude that the effective indirect exchange interaction J_{ij} between a single pair of magnetic ions, located at sites R_i and R_j, in a nonmagnetic host can be calculated with reasonable accuracy and that comparisons with experiment can be made with some confidence. The short distance behaviour shows structure which can be correlated with short range magnetic order while the large distance behaviour is of the Caroli-Blandin form, oscillatory and decaying slowly as $1/R_{ij}^3$. Although some aspects of the indicated derivations apply strictly only to a single pair of magnetic ions, we will intially take the same J_{ij} to apply to all such pairs in an alloy having a finite concentration of magnetic ions. A more careful discussion of this point will be given in section IV when damping of the effective interactions is considered. This will be particularly important for low concentration alloys for which most of the magnetic pairs are in the asymptotic regime of large separations.

We are now in a position to consider special features of these systems which lead to a "spin glass" state at low temperature. Suppose that a crystalline sample of a binary or ternary alloy of the above type has been prepared having a finite concentration of the magnetic component. Also suppose, just temporarily, that the magnetic ions are located in a periodic arrangement on the lattice, as in the case of Heusler alloys, such as Cu_2MnSn at its stoichiometric composition. As the temperature is lowered, the magnetic moments will tend to "freeze" into a magnetically ordered state at a critical temperature T_c. In the case of Cu_2MnSn, the ordered state is ferromagnetic with all moments tending to point, on the average, in the same direction in space. Other compounds may develop magnetically ordered states of other symmetry (antiferromagnetic, helical, *etc.*). Now suppose instead that the concentration of magnetic ions is low and that they are distributed more or less randomly over the lattice sites. There no longer is translational symmetry in the lattice and the magnetic ion sites are inequivalent. Consequently, there is no obvious reason for a global direction of broken spin symmetry as the temperature is lowered.

In our canonical example of low Mn concentration CuMn alloys, it is widely believed that the spin symmetry is instead locally broken due to the combined result of the disordered distribution of Mn sites and the implied quasirandom J_{ij} distribution arising from the oscillatory long range tail of the indirect exchange interaction. In other words, each spin is locally frozen in, with the distribution of such directions being a quasirandom variable. This picture of the essential new physical aspect of the low temperature state of these alloys was first recognized and explicitly articulated by Blandin and Friedel[17,18]. The term "metallic glass" was applied somewhat later to this picture[19]. It is important to appreciate that the disorder in the distribution of ionic sites need not be, in itself, sufficient to yield a spin glass state. The site disorder in combination with the oscillatory character of the spin-spin interaction leads to a quasirandom distribution of J_{ij} coupling of the Mn ions at their fixed sites. Given this distribution of "bonds" $\{J_{ij}\}$, there is no configuration of spin orientations which can minimize all of the terms in the effective Hamiltonian Eq. (1). The spin system is said to be "frustrated" [20]. Under these circumstances, there will generally be very many configurations of spin directions which are extremely close in energy. As the temperature is lowered, the system develops a spin glass state.

There are immediately many important questions which can be asked concerning the low temperature spin glass state. Is it a state of thermodynamical equilibrium or just a metastable state? What determines the temperature, T_g, at which the spin glass freezes? That is, how does T_g depend on the parameters of the material? How can T_g be monitored experimentally? Is the transition to the spin glass state a "normal" phase transition? What is the order parameter? How many different kinds of spin glasses are there? In other words, how can spin glass universality classes be decided? How can the important correlation functions which characterize the spin glass be calculated theoretically and measured experimentally? How can what we learn be extended to other types of disordered materials? And so on.

The intriguing challenges of spin glass systems have attracted the attention of many experimental and theoretical workers. There exist excellent review articles[21,22] and books[4,23] to which the reader is referred to for extensive discussion and comprehensive references. We indicate here only a few highly pertinent facts. The response of a spin glass system to an external magnetic field, as measured by the longitudinal linear magnetic susceptibility, shows a cusp at a temperature generally taken to determine T_g as illustrated schematically in Fig. 2. It is remarkable that the corresponding ac susceptibility $\chi_\omega(T)$ shows structure even at frequencies well below 1 Hz[24]. This is most unlike the magnetic response at a more conventional phase transition, such as a simple ferromagnet or antiferromagnet. Furthermore, the heat capacity shows no discernable nonanalytic or singular behaviour at T_g, which is also unlike a conventional phase transition.

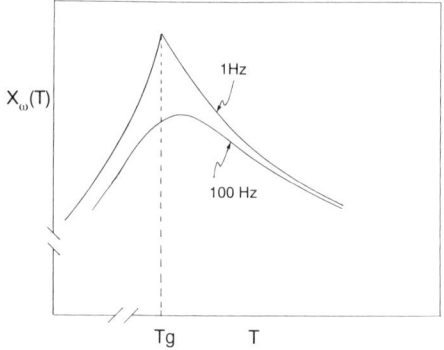

Figure 2. Schematic view of the ac susceptibility of a spin glass as a function of temperature near the spin glass freezing temperature T_g.

Metallic spin glasses present additional special interest for several reasons. The effective pair interactions due to the indirect coupling via the conduction electrons is of rather long range, although oscillatory. The range of the effective interaction is expected to play an important role in determining the universality class of a system, just as in the case of conventional phase transitions. The presence of magnetic anisotropy is another essential factor. The effective Hamiltonians considered thus far have been rotationally invariant in spin space. In practice, there may be various sources of spin anisotropy, such as uniaxial anisotropy or magnetic dipole-dipole interactions. For the present low symmetry spin systems, the Dzyaloshinskii-Moriya form of pair interaction[25,26],

$$H_{DM} = \sum_{i,j} I(\vec{R}_i, \vec{R}_j) \, (\vec{R}_i \times \vec{R}_j) \cdot (\vec{S}_i \times \vec{S}_j) \qquad (4)$$

is particularly important when the sample contains impurities with substantial spin-orbit coupling[27]. Anisotropic spin interactions such as Eq. (4) can influence strongly the lower critical dimension, d_ℓ^*, of the system, that is the spatial dimension below which fluctuations are expected to prevent the formation of a spin glass state. In fact, the value of the lower critical dimension has been an important and somewhat unsettled question for three-dimensional systems with oscillatory interactions which fall off in space as $1/R_{ij}^3$. It is partly for this reason that considerable attention has been devoted to discussion of the essential physics of the spin-spin interactions in the canonical spin glasses. Even so, only a part of the story has been given thus far.

A great deal of information on spin glass systems of a variety of types, both with and without spin anisotropies, has been generated theoretically by numerical studies, especially by Monte Carlo methods. The main results are reviewed briefly in the following section. We will then return in section IV to complete the physical picture of the effective interactions in the canonical noble metal-transition metal alloys, and to determine the implications for their spin glass phase transitions.

III MODEL SPIN GLASS SYSTEMS AND NUMERICAL RESULTS

In order that a particular model be useful for the elucidation of properties of spin glasses, the model must contain some essential features of the disorder leading to a spin glass state while at the same time being sufficiently simple that extensive theoretical and numerical analysis is posssible. Model systems which exhibit "sufficient" frustration, in the sense of Toulouse[20], have been very useful for this purpose. Following the discussion of the previous section, we define one important class of models corresponding to a random distribution of a fixed concentration of spins over the sites of a regular lattice; the interaction between the spin pairs is taken to be oscillatory as a function of the separation between each pair and specified by a function $J_{ij} = J(R_{ij})$. We refer to these as site-diluted models. An alternative model, simpler in some respects, can be defined in which each site of a regular lattice is occupied by a spin; the interaction between each pair of spins is specified by a probability distribution with interactions J_{ij} of either sign being possible. We refer to such models as bond-disordered models.

It is important to observe that both site-diluted and bond-disordered models share a feature that is common to all models of systems with quenched disorder. The system is not fully described by giving the Hamiltonian (together with the appropriate global boundary conditions). In addition, it is necessary to specify also the probability distribution of the quenched random variables (sites or bonds). The thermodynamics of the system is determined by the combination of the Hamiltonian and the probability distribution, taken together. In the case of a model for spin glass behaviour, a sufficient degree of frustration must also be present.

It is clear from previous discussion just what is meant by a site-disordered model for a system with RKKY interactions, or more generally, indirect exchange interactions. We now consider some bond-disordered models, following the seminal paper of Edwards and Anderson[28]. We consider first the case of short range interactions which, for simplicity, will be taken to act only between nearest neighbours on the lattice (additional short range interactions are not expected to be relevant for the critical properties of the model). Two procedures to specify the probability distribution of the nearest neighbour interaction have been widely used. The Gaussian probability distribution is given by

$$P(J_{ij}) = (2\pi\Delta^2)^{-1/2} \exp(-J_{ij}^2/2\Delta^2) \qquad (5)$$

For some purposes, a simpler distribution is

$$P(J_{ij}) = \frac{1}{2}\Big[\delta(J_{ij} + J) + \delta(J_{ij} - J)\Big] \tag{6}$$

This distribution is referred to as the $\pm J$ distribution. In both cases, the specification of the model is completed by giving the Hamiltonian which will be taken to be bilinear in the vector spin variables having n components. Ising, XY and Heisenberg systems correspond to $n = 1$, 2 and 3, respectively.

$$H = -\sum_{<ij>} J_{ij} \vec{S}_i \cdot \vec{S}_j \tag{7}$$

where the sum is taken once over each nearest neighbour pair on the lattice (so the bond strength is now J instead of $2J$ of Eq. (1)) and the lattice itself is in a space of d dimensions.

In the case of conventional phase transitions, not having quenched disorder, it has been instructive to consider, in additional to short range interaction models, a rather special type of long range interaction model in which every spin on the lattice interacts with equal strength with every other spin on the lattice. The bond strength of the interaction is taken to scale as $1/N$ where the additional factor of $1/N$ is necessary to obtain properly defined extensive quantities in the thermodynamic limit, $N \to \infty$. Exact analytical results are available for this model.

The strategy of this "infinite range" interaction model was adapted to the spin glass problem by Sherrington and Kirkpatrick[29]. Of course, the probability distribution of the interaction bonds must be specified for the disordered problem. Both the Gaussian and the $\pm J$ distributions have been studied extensively, both analytically and numerically. The study of this disordered infinite range interaction model has been much more challenging that the corresponding "pure" case. The low temperature spin glass phase is particularly complex but it is generally believed to be described by the solution of Parisi[4]. Although the strict applicability of an infinite range interaction model to a physical system is debatable, some ideas of the Parisi solution are still useful in realistic finite range interaction systems. The reader is referred to the literature for further discussion and details of the analysis[4]. We will now consider numerical results which have been obtained for the above models.

A wide variety of site and bond disordered spin glass models have been studied; Ising, XY and Heisenberg systems having short range, RKKY-like or infinite range (Sherrington-Kirkpatrick) interactions, with possible anisotropies. Numerical results have been obtained by an array of methods, but Monte Carlo procedures have been predominant. Only a brief review of the most pertinent (for present purposes) results will be given here. The reader is referred to the extensive literature for further discussion[21-23].

It must first be emphasized that, although Monte Carlo methods are naturally well suited to treatment of disordered systems, the simulations of disordered systems are much more demanding than are those of corresponding pure systems. The essential point is that in addition to the usual thermal averaging, it is also necessary to perform a configuration average, often over many hundreds of samples having different realizations of the quenched disorder. This is in contrast to an experiment on a truly macroscopic sample which is large enough that essentially all configurations of the quenched disorder are averaged over automatically; such systems are said to be "self-averaging". The sizes of systems which can currently be simulated in practice are not large enough to be self-averaging for a single sample, hence the need to average explicitly over many samples. Furthermore, as larger samples are used, relaxation times required to achieve equilibrium increase. Longer times to reach equilibrium are also required as the temperature is lowered. Critical slowing down in the vicinity of a phase transition is also serious. The net result is that the size of the systems which are simulated is rather limited in practice. In addition to the configuration averaging, this also necessitates extensive use of finite-size scaling methods in an attempt to extract asymptotic limiting values for the various quantities.

Three-dimensional Ising spin glasses have been extensively simulated by Ogielski[30] and by Bhatt and Young[32,33] using powerful special purpose machines. Various correlation functions and both static and dynamic critical exponents have been obtained. On the basis of results from very large simulations, it was concluded that a transition from a high temperature phase to a spin glass phase does indeed occur at a finite temperature T_g. It is useful to compare some of the correlation functions for the spin glass system with those of a simple ferromagnet. In the latter case,

$$\chi_{fm}(T) = 1/N \sum_{i,j} \langle S_i S_j \rangle \tag{8}$$

diverges as the Curie temperature, T_c, is approached from above T_c, $\langle S_i S_j \rangle \propto \exp(-R_{ij}/\xi_{fm})$ with $\xi_{fm} = \xi^o_{fm}(T - T_c)^{-\nu}$. In contrast, in the spin glass case, this quantity does not exhibit long range correlations. Instead, the configuration average, denoted by $[\langle S_i S_j \rangle]_{av}$, taken over the sample of all spin pairs having a fixed separation R_{ij}, is expected to be zero due to the quasirandom nature of the relative spin orientations. Since the vanishing of the configuration average results from the fluctuating signs, a long range correlation will appear in $[\langle S_i S_j \rangle^2]_{av} \propto \exp(-R_{ij}/\xi_{sg})$ and a divergence then is present in the "spin glass susceptibility"[28]

$$\chi_{sg}(T) = \frac{1}{N} \sum_{i,j} [\langle S_i S_j \rangle^2]_{av} = C/(T - T_g)^{\gamma_{sg}} \tag{9}$$

This permits, following Edwards and Anderson[28], the identification of a spin glass order parameter and various related correlation functions. Of course, this susceptibility is not the linear magnetic response of the spin glass to an external magnetic field. However, it is closely related to the leading nonlinear magnetic response function.

At this point, a variety of developments is possible. The question of the existence of a thermodynamic phase transition for a given pair interaction, disorder distribution, and spatial dimension can be pursued following the analogy with conventional phase transitions. In particular, it is important to determine the lower critical dimension, d^*_ℓ, below which fluctuations destroy the spin glass state. We refer to the literature for discussions of the various analytical approaches to these problems and will now turn to the numerical studies which have been performed. We limit attention to three dimensional systems and only some of the main results will be given here.

Following the discussion of Edwards and Anderson[28], the order parameter for the system of N spins in a cubical cell of edge length L can be specified as

$$q_N(t) = \frac{1}{N} \sum_{i=1}^{N} \Big(S_i(0) S_i(t) \Big)_{thermalized} \tag{10}$$

and limiting cases of large N or long time interval, t, can be studied as required. As was emphasized previously, only rather small values of N or L can be studied in the Monte Carlo simulations. It is then absolutely essential to exploit finite size scaling in order to isolate asymptotic limits. To illustrate what is involved, suppose that a physical quantity $f(L,T) = f_L(T)$ having dimension L^x has been simulated. We make the hypothesis that

$$f_L(T) = L^x \bar{f}\left((L/\xi_{sg})^{1/\nu_{sg}}\right) \qquad (11)$$

provided that both L and the correlation length $\xi_{sg}(T) = \xi_{sg}^o(T_c - T)^{-\nu_{sg}}$ are large relative to all microscopic lengths. For example. the spin glass susceptibility is then given by the scaling form

$$\chi_{sg}(T) = L^{\gamma_{sg}/\nu_{sg}} \bar{\chi}_{sg}\left((L/\xi_{sg}^o)^{1/\nu_{sg}}(T - T_g)\right) \qquad (12)$$

In simulations, as in real experiments, the analysis of data in the critical regime is complicated by the fact that the critical temperature itself is also an unknown parameter. An argument, originally due to Binder[34], is very helpful in locating T_g. From the second and the fourth moments of the order parameter, form the dimensionless combination

$$g_L(T) = \left[3 - \langle q^4 \rangle / \langle q^2 \rangle^2\right]/2 \qquad (13)$$

which, being dimensionless, has the finite size scaling form

$$g_L(T) = \bar{g}\left((L/\xi_{sg}^o)^{1/\nu_{sg}}(T - T_g)\right) \qquad (14)$$

This quantity has the important property that it is independent of the cell size L provided $T = T_g$. Thus plots of $g_L(T)$ versus T, must all intersect at T_g for the different cell sizes. This is illustrated schematically in Fig. 3. Subject to the scaling hypothesis, this procedure yields a relatively unbiased estimate of T_g and this estimate can then be used in the subsequent analysis of data. On the basis of these procedures, and very large simulations, it was concluded that the short range interaction Ising model, appropriately disordered, has a spin glass transition (at a finite T_g) and that the lower critical dimension is close to but below $d = 3$. Correlation functions have been measured in these simulations and static and dynamic critical exponents have been determined.

It has been emphasized that these simulations are computationally very demanding. It is therefore most reassuring that the existence of a phase transition (i.e. $d_\ell^* < d = 3$) has been established, independently of the simulations, by the domain wall renormalization group method which is particularly well suited to determination of the lower critical dimension [35,36].

The Ising spin glass transition has also been studied extensively for the infinite range $\pm J$ Sherrington-Kirkpatrick model. This model has the potential for providing a benchmark for the numerical work, provided the critical properties of this model are indeed correctly described by the analysis of Parisi[4], as is widely believed. Some numerical studies have suggested contrary conclusions, with the possibility of a complex two-stage ordering process[37,38]. This behaviour has now been traced to use of an insufficient number of samples when performing an average over configurations[39,40].

It has been found necessary to use hundreds, and sometimes even several thousands of configurations, in order to obtain a correct statistical description of the system. It is also worth noting that these recent studies have shown that the finite size scaling forms, for $T = T_g$, are rather accurate except for the smallest values of the cell size[39]. The Parisi solution is thus supported by these most recent numerical results.

The Ising model may be thought of as the limit of large uniaxial anisotropy of a more general vector spin model. Important numerical studies have also been carried out on short range interaction XY and Heisenberg models, sometimes with anisotropies included, using Monte Carlo methods[41-43], hybrid Monte Carlo molecular dynamics[44], and domain wall renormalization group methods[35,36]. One of the major conclusions is that the lower critical dimension of the isotropic Heisenberg model is above $d = 3$. This model therefore does not have a spin glass transition at a finite T_g. However, a transition at a finite T_g is possible if $O(N)$ symmetry breaking terms are added to the isotropic Hamiltonian. We refer to the literature for discussion.

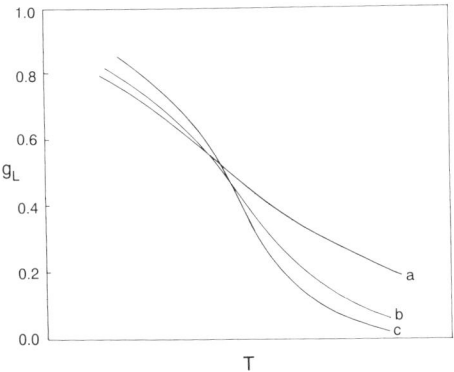

Figure 3. Schematic representation of the dimensionless combination of moments of the spin glass order parameter, $g_L(T)$ of Eq. (13), for three samples (a,b,c) of different size L, as a function of temperature.

The class of physical systems with effective spin pair interactions which are mediated by the conduction electrons and are of long range (at least potentially, subject to qualifications given in the following section) is of great interest, both theoretically and for applications. Chakrabarti and Dasgupta[45] have made a Monte Carlo simulation of a site-diluted system with an oscillatory interaction falling off as $1/R^3$ and having parameters appropriate to CuMn with 5 atomic % Mn. Care was taken to avoid the unphysical consequences of uniform rotation modes which had troubled some earlier numerical studies of such systems[46,47]. Finite size scaling methods were also used. It was found that $T_g = 0$ for this case of isotropic RKKY interaction[45]. However, a finite T_g resulted when the anisotropic Dzyaloshinskii-Moriya interaction, Eq. (4), was added. In addition, the range of values of T_g was in good correspondence with laboratory measurements on real samples. A Monte Carlo study of a RKKY-like random bond model of spins on a lattice was carried out by Reger and Young[48]. It was concluded that the model was not in the same universality class as short range interaction models and that its lower critical dimension was consistent with $d_\ell^\star = 3$ with $T_g = 0$. It is clear that the range of the interaction, as indicated by the power law at large distances, is very relevant for the spin glass phase transition. We will return to this point in the next section.

It is abundantly clear from the extensive literature on Monte Carlo and other numerical studies of these spin glass systems that the presence of the quenched disorder, with the need for averaging over a large number of configurations of the disorder, increases dramatically the computational challenges. It is natural to seek alternative approaches. For conventional second order phase transition, the introduction of replicas allows averaging over the disorder, resulting in a new effective Hamiltonian which has the advantage of being translationally invariant. This procedure has been extended to spin glasses[49]. Applications to Ising systems have been given[50-53]. This approach appears to have advantages in some cases. The development of similar methods for vector spin problems could be useful but there is no doubt that very substantial challenges will still be present.

IV PHYSICAL SPIN GLASSES REVISITED

The Monte Carlo simulations described in the previous section have been particularly useful in determining the universality classes of various spin glass model systems. In conjunction with domain wall renormalization group calculations, their lower critical dimensions have been clearly identified and the question of whether or not an equilibrium phase transition to a spin glass ordered state occurs has been largely settled, at least for the short range interaction models and also the very long range Sherrington-Kirkpatrick models. The case of systems with spin pair interactions of the long range oscillatory (RKKY) type has also been elucidated, even though these materials are somewhat more delicate in some respects. Specifically, it is believed that the lower critical dimension of these systems is $d_\ell^\star = 3$, which coincides exactly with the physical spatial dimension, leading to $T_g = 0$[54]. Such systems are then extremely sensitive to any symmetry breaking anisotropies. This sensitivity has been observed in the simulations[45]. The fact that RKKY and related long range interaction model systems are at (or at least very near to) their lower critical dimension implies that considerable care is required to specify the relevant details of the interactions in the physical systems. In fact, it was primarily for this reason that the origin of the effective interactions was examined closely in section II. The examination is continued in this section.

Since the indirect interactions are mediated by the conduction electrons, it seems reasonable that these interactions may be sensitive to the disorder which exists in the electronic background of the medium. The theoretical description of this disorder can be delicate. The initial work of Ruderman and Kittel[5] related to hyperfine interaction between nuclear spin and conduction electrons. Kayusa[6] and Yosida[7] then considered the sd exchange interaction between the itinerant s electrons and the localized d electrons. Second order perturbation theory was used to obtain the net interaction between the localized spins, with the expansion parameter being the hyperfine interaction or the sd exchange interaction. The effective interactions determined by Caroli[9] and Caroli and Blandin[10] and the extensions by Malmstrom and coworkers[13,14] were based on the Friedel-Anderson[8] resonance picture of sd mixing. In all of these cases, the calculated effective interaction is most appropriate to a single pair of magnetic moments in an otherwise pure host material (with neglect of the Kondo effect). In addition, these calculations were carried out at $T = 0$ on the assumption that the only relevant temperature scale was the Fermi temperature, which is certainly very high in metallic spin glasses.

Real indirect interaction materials in nature are much more complex, especially when account is taken of the special features of the samples prepared for laboratory experiments. A good case in point is provided by the work of Vier and Schultz[55] who made an extensive study of a family of AgMn spin glasses with added impurities. The freezing temperature and the resistivity of the ternary alloys were monitored as functions of the concentration of the added impurities, which were either nonmagnetic impurities, impurities with strong spin-orbit scattering, or magnetic transition metal impurities of another species. The work of Vier and Schultz has provided an extremely important benchmark set of data for canonical spin glass systems and has been most useful in the development of our understanding of these systems. This is discussed further below.

The ternary alloys which must be understood in order to extract the implications of the work of Vier and Schultz[55] are very different from the idealized simple RKKY type of system. Of course, there is a finite concentration of Mn ions, rather than an idealized single pair. Also, these Mn ions are in an electronic background in which a variety of other chemical species are immersed. Both of these facts have a strong influence on the effective pair interactions. In addition, although the temperature range of interest may well be small relative to T_F, additional temperature scales which reflect the nature of the disorder are present. Thus strictly finite temperature effects must also be allowed for at the outset. All of these complications do occur, in fact, in practice. We will indicate their consequences for the universality class of the spin glass transition of the canonical metallic spin glasses. The problem is somewhat complex so it is useful to approach it in stages.

Consider first the simple problem of a pair of Mn ions in a system of itinerant electrons with a finite concentration of nonmagnetic impurities at fixed but randomly distributed positions. For simplicity, also take the coupling between the electrons and the Mn ions to be of sd exchange origin. It is not practical to keep track of all of the impurity positions in determining the effective interaction J_{ij} between the Mn ions. However, the average over all configurations of the nonmagnetic impurities is accessible theoretically. It was shown by de Gennes[56] that the configuration average $[J_{ij}]_{av}$ behaves at em large R_{ij} approximately as

$$\left[J_{ij}\right]_{av} \sim J_{ij}^{(o)} \exp(-R_{ij}/\lambda) \qquad (15)$$

where λ is the conduction electron mean free path and $J_{ij}^{(o)}$ denotes the pair coupling in the absence of the nonmagnetic scatterers. That is, $J_{ij}^{(o)}$ refers to the RKKY limit and $[J_{ij}]_{av}$ has been evaluated only to second order in the sd exchange coupling. Note that Eq. (15) is intended to be, at most, a good approximation only for $R_{ij} \gg \lambda$. In a more precise treatment, the nodes of $[J_{ij}]$ as well as its amplitude are mean free path dependent[56].

It was thought for many years that Eq. (15), or variations thereof, provided a basis for a good representation of the effective pair interactions in an impure material. However, it was pointed out by de Chatel that $[J_{ij}]_{av}$ is actually a poor representation for the interaction between a pair of spins, separated by a distance R_{ij}, in a typical particular configuration of a disordered system[57]. A great deal of work has been done recently extending these ideas[58-65] for disordered electronic systems. Restricting attention to the stated case of a pair of magnetic ions and a background of nonmagnetic impurities, it has been shown that $[J_{ij}^2]_{av}$, when also calculated to only second order in the sd exchange interaction but in a self-consistent Born approximation with respect to the nonmagnetic impurity scattering, does not have the exponential mean free path damping factor at large separation. This means that the probability distribution for the quasirandom effective pair interaction is very broad and not at all sharply peaked about the mean value $[J_{ij}]_{av}$, so this latter quantity is altogether useless for describing typical spin pairs in a particular configuration. Since the mean free path damping is absent from $[J_{ij}^2]_{av}$, we expect typical interactions to fall off at large separation only as $1/R_{ij}^3$. Rather than the first moment of the probability distribution, a more suitable approximation to the *magnitude* of the typical pair interaction is the square root of the second moment

$$\left| J_{ij} \right|^{typical} = \left[J_{ij}^2 \right]_{av}^{1/2} \tag{16}$$

All of the above considerations have been based on simple second order perturbation theory expansions with respect to the sd exchange interaction and are taken in the low temperature limit. Of course, in a real system in the thermodynamic limit, there is a finite concentration of magnetic ions, and the applicability of elementary forms of perturbation expansion must be reconsidered. It is appropriate to consider the Hamiltonian

$$H = H_o + V_{imp} + V_{sd} \tag{17}$$

where H_o is the Hamiltonian for free conduction electrons, and V_{imp} and V_{sd} denote the perturbations due to the nonmagnetic impurities and the intrinsic sd exchange scattering of the magnetic ions, respectively. It was shown by Shegelski and Geldart[63-65] that a perturbation expansion with respect to V_{sd} is not of uniform validity for all R_{ij} and that a nonperturbative treatment is necessary at large R_{ij}, just as in the usual diffusion problems. Both the nonmagnetic scattering and the sd exchange scattering were treated in a self-consistent Born approximation. It is then necessary to distinguish several length scales at this point.

The dc conductivity is given by $\sigma = (ne^2/m)\tau$ where $\tau = \lambda/v_F$, with v_F being the group velocity of electrons at the Fermi energy and the transport mean free path is given by

$$1/\lambda = 1/\lambda_{in} + 1/\lambda_{non} + 1/\lambda_{sd} \tag{18}$$

with the three terms representing contributions due to inelastic phonon scattering (which will be neglected at low T), nonmagnetic and sd scattering, respectively, to the total inverse mean free path. Shegelski and Geldart found that the total inverse mean free path enters the first moment at large distance, as expected,

$$[J_{ij}]_{av} \sim J_{ij}^{(o)} \exp(-R_{ij}/\lambda) \tag{19}$$

However, the large distance benaviour of $[J_{ij}^2]_{av}^{1/2}$ is given by

$$[J_{ij}^2]_{av}^{1/2} \propto \exp\left(-R_{ij}/\Lambda_T\right), \quad R_{ij} > \Lambda_T \tag{20}$$

where $\Lambda_T = (\lambda_T \lambda/3)^{1/2}$ and $\lambda_T = T_F/\pi k_F T$. At intermediate distances, $\lambda < R_{ij} < \Lambda_T$, the second moment of the probability distribution contains additional length scales $\Lambda_{sd} = \lambda\left(\lambda_{sd}/2n\lambda - 1/3\right)^{1/2}$, where $(n = 1, 2, 3)$, when λ_{sd}/λ is large. The appearance of all of these length scales, rather than the simple total mean free path, in

$[J_{ij}^2]_{av}$ but not in $[J_{ij}]_{av}$ implies that the first moment $[J_{ij}]_{av}$ is a poor representation at large R_{ij} of the effective interaction, just as in the previous example. A better approximation to the magnitude of the typical pair interaction is given by Eq. (17) in this present case as well.

It is useful to compare Λ_{sd} to the important T dependent length scales in this spin glass problem. The distance travelled by an electron at the Fermi energy before loss of phase coherence due to thermal smearing is $\lambda_T = T_F/\pi k_F T$ in an idealized pure sample. In the presence of disorder, the long time displacements show diffusive rather that ballistic character, with the relevant length scale being $\Lambda_T = \{\lambda_T \lambda/3\}^{1/2}$. A further length scale of interest is $T_o = T_F/\pi k_F \lambda$. For the canonical spin glasses of present interest, T_g is always much smaller than T_F and T_o. However, Λ_{sd} and Λ_T are of the same magnitude in practice.

Taking all of these facts into account leads to the important conclusion that a correct description of the effective interactions in these spin glasses, in particular, requires not only that the sd exchange interaction be treated in a self-consistent, nonperturbative manner but also that the important finite temperature effects be explicitly included. That is, nontrivial temperature scales enter in a correct treatment of the problem. A fairly detailed discussion of the implications of these ideas to canonical spin glasses was given by Shegelski and Geldart[65] and comparisons were made to the data of Vier and Schultz[55]. An approximate but intuitively appealing description of T_g in terms of the magnitude of the typical pair interactions was given. The dependence of this T_g on the concentration of an added nonmagnetic impurity, as measured by the total electrical resistivity, was calculated. The results are shown in Fig. 4. The agreement with the experimental data is good. Of course there are parameters in the theory but realistic procedures have been used to obtain estimates. In the case of the spin glass AgMn without any added impurities, the dependence of T_g on the concentration of Mn was then calculated. The result is shown in Fig. 5 together with experimental data taken from the literature. In contrast to Fig. 4, it should be pointed out that there are no free parameters used in Fig. 5 to obtain fits, so this is a good test of the theory. For details and discussion of other calculated quantities, we refer to the literature.

We are now in a position to consider the implications of these results for the fundamental question of the universality class of real metallic spin glasses. Provided we are willing to describe the spin glass transition in terms of localized spins with an effective interaction given by averaging over electronic configurations, we conclude from the above that the effective interactions between spins are of strictly finite range in the sense that there is always an exponential damping factor at large separations as a result of finite temperature and the intrinsic self-damping of the sd exchange interactions. Previous treatments of the spin glass problem did not take the thermal smearing and the quenched disorder into account. Consequently, previous conclusions based on undamped interactions of the RKKY type, as in the previous section, do not strictly apply to real metallic spin glasses. Furthermore, treatments of this type of disordered electronic sytems which take the effective interaction to be exponentially damped at large distance by the mean free path given simply by the conductivity, as is the case for the first moment in Eqs. (15) or (19) are physically wrong. The damping is not of this simple physical origin.

Even though the range of the interaction is concentration and temperature dependent, we conclude that a given sample will undergo a spin glass transition much like a common short range fixed interaction system. That is, the lower critical dimension of a real metallic spin glass is indeed *less* than the physical dimension; $d_\ell^* < 3$.

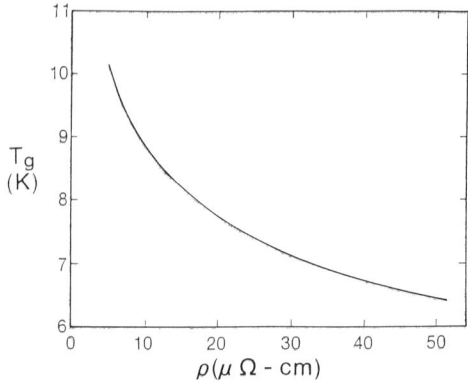

Figure 4. Calculated T_g for AgMn (2.2 at. %) as a function of the total resistivity from Ref. 64.

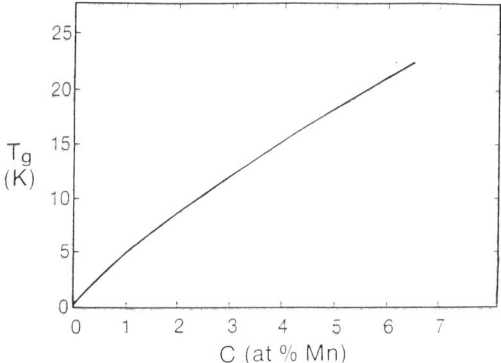

Figure 5. Calculated T_g for AgMn as a function of Mn concentration c from Ref. 64.

It is also important to emphasize that the transition occurs even in the absence of anisotropies induced by spin-orbit interaction impurities. It is not necessary to invoke Dzyaloshinskii-Moriya interactions, in contrast to the undamped RKKY case. In this context, it is interesting to note that Fert, de Courtenay and Bouchiat[66] have analysed the data of Vier and Schultz[55] taken on samples prepared to monitor the effect of added spin-orbit impurities. It was found that T_g tended to a finite nonzero value in the limit of zero concentration of spin-orbit scatterers.

The conclusions of this section have been obtained by a self-consistent application of finite temperature many-body theory. More extensive calculations would provide additional details of correlation functions and thermodynamic properties. However, it would be of very great interest to develop Monte Carlo procedures to deal with these coupled conduction and localized quantum spin systems. New numerical and theoretical developments will probably be required for this purpose.

Finally, it must be pointed out that neutron scattering data on CuMn and AgMn dilute alloys has been interpreted in a picture of spin density waves and domain structure[67], which is very different from the picture which we have described here. Much work has been done in an attempt to obtain further information on these systems at mesoscopic scales. We refer to reviews for detail[68,69]. A coherent complete picture requires consideration of all of the relevant physical length scales.

IV SUMMARY

We have given a brief review of some of the physics of disordered materials for which there are important technical applications as well as theoretical challenges. The emphasis has been placed on spin glasses. The main results of numerical studies, mainly by Monte Carlo simulations, have been indicated. The physical basis of the effective indirect pair interactions, which are mediated by the conduction electrons in metallic spin glasses, has been emphasized. As a result of disorder-induced correlations (the same type which lead also to weak localization and universal conductance fluctuations), the effective pair interactions are found to be of strictly finite range when self-consistent sd exchange scattering and finite temperature are taken into account. The lower critical dimension of the canonical spin glass systems is consequently predicted to be below the physical dimension $d = 3$ in contrast to the highly simplified case of undamped RKKY interactions.

ACKNOWLEDGEMENTS

This research was supported by the Natural Sciences and Engineering Research Council of Canada and by the Gordon Godfrey Foundation at The University of New South Wales. It is a pleasure to acknowledge many informative discussions of disordered materials with R.A. Dunlap.

REFERENCES

1. A collection of short articles on "Disordered Solids" is given *in*: Physics Today **41** Dec. (1988).
2. A series of interesting lectures and reviews is given *in*: "Les Houches, Session XXXI, 1978– La matiere mal condensee/Ill-condensed matter," R. Balian, R. Maynard, and G. Toulouse, eds., North Holland/World Scientific, Singapore (1983).
3. Various sample preparation and characterization methods are described by H. H. Liebermann *in*: "Amorphous Metallic Alloys", F. E. Luborsky, ed., Butterworths, London (1983), p. 26.
4. M. Mezard, G. Parisi, M. A. Virasoro. "Spin Glass Theory and Beyond," World Scientific, Singapore (1987).
5. M. A. Ruderman and C. Kittel, Phys. Rev. **96**, 99 (1960).
6. T. Kasuya, Prog. Theor. Phys. **16**, 45 (1956).
7. K. Yosida, Phys. Rev. **106**, 893 (1957).
8. A review of this picture is given by A. Blandin, J. Appl. Phys. **39**, 1285 (1968).
9. B. Caroli, J. Phys. Chem. Solids **28**, 1427 (1967).

10. B. Caroli and A. Blandin, J. Phys. Chem. Solids **27**, 503 (1966).
11. D. J. W. Geldart, Physics Letters **38A**, 25 (1972).
12. P. Jena and D. J. W. Geldart, Phys. Rev. B **7**, 439 (1973).
13. G. Malmstrom, D. J. W. Geldart, and C. Blomberg, J. Phys. F **6**, 233 (1976).
14. G. Malmstrom, D. J. W. Geldart, and C. Blomberg, J. Phys. F **6**, 1953 (1976).
15. D. C. Price, J. Phys. F **8**, 933 (1978).
16. P. M. Levy and Q. Zhang, Phys. Rev. B **33**, 665 (1986).
17. A. Blandin and J. Friedel, J. Phys. Radium **20**, 160 (1959).
18. A. Blandin, J. Physique **39**, C6-1499 (1978).
19. See reference 1 *in* the experimental review by G. Williams, Can. J. Phys. **65**, 1251 (1987).
20. G. Toulouse, Comm. on Phys. **3**, 115-119 (1977),
21. K. Binder and A. P. Young, Rev. Mod. Phys. **58** 801 (1986).
22. A. P. Young, J. D. Reger, and K. Binder, Spin glasses, orientational glasses and random field systems, *in*: "The Monte Carlo Method in Condensed Matter Physics," K. Binder, ed., Springer-Verlag, Berlin, Heidelberg (1992).
23. K. H. Fischer and J. A. Hertz. "Spin Glasses," Cambridge University Press, Cambridge (1991).
24. L. P. Levy, Phys. Rev. B **38**, 4963 (1988).
25. I. Dzyaloshinsky, J. Phys. Chem. Solids 4, 241 (1958).
26. T. Moriya, Phys. Rev. Lett. **4**, 5 (1960).
27. A. Fert and P. M. Levy, Phys. Rev. Lett. **44**, 1538 (1980).
28. S. F. Edwards and P. W. Anderson, J. Phys. F **5**, 89 (1975).
29. D. Sherrington and S. Kirkpatrick, Phys. Rev. Lett. **35**, 1792 (1975).
30. A. T. Ogielski, Phys. Rev. B **32**, 7384 (1985).
31. A. T. Ogielski and I. Morgenstern, Phys. Rev. Lett. **54** 928 (1985).
32. R. N. Bhatt and A. P. Young, Phys. Rev. Lett. **54** 924 (1985).
33. R. N. Bhatt and A. P. Young, Phys. Rev. B **37**, 5606 (1988).
34. K. Binder, Z. Phys. B **43** 119 (1981).
35. B. W. Morris, S. G. Colborne, M. A. Moore, A. J. Bray, and J. Canisius, J. Phys. C **19**
36. M. P. Gingras, Phys. Rev. Lett. **71** 1637 (1993).
37. B. Derrida, Physics Reports **184** 207 (1989).
38. I. A. Campbell, Phys. Rev. Lett. **68** 3351 (1992).
39. R. N. Bhatt and A. P. Young, Phys. Rev. Lett. **69** 3130 (1992).
40. I. A. Campbell, Phys. Rev. Lett. **69** 3131 (1992).
41. B. W. Morris and A. J. Bray, J. Phys. C **17** 1717 (1984).
42. S. Jain and A. P. Young, J. Phys. C **19** 3913 (1986).
43. J. A. Olive, A. P. Young and D. Sherrington, Phys. Rev. B **34** 6341 (1986).
44. F. Matsubara, T. Iyota, and S. Inawashiro, Phys. Rev. Lett. **67**, 1458 (1991).
45. A. Chakrabarti and C. Dasgupta, J. Phys. C **21** 1613 (1985).
46. L. R. Walker and R. E. Waldstedt, Phys. Rev. Lett. **38** 514 (1977).
47. L. R. Walker and R. E. Waldstedt, Phys. Rev. B **22** 3816 (1980).
48. J. D. Reger and A. P. Young, Phys. Rev. B **37** 5493 (1988).
49. F. Haake, M. Lewenstein, and M. Wilkens, Phys. Rev. Lett. **55** 2606 (1985).
50. R. H. Swendsen and J.-S. Wang, Phys. Rev. Lett. **57** 2607 (1986).
51. J.-S. Wang and R. H. Swendsen, Phys. Rev. B **37** 7745 (1988).
52. J.-S. Wang and R. H. Swendsen, Phys. Rev. B **38** 4840 (1988).
53. J.-S. Wang and R. H. Swendsen, Phys. Rev. B **38** 9086 (1988).
54. A. J. Bray, M. A. Moore, and A. P. Young, Phys. Rev. Lett. **56**, 2641 (1986).
55. D. C. Vier and S. Schultz, Phys. Rev. Lett. **54**, 150 (1985).
56. P. G. de Gennes, J. Phys. Radium **23**, 630 (1962).
57. P. F. de Chatel, J. Magn. Magn. Mater. **23**, 28 (1981).
58. A. Yu. Zyuzin and B. Z. Spivak, Pis'ma Zh. Eksp. Teor. Fiz. **43**, 185 (1986); [JETP Lett. **43**, 234 (1986)].
59. L. N. Bulaevskii and S. V. Panyukov, Pis'ma Zh. Eksp. Teor. Fiz. **43**, 190 (1986); [JETP Lett. **43**, 240 (1986)].

60. G. Bergmann, Phys. Rev. B **36**, 2469 (1987).
61. M. J. Stephen and E. Abrahams, Sol. St. Comm. **11**, 1423 (1988).
62. A. Jagannathan, E. Abrahams, and M. Stephen, Phys. Rev. B **37**, 436 (1988).
63. M. R. A. Shegelski and D. J. W. Geldart, Sol. St. Comm. **79**, 769 (1991).
64. M. R. A. Shegelski and D. J. W. Geldart, Phys. Rev. B **46** 2853 (1992).
65. M. R. A. Shegelski and D. J. W. Geldart, Phys. Rev. B **46** 5318 (1992).
66. A. Fert, N. de Courtenay and H. Bouchiat, J. Phys. France **49**, 1173-1178 (1988).
67. S. A. Werner, Comments Cond. Mat. Phys. **15** 55 (1990).
68. M. B. Weissman, N. E. Israeloff and G. B. Alers, J. Magn. Magn. Mater. **114** 87 (1992).
69. M. B. Weissman, Rev. Mod. Phys. **65** 829 (1993).

FREEZING: DENSITY FUNCTIONAL THEORY

A.D.J. Haymet

School of Chemistry
University of Sydney
NSW 2006 Australia

WHY DO LIQUIDS FREEZE?

At fixed pressure, as the temperature is decreased all liquids freeze. [Liquid helium at pressures below about 25 atm appears to be the only exception!] Why do liquids abandoned their random, 'disordered' structure, and form periodic arrays? As surprising as it seems, at present there is no molecular level, first principles theory of freezing or melting, even for the simplest materials. The prediction of phase diagrams is an important first step in understanding the crystal/melt interface, crystallization near equilibrium, and nucleation. Recently a new approximate theory for the freezing of classical liquids, known as the density functional (DF) theory, has been developed.[1] The mathematical structure of the theory is simple enough that it provides an attractive starting point for theories of more complex, dynamical phenomena.

Here we will discuss simple liquids, where by 'simple' we mean liquids such as pure methane, sodium or water. The crystallization of huge molecules, such as proteins, is an essential first step in the determination of structure from scattering experiments. Even for the simplest, classical liquids there is no universal (or universally accepted) theory of freezing, or indeed of first order phase transitions in general. Theories for the behavior of glass formers, such as many polymeric systems, are a separate issue not discussed here, but start from ideas developed for materials which crystallise. [Even glass formers have a lower free energy crystalline state, which may be kinetically inaccessible under many conditions.]

The lack of theory might seem puzzling, since the thermodynamic conditions for phase equilibrium stated by Gibbs are well known. At a given temperature T and pressure p, the laws of Thermodynamics prove that the phase with the lowest free energy is the stable phase. For two coexisting phases, denoted here by the subscripts 'S' for solid and 'L' for liquid, the temperatures, pressures and chemical potentials μ_j of all components 'j' must be equal:

$$T_L = T_S \qquad p_L = p_S \qquad \mu_L^{(j)} = \mu_S^{(j)}. \tag{1}$$

From a microscopic point of view, the prediction of freezing, and phase diagrams in general, is straightforward in principle. One should use the techniques of statistical mechanics to predict the thermodynamic properties of the material under study, and use equation (1) to determine the phase boundaries. In practice, the calculation of reliable values for the free energy has proven extremely difficult, and hence the phenomenon of freezing/melting has attracted the attention of many scientists and generated a huge literature.

TRANSLATIONAL SYMMETRY BREAKING

There are additional, much deeper questions concerning freezing. Why do materials adopt a specific symmetric crystal structure at all? This phenomenon is called 'spontaneous translational symmetry breaking' and occurs in many branches of science, including particle physics. The short answer is: we do not know. Even within classical mechanics, we have no theory which can analyse a given Hamiltonian and predict the symmetry of the crystal to which it will freeze. At present, the symmetry breaking has to be put into the theory 'by hand'. By this we mean that a particular symmetry is assumed and its free energy evaluated relative to other candidate crystal symmetries.

Most substances contract when they freeze, but water and a few other materials (silicon, gallium arsenide, gallium and bismuth, to name a few) expand. The degree of expansion or contraction varies widely, even for simple materials at one atmosphere pressure. Liquid sodium metal contracts 3% when it freezes at 98 °C, molten sodium chloride contracts 39% at 801 °C, and gallium expands by 3% at 30 °C! The fractional density change on freezing, denoted η, may be calculated from the liquid (number) density ρ_L and the average crystal density ρ_S, using the definition

$$\eta = \frac{\rho_S - \rho_L}{\rho_L}. \tag{2}$$

This 'non-universal' property of the freezing transition is displayed in Table 2 for a variety of materials. Prediction of this property constitutes a major challenge to any theory.

THE DENSITY WAVE PICTURE

Both qualitative and quantitative predictions for freezing arise from a relatively new, approximate theory of freezing, known from its mathematical structure as the 'density functional' (DF) theory. Although it side-steps the most fundamental mathematical question of crystallization, namely the spontaneous symmetry breaking discussed above, the density functional theory is proving to be a useful, numerically simple tool for treating practical problems of phase coexistence. It has been used to construct approximate theories of the crystal/melt interface, nucleation, glasses, the stability of quasicrystals, the prediction of vacancy concentrations in the crystal at melting, and the freezing of quantum liquids, as well as the standard mathematical models of freezing such as hard spheres, the Lennard-Jones system and mixtures. There are a number of review articles which cover these developments.[2,3,1,4]

Here we focus on the qualitative ideas, which should turn out to be helpful in other areas. At fixed pressure, the determination of the freezing points amounts to finding the temperature of at which the free energy of the random liquid exactly equals the free energy of the spatially ordered crystal. The density wave point of view, which originates with Ramakrishnan, views the crystal as a liquid permeated with standing waves in the density. These waves measure the displacement of the particles from their perfect lattice sites, and have wavelengths corresponding to all possible reciprocal lattice vectors (RLV's) in the crystal. The amplitudes of the waves are related to the (wavevector dependent) Debye-Waller factor in the equilibrium crystal. The freezing temperature is that unique temperature at which the free energy penalty for creating such an infinite network of density waves is exactly zero.

MATHEMATICAL THEORY

The complete mathematical theory has been reviewed in detail very recently.[1] Here we focus on the concepts. The central mathematical quantity in the DF freezing theory is the equilibrium average, single particle density $\rho(\mathbf{r})$. In the isotropic liquid phase this quantity is simply a constant, ρ_L, the number density of the material. In the solid phase the density is spatially varying, with a symmetry determined by the crystal type and a period determined by the average (over a unit cell) crystal density ρ_S. It is convenient to write the crystal density as a Fourier sum

$$\rho(\mathbf{r}) = \rho_L \left[1 + \eta + \sum_n \mu_n \exp(i\mathbf{k}_n \cdot \mathbf{r}) \right] , \qquad (3)$$

where η is the fractional density change on freezing defined above, $\{k_n\}$ is the set of reciprocal lattice vectors which defines the lattice, and μ_n are order parameters which measure the degree of periodic order of wavevector k_n in the crystal.

The goal of the freezing theory is simply to predict the temperature dependence of these order parameters. At high temperature all will be zero. At the freezing point they will assume some non-zero, finite value, which will gradually increase as the temperature is lowered further, until absolute zero temperature is reached, or a second (crystal-crystal) phase transition intervenes. In some more recent work the crystal density is expanded as a sum of Gaussians centered at each lattice site; this additional approximation simply imposes a relation which fixes all the order parameters μ_n given just one of them, say μ_1. Such an approximation is often useful and accurate for close-packed crystals.

The free energy $\mathcal{F}$ of the liquid or crystal, along with the other thermodynamic quantities, can be expressed as a functional $\mathcal{F}[\rho(\mathbf{r})]$ of the density $\rho(\mathbf{r})$. This is the origin of both the name 'density functional theory', and the power of the technique. Since we are seeking the temperature at which phases coexist at the same pressure and chemical potential, it is convenient to express the theory in terms of the grand potential difference between the liquid and crystal, $\Delta\beta\Omega$, where $\beta^{-1} = kT$. The exact expression for this difference, which will vanish exactly at the freezing point, is given, for example, in a paper by Laird, McCoy and Haymet,[5] equation (4.13). Simplifying that equation we obtain the form,

$$\frac{\Delta\beta\Omega}{\rho_L V} = c_0\eta + \tfrac{1}{2} \sum_n c_n \mu_n^2 + \text{higher order terms} , \qquad (4)$$

where the sum is over all RLV's of the crystal $\{k_n\}$. The first term is the change in grand potential due to the overall contraction (or expansion) of the liquid on freezing. The sum is nothing but the free energy of setting up standing waves of wavelength $\{k_n\}$.

The theory is completed by specifying the coefficients c_0 and c_n. To first order in thermodynamic perturbation theory, these coefficients are related to the structure factor of the equilibrium liquid, a quantity which is accessible via X-ray or neutron elastic scattering. Specifically, $c_n = c(k_n) = 1 - 1/S(k_n)$. These coefficients c_n of course depend on the temperature and pressure of the liquid. In the years since the original calculations, a host of other methods have been devised for relating the coefficients c_n to known properties, but in each case knowledge of the equilibrium liquid is required. Since we focus only on qualitative concepts here, we will not review the strengths and weaknesses of each method.

The freezing point is then located by lowering the temperature of the liquid. At a certain unique temperature, the terms in equation (4) will exactly balance each other, and the grand potential difference between liquid and crystal will be zero. That is, the temperatures, pressures and chemical potentials of the liquid and crystal will be exactly equal. Although it may not be obvious from the above discussion, this process is simply the Maxwell construction in the grand ensemble.

Note that some density waves lower the free energy of the crystal relative to the liquid, and other raise it (depending on the sign of c_n for the wave vector k_n of the density wave). But the crystal cannot just select a subset of density waves: it has to take them all. The crystal symmetry dictates that if the crystal has one density wave, it has all the overtones, sums and differences; that is, all the symmetry related density waves consistent with the specified symmetry. Hence the process of crystallization is seen to be an incredibly delicate balance of three contributions, (i) the overall expansion or contraction of the crystal (which may raise or lower the free energy, depending on the crystal), (ii) those standing waves in the singlet density which increase the free energy, and (iii) those standing waves which lower the free energy. Only when all three contributions cancel does the free energy of the periodic crystal exactly equal the free energy of the liquid. Perhaps this density wave point of view can be transferred to other, even more complex problems.

SUMMARY

The radical feature of the DF theory is that the free energy is assumed to be an analytic functional of the singlet density. For a second order transition, such as the gas-liquid critical point, this would be a poor approximation (and lead, among other things, to incorrect 'classical' critical exponents). However, for certain first order phase transitions, the empirical evidence is that the truncation of this expansion at first order is useful. The mathematical approximations mean that the density functional theory is far from a rigorous solution to the freezing problem. Nevertheless, by building upon the advances in liquid theory, and using the structure of the liquid as a starting point for perturbation theory, the density functional theory does constitute a complete theory of freezing. It starts from the laws of statistical mechanics and a knowledge of the forces between the molecules, and by making a series of well defined (and relatively well tested) approximations, the theory predicts the phase diagram.

Acknowledgments

Our research in freezing is supported by the Australian Research Council (ARC) (grant No. A29131271), to whom grateful acknowledgment is made.

References

(1) A.D.J. Haymet, Freezing in *Fundamentals of Inhomogeneous Fluids*, edited by D. Henderson, (Marcel Dekker, New York, 1992), Chap. 9, pages 363–405.

(2) R. Evans, Density functionals in the theory of nonuniform fluids in *Fundamentals of Inhomogeneous Fluids*, edited by D. Henderson, (Marcel Dekker, New York, 1992), Chap. 3, pages 85–175.

(3) D.W. Oxtoby, Nucleation in *Fundamentals of Inhomogeneous Fluids*, edited by D. Henderson, (Marcel Dekker, New York, 1992), Chap. 10, pages 407–442.

(4) B.B. Laird and A.D.J. Haymet, The crystal / liquid interface: Recent computer simulations, Chemical Reviews **92**, 1819–38 (1992).

(5) B.B. Laird, J.D. McCoy, and A.D.J. Haymet, Density functional theory of freezing: Analysis of crystal density, J. Chem. Phys. **87**, 5449 (1987).

APPLICATION OF THE LOCAL CHEMICAL POTENTIAL TO THE QUANTUM HALL EFFECT IN A BALLISTIC QUANTUM WIRE

P. N. Butcher[1] and D. P. Chu[1]

[1] Department of Physics, University of Warwick, Coventry CV4 7AL, UK

1. INTRODUCTION

We have recently made calculations of the Hall resistance of a ballistic two-dimensional electron gas confined in a quantum wire by hard walls (Chu and Butcher 1993 a and b). The calculations are made in the linear transport regime at low temperatures and are based on a formula for the local chemical potential given by Imry (1989). Other authors have sometimes used a different formula which gives different results (Peeters 1988, Akera and Ando 1989 and 1990). Here we outline our calculations and present some of our results. We also give a new definition of the local chemical potential which can be evaluated exactly in the linear transport regime at low temperatures. The outcome is identical to Imry's formula. An introduction to the theory of electron transport in low-dimensional semiconductor structures is given by Butcher (1993).

2. THE CALCULATIONS

A schematic diagram of the system under consideration is given in figure 1. It is a two-dimensional electron gas with electron density n_s which is confined in a space of width W in the x-y plane by infinite potential barriers at $y = \pm W/2$. A uniform magnetic field B is applied in the z direction and described in the Landau gauge by writing the vector potential as $\mathbf{A} = (-By, 0, 0)$. The behaviour of the system is discussed here within the framework of the self-consistent, zero temperature model introduced by Li and Thouless (1990).

We calculate both V(y), the electrostatic Hall potential (EHP), which arises in the presence of the magnetic field, and $\mu(y)$, the local chemical potential (LCP). These potential energies are used to calculate the corresponding Hall resistances, R_{EHP} and R_{LCP} respectively, from the potential differences across the waveguide divided by -e times the total current. As a check on our numerical accuracy we also calculate a longitudinal resistance R_L from the chemical potential difference $\mu_1-\mu_2$ between the reservoirs feeding the two ends of the waveguide (see figure 1). We neglect spin splitting and suppose that the reservoirs are non-reflecting. Then R_L is quantized in units of $2e^2/h$.

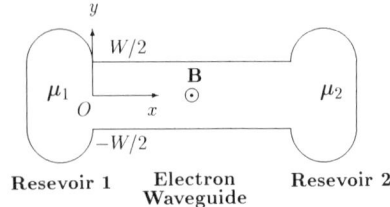

Figure 1. Schematic diagram of a 2D electron waveguide with reservoirs.

The electron wave function $\psi_{\alpha k}(x,y)$ satisfies the Schrodinger equation

$$\left[\frac{1}{2m^*}(p+eA)^2 + V(y)\right]\psi_{\alpha k}(x,y) = E_{\alpha k}\psi_{\alpha k}(x,y) \tag{1}$$

where m^* is the effective mass and $\alpha = 1,2,\ldots$ labels the subbands. The normalised eigenfunctions take the form

$$\psi_{\alpha k}(x,y) = \ell^{-1/2}\exp(ik_\alpha x)\chi_{\alpha k}(y) \tag{2}$$

where ℓ is the length of the wire. The EHP, which must be determined self-consistently, can be expressed as

$$V(y) = \frac{e}{4\pi\epsilon_o\epsilon}\int_{-w/2}^{w/2}dy'\, 2\ln|y-y'|\,\delta\sigma(y) \tag{3}$$

In this equation the increment of electron charge density produced by the magnetic field is

$$\delta\sigma(y) = \frac{-e}{\pi}\sum_\alpha \int_{-k_{\alpha 2}}^{k_{\alpha 1}} dk\left[|\chi_{\alpha k}(y)|^2 - |\chi^0_{\alpha k}(y)|^2\right] \tag{4}$$

where we have allowed for spin degeneracy. The functions $\chi^0_{\alpha k}(y)$ are the eigenfunctions when B=0 (Li and Thouless 1990) and $k_{\alpha 1}$ and $k_{\alpha 2}$ are the magnitudes of the maximum wave numbers of the electrons emerging in subband α from reservoirs 1 and 2 respectively. They are evaluated by setting $\mu_1 = \epsilon_F + \Delta/2$ and $\mu_2 = \epsilon_F - \Delta/2$ where ϵ_F is the equilibrium Fermi level and Δ is a small chemical potential difference between the reservoirs which is introduced to drive the current.

To complete the calculation we have to constrain ϵ_F so as to yield a given electron density n_s for a fixed value of Δ, i.e. so that

$$n_s = \frac{1}{\pi w}\sum_\alpha [k_{\alpha 1} + k_{\alpha 2}] \tag{5}$$

The current distribution in the wire can then be calculated from

$$j_x(y) = \frac{-e\hbar}{4\pi m^*}\sum_\alpha \int_{-k_{\alpha 2}}^{k_{\alpha 1}} dk\,(k - y/\ell_B^2)\,|\chi_{\alpha k}(y)|^2 \tag{6}$$

where $\ell_B = (\hbar/eB)^{1/2}$ is the magnetic length. The total current is the integral of $j_x(y)$

across the waveguide from which the longitudinal resistance R_L can be calculated. The resistance R_{EHP} can be similarly derived from $V(y)$. Finally, R_{LCP} is obtained from the formula for the local chemical potential given by Imry (1989):

$$\mu(y) = (p_1\mu_1 + p_2\mu_2)/(p_1 + p_2) \qquad (7)$$

where

$$p_1 = \sum_\alpha |X_{\alpha,k_{\alpha F}}|^2 / v_{\alpha k_{\alpha F}} \qquad (8a)$$

and

$$p_2 = \sum_\alpha |X_{\alpha,-k_{\alpha F}}|^2 / v_{\alpha k_{\alpha F}} \qquad (8b)$$

In these equations $k_{\alpha F}$ is the Fermi wave number in subband α when $\Delta=0$ and $v_{\alpha k_{\alpha F}}$ is the group velocity of subband α at the Fermi level. (In practice we replace $k_{\alpha F}$ by $k_{\alpha 1}$ in p_1 and by $k_{\alpha 2}$ in p_2. This has no effect on $\mu(y)$ in the linear regime.)

3. NUMERICAL RESULTS

When the above equations are solved we obtain self-consistent electron wave functions, the distributions of $V(y)$ and $\mu(y)$ across the wire and the two kinds of Hall resistance, R_{EHP} and R_{LCP}. To interpret the physics which is revealed in the results it is useful to have representations of the electron distributions in the subband wave functions at the equilibrium Fermi level. We therefore define a convenient mean position $<y_\alpha>$ and a half-width $<(\Delta y_\alpha^2)>^{1/2}$ in subband α and cross hatch the region lying between $<y_\alpha>-<(\Delta y_\alpha^2)>^{1/2}$ and $<y_\alpha>+<(\Delta y_\alpha^2)>^{1/2}$ to indicate the spread of the wave functions. We take $<y_\alpha> = <\alpha|y|\alpha>$ and $<(\Delta y_\alpha^2)> = <\alpha|y^2|\alpha>-<y_\alpha>^2$. However, when $\alpha>1$, in evaluating $<y_\alpha>$ and $<(\Delta y_\alpha^2)>$ we keep only the renormalised part of the wave function lying between a side wall and the node closest to it. This procedure has the merit of producing useful pictorial representations of the electron distributions for the excited subband wave functions while avoiding unhelpful complications due to the nodes and gives a better description of the behaviour of the electron wave functions near the edges which is what determines R_{LCP}. In our numerical calculations, we always use the parameters of a GaAs wire (Li and Thouless 1990) of width $w = 100$ nm.

Figure 2(a) shows plots against B which indicate the degree of overlap of the oppositely propagating wave functions in the ground subband when $n_s = 2 \times 10^{14}$ m^{-2}. For this electron density only the ground subband is occupied. In this case, the formula for the local chemical potential becomes very simple and the density of states factor $v_{1k_F}^{-1}$ cancels out. The departure of R_{LCP} from its quantised value is therefore, entirely due to the overlap of opposite-going wave functions. Data for $<y_1>$ and the wave function spread is given by the up(down) triangles and the cross-hatch lines sloping down to the right(left) for right(left)-going wave function respectively. Figure 2(b) gives data for R_L (crosses), R_{EHP} (Squares), and R_{LCP} (circles) for the same value of n_s. We see that R_{LCP} increases rapidly as the opposite-going functions begin to separate and stays at the quantised value when they are well separated at the two edges of the wire.

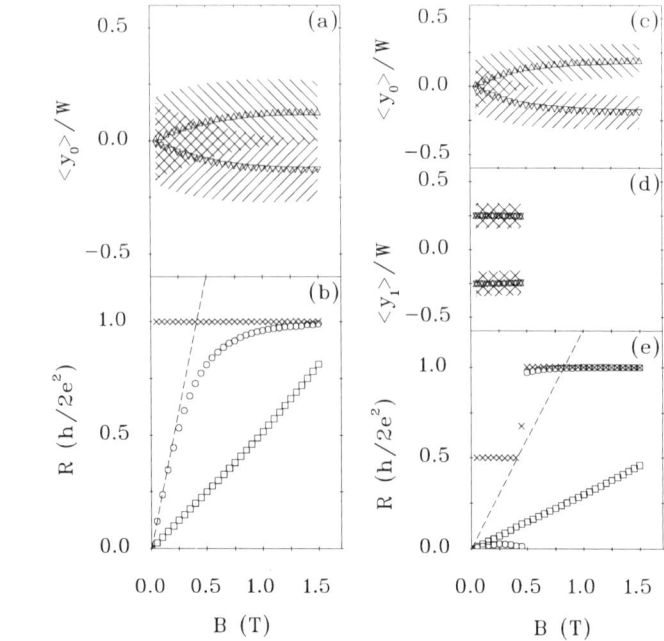

Figure 2. Plots showing overlap of the opposite-going wave functions in different subbands as a function of B and the corresponding R_L (crosses) and Hall resistances: R_{EHP} (squares) and R_{LCP} (circles). Up(down)-triangle show the average position of right(left)-going wave functions and the cross hatch lines sloping down to the right(left) mark the corresponding wave function spread. The width of the wire is $W = 100$ nm and the electron densities are $n_s = 2 \times 10^{14}$ m^{-2} in (a) and (b) and $n_s = 4 \times 10^{14}$ m^{-2} in (c), (d), and (e). The dashed lines show the classical Hall resistances.

Figures 2(c) and (e) show corresponding plots when $n_s = 4 \times 10^{14}$ m^{-2}. For this density subbands 1 and 2 are occupied when B < 0.5 T but only the ground subband is occupied when B > 0.5 T. We chose this n_s to make the Fermi wave number of subband 2 much smaller than that of subband 1 when B → 0 so that the Fermi level is very close to the bottom of the dispersion curve which is flat. In this situation, the local chemical potential difference between the two edges of the wire is greatly reduced by both the large overlap of the opposite-going wave functions at low B and the significant increase of the density of states subband 2. Figures 2(c) and 2(d) show the degree of overlap in the $\alpha=1$ and $\alpha=2$ subbands respectively. Figure 2(e) shows that R_{LCP} is quenched when B <0.5 T and Figures 2(c) and (d) confirm that the quenching is associated with severe overlap of the $\alpha=2$ wave functions while the $\alpha=1$ wave functions are separating as they did in Figure 2(a). As soon as the second subband is depopulated, R_{LCP} jumps to the quantised value because the opposite-going $\alpha=1$ wave functions are separated at the two edges. Finally, we note that R_{EHP} (squares) is linear in B and remains linear when the second subband depopulates. Linear behaviour is also found for the classical Hall resistance which is shown by the dashed lines in figures 2(b) and (e).

We see from equations (7) and (8) that $\mu(y)$ at a wire edge is determined by the values of the opposite-going electron wave functions of all of the occupied subbands. Increasing the overlap at the edges decreases the difference of $\mu(y)$ between the edges. The singularity of the density of states at the bottom of a subband enhances this effect enormously. In the limit, B → 0, the left- and right-going wave functions coincide and the difference of $\mu(y)$ across the wire is zero. For small B, there will consequently be almost complete quenching of R_{LCP} because the opposite-going wave functions of all the occupied subbands overlap heavily at the edges. When the opposite-going wave functions of any one subband are separated, the level of quenching of R_{LCP} is reduced. Finally, when every pair of opposite-going wave functions are well separated, the LCP difference approaches the chemical potential difference between the two ends of wire and the R_{LCP} is almost exactly quantised.

We always suppose in the calculations described above that n_s is fixed. Consequently, when only the ground subband is occupied, the corresponding Fermi wave number k_{1F} is also fixed. In that case, increasing B from zero simply separates the wave functions. On the other hand, when two subbands are occupied, increasing B increases k_{1F} but reduces the Fermi wave number k_{2F}. For $n_s = 4 \times 10^{14}$ m^{-2}, $k_{2F} \ll k_{1F}$ at B=0. Hence, when B is increased, the $\alpha = 1$ wave functions separate quickly but the $\alpha = 2$ wave functions do not separate with the result that R_{LCP} is quenched.

Peeters (1988) and Akera and Ando (1989) make calculations for an alternative choice of the weighting factors in which the density of states factors are omitted from equation (8). No quenching is found in this case. Akera and Ando (1990) find similar results to those shown in Figure 2 in calculations of R_{LCP} by using equations (7) and (8) and wave functions which are not self-consistent. They also show that $R_{LCP} = 0$ when ϵ_F coincides with the bottom of an excited subband. This comes about because the density of states factor $v_{\alpha k_{\alpha F}}^{-1}$ in equation (8) ensures that the subband in question dominates the weighting factors p_1 and p_2 and the magnitude of the subband wave function becomes symmetrical about the centre of the waveguide. We compare results obtained by using equation (8) as it stands ($\bigcirc$) and with $v_{\alpha k_{\alpha F}}^{-1}$ removed (+) in figure 3. It is important to discriminate between these two ways of calculating $\mu(y)$. We turn to this question in the next section.

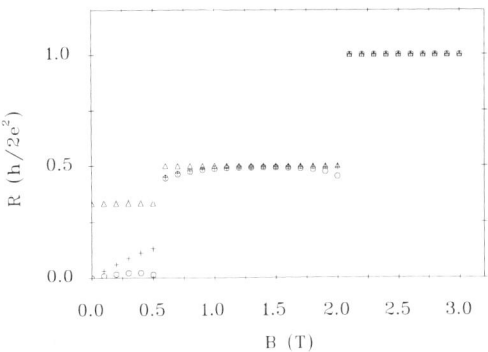

Figure 3. Longitudinal and Hall resistances calculated as a function of B. The longitudinal resistance is denoted by Δ. The Hall resistances calculated from Eq.(7) using the weighting factors (8) and also (8) with $v_{\alpha k_{\alpha F}}^{-1}$ removed are denoted by $\bigcirc$ and + respectively. (w = 100 nm and $n_{1D} = 1.1 \times 10^6$ cm^{-1}).

4. A NEW DEFINITION OF THE LOCAL CHEMICAL POTENTIAL

The formulae (7) and (8) for the local chemical potential in a 2D waveguide structure were taken from Imry (1989). They rest on particular assumptions about the behaviour of non-invasive voltage probes. Akera and Ando (1989, 1990) sometimes use the same formula and sometimes use a different formula derived by Peeters (1988), by making different assumptions. Büttiker (1988) makes an approximate many body calculation and obtains the expression on the right-hand side of equation (7) for the self-consistent potential due to the unequal chemical potentials in the reservoirs. None of these derivations is very satisfactory because a non-invasive probe has yet to be developed and the Landauer-Büttiker formalism is a one-electron theory (Büttiker 1988). In this section we consider a general 3D nanostructure fed by ideal reservoirs. Then the local chemical potential $\mu(\mathbf{r})$ depends of on $\mathbf{r} = (x,y,z)$. A new definition of $\mu(\mathbf{r})$ is given and a formula for it is developed without making any assumptions other than those involved in the Landauer-Büttiker formalism.

The statistics of the incident electrons are described by Fermi-Dirac functions with temperatures and chemical potentials which are determined in the reservoirs. Within the terminals and the nanostructure the behaviour of the electrons is determined by a one-electron Schrödinger equation containing a static potential energy function and a static vector potential describing a constant magnetic induction field with magnitude B. It is convenient to label the terminals by an integer $t = 1,2,...N$ and the transport channels within a terminal by another integer $\alpha = 1,2...$.

The most general configuration of the nanostructure which we consider is a non-equilibrium one in which the chemical potential in the reservoir feeding terminal t takes an arbitrary value μ_t. We write $n(\mathbf{r}, \mu_1, \mu_2,... \mu_N)$ for the electron density at $\mathbf{r}$ in this case. We also consider an equilibrium situation in which the chemical potential in every terminal takes a common value μ_o. The electron density in this case is denoted by

$$n_o(\mathbf{r}, \mu_o) \equiv n(\mathbf{r}, \mu_o, \mu_o,... \mu_o) \qquad (9)$$

The local chemical potential in the non-equilibrium case, $\mu(\mathbf{r})$, is defined by

$$\mu(r) \equiv \mu_L \qquad (10)$$

where μ_L is the solution of the equation

$$n_o(\mathbf{r}, \mu_L) = n(\mathbf{r}, \mu_1, \mu_2,... \mu_N) \qquad (11)$$

We see that $\mu(\mathbf{r})$ is the chemical potential in the reservoirs of the equilibrium system which creates the same electron density at $\mathbf{r}$ as do the actual chemical potentials $\mu_1, \mu_2,... \mu_N$ in the non-equilibrium system.

To begin the calculation of $\mu(\mathbf{r})$, as defined above, we suppose that an incident wave in channel α in terminal t is described by a known wave function

$$\psi_{t\alpha} = \ell^{-\frac{1}{2}} \exp(ik_{t\alpha}x) \chi_{t\alpha}(y,z) \qquad (12)$$

where ℓ is the length of the terminal region in which $\psi_{t\alpha}$ is normalised, $k_{t\alpha}$ is a positive wave number and $\chi_{t\alpha}(y,z)$ is normalised over the cross-section of the terminal. The notation used here is similar to that in Butcher (1993). The Cartesian coordinate axes Oxyz in each terminal are oriented with Ox pointing towards the nanostructure and the local gauge of the vector potential is chosen so that x does not appear in the Hamiltonian which makes $k_{t\alpha}$ a good quantum number. Both $k_{t\alpha}$ and $\psi_{t\alpha}$ are functions of the electron energy ϵ. In the calculation to follow we make use of the expectation value of the electron velocity along Ox: $v_{t\alpha} = \hbar^{-1} d\epsilon/dk_{t\alpha}$ and the associated density of incident states per unit length of the terminal $N_{t\alpha} = 2/hv_{t\alpha}$. We note that $v_{t\alpha}$ is positive, by definition, because it relates to an incident wave.

To continue with the calculation of $n(\mathbf{r}, \mu_1, \mu_2,... \mu_N)$ we suppose that the scattering state $\theta_{t\alpha}(\mathbf{r})$ generated by the incident wave $\psi_{t\alpha}$ is known. The phases of the scattering states are random because the phases of the incident waves are randomised by the reservoirs. Consequently, we have simply to add the contributions to $n(r, \mu_1, \mu_2, \ldots \mu_N)$ produced by every incident wave in every terminal. The result is

$$n(r, \mu_1, \mu_2, \ldots \mu_N) = \sum_t \sum_\alpha \int d\epsilon\, f_t(\epsilon)\, \ell N_{t\alpha}\, |\theta_{t\alpha}(r)|^2 \qquad (13)$$

where $f_t(\epsilon)$ is the Fermi-Dirac function of ϵ with chemical potential μ_t which specifies the occupancy of $\psi_{t\alpha}$. To calculate $n_o(r,\mu_L)$ we set all the chemical potentials in equation (13) equal to μ_L:

$$n_o(r, \mu_L) = \sum_t \sum_\alpha \int d\epsilon\, f_L(\epsilon)\, \ell N_{t\alpha}\, |\theta_{t\alpha}(r)|^2 \qquad (14)$$

where the local chemical potential μ_L appears in the Fermi-Dirac function $f_L(\epsilon)$. Inspection of equations (6), (13) and (14) shows that Eq.(11) for μ_L reduces to

$$\sum_t \sum_\alpha \int d\epsilon\, [f_t(\epsilon) - f_L(\epsilon)]\, |\theta_{t\alpha}(r)|^2 / v_{t\alpha} = 0. \qquad (15)$$

It only remains for us to solve equation (15) for μ_L. The solution can be obtained easily in the linear regime when μ_t is close to the common chemical potential ϵ_F which existed in each reservoir prior to the establishment of the chemical potentials $\mu_1, \mu_2, \ldots \mu_N$. Then μ_L is also close to ϵ_F and we may write

$$f_t(\epsilon) - f_L(\epsilon) = -(\mu_t - \mu_L)\, df_o(\epsilon)/d\epsilon \qquad (16)$$

where the Fermi-Dirac function $f_o(\epsilon)$ involves the chemical potential ϵ_F. When equation (16) is used in equation (15) we immediately obtain the result

$$\mu_L = \sum_t p_t \mu_t / \sum_t p_t \qquad (17)$$

where the weighting factors p_t are given by

$$p_t = -\sum_\alpha \int d\epsilon\, df_o(\epsilon)/d\epsilon\, |\theta_{t\alpha}(r)|^2 / v_{t\alpha} \qquad (18)$$

At low temperature we may write $df_o(\epsilon)d\epsilon \simeq -\delta(\epsilon - \epsilon_F)$ so that

$$p_t = \sum_\alpha |\theta_{t\alpha}(r)|^2 / v_{t\alpha} \qquad (19)$$

where it is left understood that $\theta_{t\alpha}(r)$ and $v_{t\alpha}$ are to be evaluated at the Fermi level. Equations (18) and (19) are our final results for a general nanostructure. For the 2D waveguide geometry considered in Section 2 they are identical to equations (7) and (8) because the scattering state $\theta_{t\alpha}(r)$ is identical to the incident wave in equation (12) and the coordinate z is irrelevant.

5. CONCLUSION

We have seen that the resistance R_{EHP} is never quantized. Consequently it has a negligible part to play in theoretical studies of the quantum Hall effect. On the other hand, the resistance R_{LCP} is quantised in appropriate circumstances and shows intermittent quenching at values of n_s for which the Fermi level approaches a subband minimum.

The advantage of the LCP concept is that it permits calculations to be made on the quantum Hall effect in a simple waveguide geometry. The new definition of the LCP given in Section 4 avoids the approximations made in previous discussions. The outstanding problem is how to measure the LCP. Theoretical studies of probes specifically designed to be non-invasive would be useful in pointing the way towards a reliable measurement procedure or, alternatively, suggesting a more useful definition.

Numerical studies of semiconductor nanostructures is a relatively new and challenging field. Many aspects of the physics are essentially two-dimensional. We have described a particularly simple problem in which the differential equations to be solved are one-dimensional. More complicated structures, e.g. a waveguide cross, are necessarily two-dimensional in character. The third dimension, perpendicular to the plane of the two-dimensional electron gas, is by no means irrelevant to the real behaviour of the system. It is usually ignored in numerical studies at the present time because the available computing power is inadequate. That situation is changing very rapidly. A great deal remains to be done in two dimensions and three dimensional calculations are just over the horizon.

REFERENCES

Akera, H. and Ando, T., 1989, Hall effect in quantum wires, *Phys. Rev. B* 39: 5508.

Akera, H. and Ando, T., 1990, Theory of the Hall effect in quantum wires: effects of scattering, *Phys. Rev. B* 41: 11967.

Baranger, H.U., Di Vincenzo, D.P., Jalabert, R.A. and Stone, A.D., 1991, Classical and quantum ballistic-transport anomalies in microjunctions, *Phys. Rev. B* 44: 10,637.

Butcher, P.N., 1993, Theory of electron transport in low-dimensional semiconductor structures, *in* "Physics of Low-Dimensional Semiconductor Structures", P. Butcher, N.H. March and M. Tosi, eds., Plenum Press, New York.

Büttiker, M., 1988, Symmetry in electrical condition, *IBM Journ. Res. Dev.* 32: 317.

Chu, D.P. and Butcher, P.N., 1993a, The integer quantum Hall effect in a quantum wire, *Phys. Rev. B* 47: 10008.

Chu, D.P. and Butcher, P.N., 1993b, Quenching of the Hall effect in a uniform ballistic quantum wire, *J. Phys.: Condens. Matter* 5: L397.

Imry, Y., 1989, Theoretical considerations for some new effects in narrow wires, *in* Nanostructure Physics and Fabrication, M.A. Reed and W.P. Kirk, eds., Academic Press, New York.

Li, Q. and Thouless, D.J., 1990, Electric potential and current distributions in a quantum wire under weak magnetic fields, *Phys. Rev. Lett.* 65: 767.

Peeters, F.M., 1988, Quantum Hall resistance in a quasi-one-dimensional electron gas, *Phys. Rev. Lett.* 61: 589.

FINITE LATTICE CALCULATIONS FOR MAGNETIC SYSTEMS

J. Oitmaa

School of Physics
University of New South Wales
Sydney 2052, Australia

INTRODUCTION

Many of the rich phenomena and properties of condensed matter physics are utterly unpredictable on a "noninteracting particle" paradigm. This implies that theory must confront the problem of a large number of strongly interacting particles. The usual strategy of theorists is to construct a model Hamiltonian which is simple enough to allow controlled calculations but not so simple that the interesting physics is lost. Computational methods, of the kinds described at this workshop, have an essential role to play.

In this lecture I want to describe a method which, while conceptually simple, can nevertheless provide important information. The basic idea is to choose a sequence of small clusters of particles, for each cluster obtain numerically exact estimates of properties of interest, and the attempt to extrapolate to the bulk limit.

The method arises naturally in lattice systems, and all of the discussion will be confined to magnetic spin models. The ideas can be generalised to more complex lattice systems and also to continuum models.

The typical Hamiltonian takes the form

$$H = \sum_i h_i + \sum_{i j} v_{ij} + ...$$

where $i,j = 1,2, ...N$ are the sites of a lattice, h_i are single-site terms, v_{ij} are pair interactions, etc. There are g quantum states per site and the dimensionality of the Hilbert space $D = g^N$ diverges exponentially in the thermodynamic limit $N \rightarrow \infty$. The questions that will need to be addressed include

- ground state energy, wavefunctions, correlations ..... . This requires the solution of the eigenvalue equation

$$H\Psi_0 = E_0\Psi_0$$

- thermodynamic properties, which are obtained from the partition function

$$Z = \text{Tr}\{e^{-\beta H}\} = \sum_i e^{-\beta E_i}$$

via standard relations.

Exact analytic calculations are only possible for a few 1-dimensional systems, and thus approximate methods must usually be employed. The finite-lattice approach starts with a cluster of M sites, for which the number of basis states g^M is finite, and follows the steps
1. Construct the $g^M \times g^M$ matrix H
2. Find the eigenvalues and eigenvectors numerically
3. Compute the quantities of interest
4. From a sequence of clusters, extrapolate to $M = \infty$

We expect that for large M
$$E_0(M) = M\varepsilon_0 + \text{corrections}$$
$$\ln Z(M) = M(-\beta f_0) + \text{corrections}$$
where ε_0 and f_0 are the bulk energy and free energy per site and the correction terms become negligible as $M \to \infty$.

The limiting factor is the exponential increase in basic size. For example, for a spin $\frac{1}{2}$ system g=2, and we have the following

M	8	12	16	20	..	32
D=2^M	256	4096	65536	1048576	..	~4 x 10^9

Thus the calculation quickly becomes prohibitive, in terms of computer time and memory requirements. There are a number of potential simplifications.
- the H matrix is often very sparse, and only the non-zero elements and the locations need be stored.
- symmetries can be exploited to reduce the size of matrices to be diagonalized
- for ground state problems only the lowest eigenvalue and eigenvector are needed and iterative methods, such as the Lanczos method, can be used.

Despite these limitations recent achievements are impressive. Schulz and Ziman [1] have obtained results for an $s=\frac{1}{2}$ spin system on a 6x6 lattice (M=36). In their work the largest matrix is of dimension 15 804 956 with approx. 1.2×10^9 nonzero elements. Lin [2] has treated a Hubbard model (g=4) with M=18 where the largest sector has dimension 16 445 304.

THE 1-DIMENSIONAL ANTIFERROMAGNET

The $s=\frac{1}{2}$ Heisenberg antiferromagnet in 1 dimension is described by the Hamiltonian

$$H = 2J \sum_{i=1}^{N} \{S_i^z S_{i+1}^z + \frac{1}{2}(S_i^+ S_{i+1}^- + S_i^- S_{i+1}^+)\}$$

with $S_{N+1} = S_1$. The ground state energy and wavefunction of this system are known exactly (see Mattis [3]) and thus it provides a nice pedagogical example to demonstrate the method.

We start with small systems for which the calculations can be done by hand. The basis states are the eigenstates of $\{S_i^z\}$

$$|m\rangle = |m_1, m_2, \ldots m_N\rangle \qquad m_i = \pm\frac{1}{2}$$

N=2

This case is identical to the standard quantum problem of the spin states of the He atom. In an obvious notation the 4 basis states are $|++\rangle, |+-\rangle, |-+\rangle, |--\rangle$.

It is a simple exercise to obtain the H matrix

$$H = J \begin{pmatrix} 1 & 0 & 0 & 0 \\ 0 & -1 & 2 & 0 \\ 0 & 2 & -1 & 0 \\ 0 & 0 & 0 & 1 \end{pmatrix}$$

and the ground state energy and eigenvector

$$E_0 = -3J, \qquad \Psi_0 = \frac{1}{\sqrt{2}} [|+-\rangle - |-+\rangle]$$

We note that Ψ_0 lies in the sector with $S_{tot}^z = 0$, which is known to be true for all even N [4].

N=4

```
1 O———O 2
  |     |
4 O———O 3
```

There are $2^4=16$ basis states, but only 6 with $S^z_{tot}= 0$. These are, in an obvious notation

$|{}^{+-}_{-+}\rangle, |{}^{-+}_{+-}\rangle, |{}^{++}_{--}\rangle, |{}^{-+}_{-+}\rangle, |{}^{--}_{++}\rangle, |{}^{+-}_{+-}\rangle$

The Hamiltonian matrix is

$$H = J \begin{pmatrix} -2 & 0 & 1 & 1 & 1 & 1 \\ 0 & -2 & 1 & 1 & 1 & 1 \\ 1 & 1 & 0 & 0 & 0 & 0 \\ 1 & 1 & 0 & 0 & 0 & 0 \\ 1 & 1 & 0 & 0 & 0 & 0 \\ 1 & 1 & 0 & 0 & 0 & 0 \end{pmatrix}$$

from which it follows that

$E_0 = -4J$

$\Psi_0 = \frac{1}{\sqrt{12}}\{ 2|{}^{+-}_{-+}\rangle + 2|{}^{-+}_{+-}\rangle - |{}^{++}_{--}\rangle - |{}^{-+}_{-+}\rangle - |{}^{--}_{++}\rangle - |{}^{+-}_{+-}\rangle \}$

Note that the classical Néel states $|{}^{+-}_{-+}\rangle$ and $|{}^{-+}_{+-}\rangle$ are not the true ground states, although they have the largest amplitude in Ψ_0. This feature remains true for larger N.

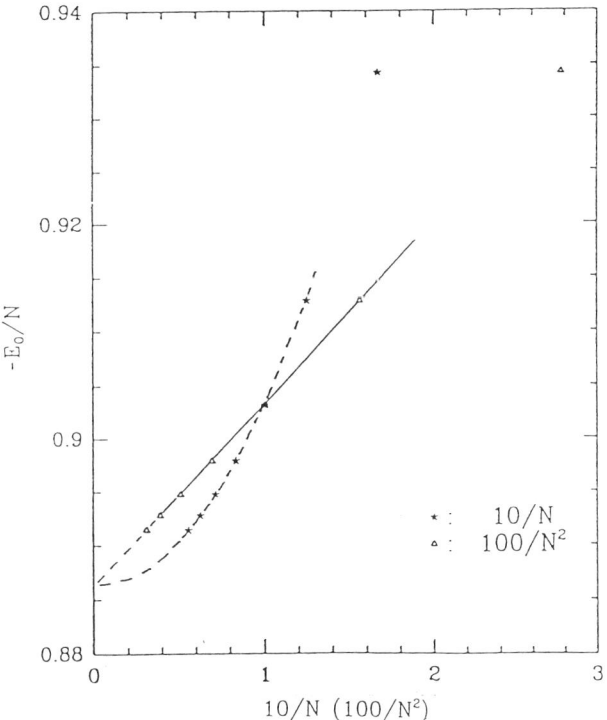

Figure 1. Ground state energy of 1-d Heisenberg antiferromagnet

For N > 4 it is no longer feasible to do these calculations by hand. The first substantial application of computers to this type of calculation was by Bonner and Fisher [5], who studied closed chains up to N=11. With present day computer technology much more can be achieved. Using the Lanczos method and modest computer time I have obtained the ground state energies for N≤18. In Figure 1 I show E_0/N plotted versus $1/N$ and versus $1/N^2$. The former shows residual curvature but the latter appears quite linear, suggesting the limiting behaviour

$$E_0/N = \varepsilon_0 + \alpha/N^2 + \ldots\ldots$$

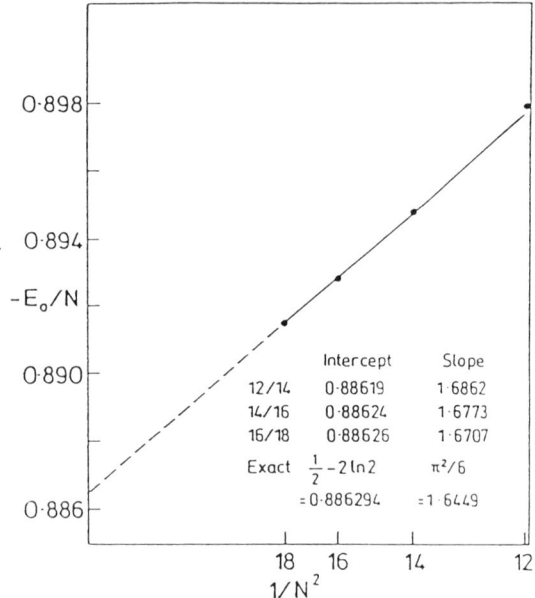

Figure 2. Ground state energy of 1-d antiferromagnet with extrapolation to N=∞

In Figure 2 an expanded plot is shown, as well as extrapolations to the intercept and slope from the pairs of points 12/14, 14/16, 16/18. For comparison the exact results for ε_0 [3] and α [6] are also shown. The correct ground state energy is obtained to better than 4 figures, and the slope to within 2%. The presence of logarithmic corrections slows the convergence of the α estimates.

To illustrate the calculation of thermodynamic properties I show, in Figure 3, the specific heat of the 1-d antiferromagnet. Curves are shown for N=4,6,8,10,12. The convergence is quite rapid. For T>1 the N=12 and N=10 results are indistinguishable in the diagram, and the bulk behaviour can be predicted with high confidence. Even at low temperatures the difference between N=10 and N=12 is small although as T→0 convergence becomes poor.

The finite lattice method is thus able to yield results for this model which are at least as accurate as other approximate methods, and can be applied equally well to other spin Hamiltonians. In the following section we consider 2-d antiferromagnets.

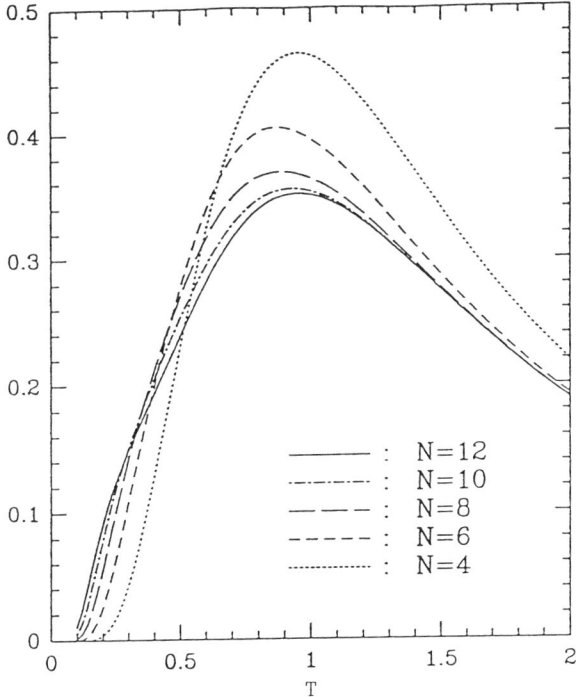

Figure 3. Specific heat of 1-d Heisenberg antiferromagnet

THE 2-DIMENSIONAL ANTIFERROMAGNET

The 2-dimensional $s=\frac{1}{2}$ Heisenberg antiferromagnet remains an important unsolved problem. Interest in this system has been stimulated recently by its relevance to high T superconductivity in the cuprates. All of these materials contain CuO_2 planes, as shown. In the undoped state the Cu^{++} ions have spin $s=\frac{1}{2}$ while the O^- ions have s=0.

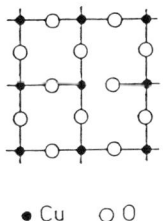

• Cu   ○ O

This system is perhaps the best realization of the ideal 2-d $s=\frac{1}{2}$ antiferro-magnet, and at low temperatures these materials do show antiferromagnetic order. Doping produces holes on the O sites and thus leaves the doped sites with residual $s=\frac{1}{2}$ spins. With increased doping the AF order is destroyed and a superconducting state is achieved.

The first serious application of finite lattice techniques to the 2-D antiferromagnet was by Oitmaa and Betts [7] who studied systems up to N=18. Subsequent calculations have extended this to N=26 [8], N=32 [9], and N=36 [1]. While the choice of finite lattices in 1-dimension is obvious, this is not so in 2-dimensions. To maintain square cells with the symmetries of the host lattice it is necessary to choose cells with $N=m^2+n^2$, with m, n integers and (m+n) even. The various possibilities are shown below

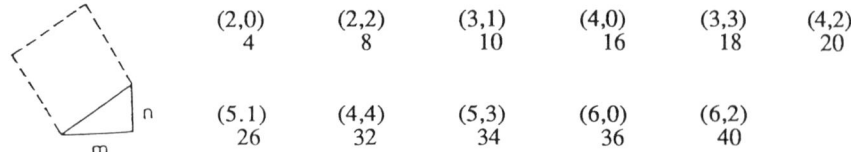

(2,0)	(2,2)	(3,1)	(4,0)	(3,3)	(4,2)
4	8	10	16	18	20
(5,1)	(4,4)	(5,3)	(6,0)	(6,2)	
26	32	34	36	40	

Figure 4 shows the ground state energy per site plotted versus N^{-1} and versus $N^{-3/2}$, the latter being a field theory prediction [10]. Neither plot shows the same regularity as the 1-d case, which may be a residual effect of the differently oriented cells or simply due to the small linear dimension of the cells. Nevertheless it seems reasonable to conclude that

$$-E_0/N = 1.345 \pm 0.005$$

It is known, from the Mermin-Wagner theorem [11] that this system can have no long range order at any T>0. The long range order parameter can be defined as

$$M_{AF} = \frac{1}{N} \langle M_{st}^2 \rangle$$

where $M_{st} = \sum_i \eta_i S_i^z$ and $\eta_i = \pm 1$ for alternate sublattices. Figure 5 shows a plot of M_{AF} versus $N^{-1/2}$ for N up to 26. An extrapolation to N→∞ gives clear evidence of a residual staggered magnetization, of magnitude $M_{AF} \sim 0.32$. The existence of long range antiferromagnetic order at T=0 in the Heisenberg antiferromagnet was first observed by Oitmaa and Betts [7]. Subsequent calculations, using a number of methods, have confirmed this feature.

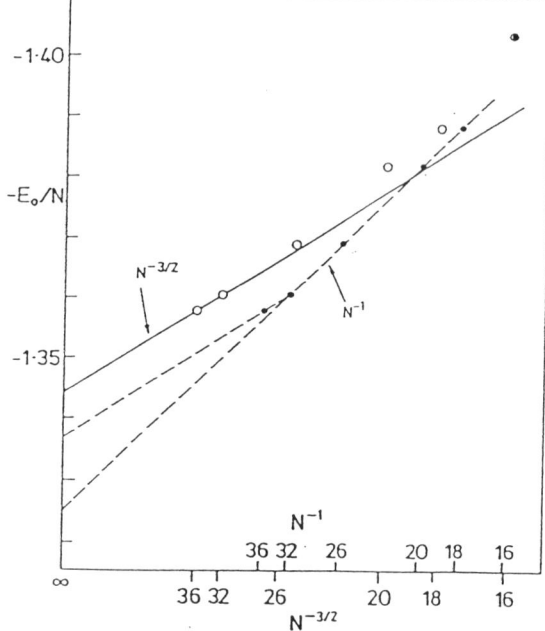

Figure 4. Ground state energy of Heisenberg antiferromagnet on square lattice.

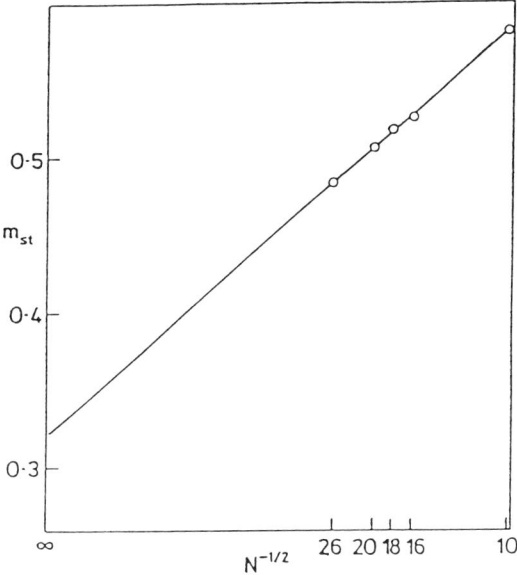

Figure 5. Ground state staggered magnetization for Heisenberg antiferromagnet on square lattice

To conclude this lecture I want to present results of some recent research carried out by Betts and myself. We have studied the Heisenberg antiferromagnet with one or more additional spins ("spin defects") at the mid-points of lattice bonds. This is intended to model static holes on oxygen atoms in the CuO_2 planes of the cuprate superconductors. The Hamiltonian is taken to be

$$H = 2J \sum_{host} S_i \cdot S_j + 2K \sum_{defect} S_k \cdot S_l$$

The following configurations have been considered

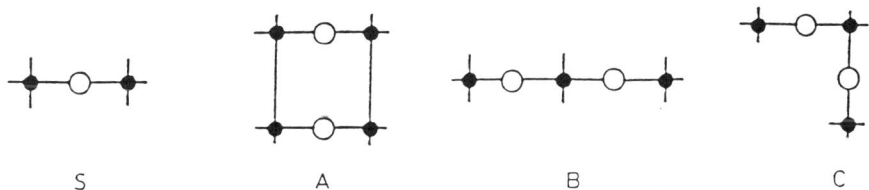

for systems of size up to 20 + 1, for the 1-defect configuration S, and up to size 18 + 2 for the 2-defect cases A,B,C.

We have computed the ground state energy and spin-spin correlations for a range of values of $x = K/J$, including both ferromagnetic and antiferromagnetic coupling of the defect spin to the host lattice. Detailed results are presented in the literature [1]; [13]. For the two defect case it is an interesting question which of the configurations A,B,C has the lowest energy. It turns out that this depends on x. For N=18 the configuration A is most stable for $-3.54 < x < 0.76$ and $x > 6.41$. Otherwise C is most stable. Another interesting question is that of binding. Is it energetically favourable for two defect spins to bind or to be separated? The binding energy can be defined as

$$-\Delta E = E_2(x) - 2E_1(x) - E_2(0) + 2E_1(0)$$

The binding energy is plotted in Figure 6 for the N=18 system versus $y = x/(|x|+5)$. It is apparent from this figure that binding is favoured for all x.

These examples should suffice to illustrate the power and generality of the method. The finite-lattice method is currently used by many research groups and can yield results not obtainable by other methods. When used in conjunction with other approaches it provides a powerful tool to study the properties of strongly interacting many body systems.

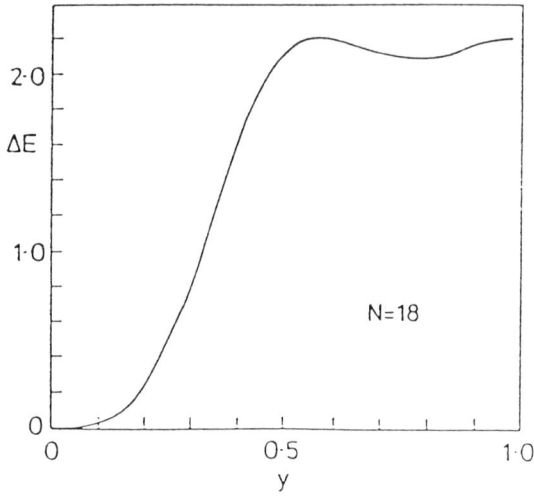

Figure 6. Binding energy for 2 defects, for the case N=18 verus y=x/(5+|x|) with x=K/J.

REFERENCES

[1] H.J. Schulz and T.A.L. Ziman, Finite-size scaling for the two-dimensional frustrated quantum Heisenberg antiferromagnet, Europhys. Lett. 18, 355 (1992).

[2] H.Q. Lin, Ground-state properties of the two-dimensional Hubbard model, Phys. Rev. B44, 7151 (1991).

[3] D.C. Mattis, "The Theory of Magnetism I", Springer, Berlin (1981).

[4] E.Lieb and D. Mattis, Ordering energy levels of interacting spin systems, J. Math. Phys. 3, 749 (1962).

[5] J.C. Bonner and M.E. Fisher, Linear magnetic chains with anisotropic coupling, Phys. Rev. 135, A640 (1964).

[6] C.J. Hamer, Finite-size corrections for ground states of the XXZ Heisenberg chain, J. Phys. A19, 3335 (1986).

[7] J. Oitmaa and D.D. Betts, The ground state of two quantum models of magnetism, Can. J. Phys. 56, 897 (1978).

[8] S. Tang and J.E. Hirsch, Long-range order without broken symmetry: two-dimensional Heisenberg antiferromagnet at zero temperature (1989).

[9] H.Q. Lin, Exact diagonalization of quantum-spin models, Phys. Rev. B42, 6561 (1990).

[10] H. Neuberger and T. Ziman, Finite-size effects in Heisenberg antiferromagnets, Phys.Rev. B39, 2608 (1989).

[11] N.D. Mermin and H. Wagner, Absence of ferromagnetism or antiferromagnetism in one or two-dimensional isotropic Heisenberg models, Phys. Rev. Lett. 17, 1133 (1966).

[12] D.D. Betts and J. Oitmaa,, Effects of perturbing spins on the properties of the $S=\frac{1}{2}$ Heisenberg antiferromagnet on the square lattices, Phys. Rev. B48, 10602 (1993).

[13] J. Oitmaa, D.D. Betts and M. Aydin, Studies of the two-dimensional Heisenberg antiferromagnet with perturbing spin defects, Phys. Rev. B. (submitted)(1994).

INDEX

Ab initio, 1,9,31,38,67,176
Antiferromagnet, 270,272
Atomic cluster, 50

Boson(s), 10,14,203,208
Branching, 16,19
Brillouin zone sampling, 75

C_{60}, 73,88,103,105,163
Car-Parrinello, 1,40,72,73,78,81
Central limit theorem, 11
Chemisorption, 175,177
Clusters, 30,50,54,56,163,269
Collective oscillation, 164
Computer simulation, 87
Configuration interaction(CI), 43
Correlation energy, 24,25
Crystal'92, 178,181,182,187

Density functional, 1,8,37,40,43,49,67,
141,169,255,258
Density wave, 257,258
Diagonalization, 77
Diffusion Monte-Carlo, 3,14,21
Direct minimization, 2,72,78,81
Disordered electronic materials, 235
Dzyaloshinskii-Moriya interaction, 245,250

Edwards-Anderson, 243
EELS, 141,164,169,170
Electron gas, 22,140,161
Electronic structure, 67,164
Energy surfaces, 41,50
Exchange-correlation, 2,44,69,95,154

Fermion(s), 10,11,21,22,31,210,220
Fermion-Boson interaction, 230
Forces, 70
Freezing, 255
Friedel oscillations, 139,143,144,156
Functional integral, 191,192,204,207,225

Gamess' 92, 178,182
Gaussian' 90, 178,182
Ginzburg-Landau equation, 232
Graphitic tubules, 88,90,103,110
Grassmann variable, 209
Green's function, 15,19,68,79,80

Hartree-Fock, 8,23,42,44,68
Hohenberg-Kohn, 43
Hubbard-Stratonovich transform, 224,231,232
Hydrodynamic model, 166

Imaginary time, 198,201,204,218
Isomers, 51,57

Kohn-Sham, 2,26,44,68,94,140,141,146

Lattice calculations, 269
Lanczos method, 270,272
Lennard-Jones, 39,40,257
Linear response, 143,145
Linear scaling, 79
Local chemical potential, 261,265
Local density approx (LDA), 2,8,25,26,28,
29,30,68,70,82
Local spin density (LSD), 44,45,56,63

Magnetic systems, 269
MasPar, 96,119,128,129,130
Metallic glass, 239
Metallic hydrogen, 24
Metal surface, 30
Metropolis algorithm, 13,14
MINDO, 178,181,182,185,187
Molecular dynamics, 1,37,40,72,73,87,
125,126,137
Molecules, 30,37,56
Monte-Carlo, 12,29,69,246
Morse, 39

Orientational disorder, 107
Ozone molecule, 38,40

Parallel molecular dymanics algorithm, 91
Parallel quantum dynamics, 94
Partition function, 193, 217
Path integral, 3,191
Phonon(s), 105,106,108,111,116,161
Plasmon(s), 148,167,229
Porous silica, 101
Pseudopotential, 10,28,29,31,49,74,82,176

Quantum field theory, 191
Quantum Hall effect, 261,267
Quantum molecular dynamics, 3, 87,97
Quantum Monte-Carlo, 1,3,7,8,9,25,68,83

Renormalization group, 245
RKKY interaction, 241,245,251

Semiconductor surfaces, 175
Sherrington-Kirkpatrick model, 242,244,246
Sign problem, 10,21,31
Simulated annealing, 39,96
SiO_2, 88,97

Spin glass(es), 235,236,239,241
Supercells, 72
Surface plasmon, 168,171
Susceptibility, 142

Tetrahedron technique, 76
TBMD, 90,103
TDLDA, 143,144

Total energy, 40

Variational Monte-Carlo, 3,12
Verlet algorithm, 50

Wick theorem, 199
Wigner crystal, 24

Zero point energy, 27